Nonlinear Physical Science

Nonlinear Physical Science focuses on recent advances of fundamental theories and principles, analytical and symbolic approaches, as well as computational techniques in nonlinear physical science and nonlinear mathematics with engineering applications.

Topics of interest in *Nonlinear Physical Science* include but are not limited to:

- New findings and discoveries in nonlinear physics and mathematics
- Nonlinearity, complexity and mathematical structures in nonlinear physics
- Nonlinear phenomena and observations in nature and engineering
- Computational methods and theories in complex systems
- Lie group analysis, new theories and principles in mathematical modeling
- Stability, bifurcation, chaos and fractals in physical science and engineering
- Discontinuity, synchronization and natural complexity in physical sciences
- Nonlinear chemical and biological physics

This book series is indexed by the SCOPUS database.

To submit a proposal or request further information, please contact Dr. Mengchu Huang(Email: mengchu.huang@springer.com).

More information about this series at https://link.springer.com/bookseries/8389

Albert C. J. Luo · Rafail K. Gazizov
Editors

Symmetries and Applications of Differential Equations

In Memory of Nail H. Ibragimov (1939–2018)

Higher Education Press

Springer

Editors
Albert C. J. Luo
Department of Mechanical
and Mechatronics Engineering
Southern Illinois University Edwardsville
Edwardsville, IL, USA

Rafail K. Gazizov
Department of High Performance
Computing
Ufa State Aviation Technical University
Ufa, Russia

ISSN 1867-8440 ISSN 1867-8459 (electronic)
Nonlinear Physical Science
ISBN 978-981-16-4682-9 ISBN 978-981-16-4683-6 (eBook)
https://doi.org/10.1007/978-981-16-4683-6

Jointly published with Higher Education Press
The print edition is not for sale in China (Mainland). Customers from China (Mainland) please order the print book from: Higher Education Press.

This Springer imprint is published by the registered company Springer Nature Singapore Pte Ltd.
The registered company address is: 152 Beach Road, #21-01/04 Gateway East, Singapore 189721, Singapore

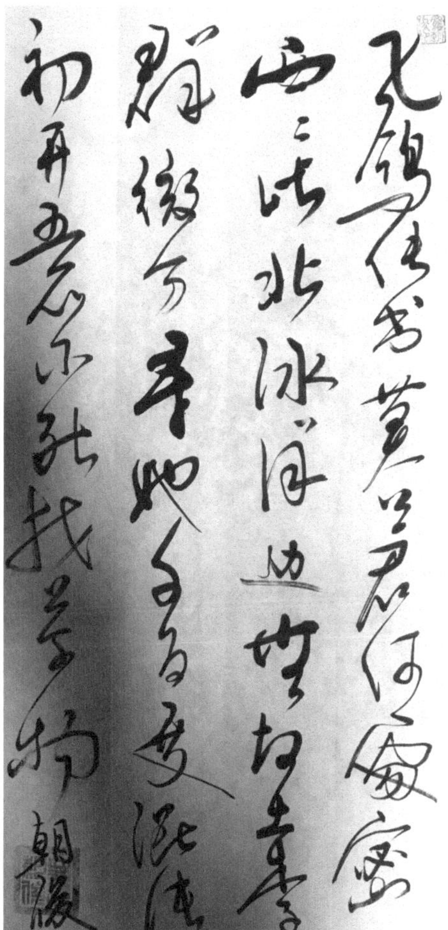

Preface

Professor Nail H. Ibragimov loved mathematics passionately. "It gives you a whole new window on the world," he used to say on his favorite subject. His work was focused on modern group analysis, "a very attractive area." This research is based on the results and ideas of Sophus Lie, who Ibragimov described as "one of the greatest mathematicians of the 19th century." According to Ibragimov, the importance of Lie group analysis cannot be overemphasized—it helps to solve very important differential equations used in mathematical models in many and diverse fields. He also believed that it was essential that education in this field be expanded.

Nail H. Ibragimov was born on January 18, 1939, in the village of Urussu, in Tatarstan, Russia. School years fell on the difficult post-war years, but it was then that the interest in physics and mathematics appeared. A lot of this was facilitated by the excellent teachers of secondary school in the town of Urussu, located 5 km from his village. Nail had always been grateful to his mathematics teacher Larisa Petrovna Barkhat, to whom he dedicated the first volume of his selected works.

While serving in the army in 1958–1961, he began to prepare for studies at university: he independently studied mathematical analysis using textbooks of N. N. Luzin, physics according to the multivolume course of O. D. Khvolson. He also studied foreign languages. He got admitted to the Moscow Institute of Physics and Technology and after the first year he moved to newly created Academgorodok and transferred to Novosibirsk University. Outstanding scientists such as M. A. Lavrentiev, S. L. Sobolev, B.Yu. Rumer, D. V. Shirkov, and A. D. Aleksandrov worked there at that time and influenced the formation of Nail as a scientist.

Ibragimov combined his studies at the university with work at the Institute of Hydrodynamics in the scientific group of his teacher Lev Vasilievich Ovsyannikov, where the methods of group analysis were actively developed in those years. The creative atmosphere that prevailed in the institutes of Academgorodok, the high efficiency, and talent of N. H. Ibragimov led to early graduation from the university, defending a Ph.D. thesis in 1967 only 2 years after graduation, and defending the Doctor of Science dissertation already in 1973. The main results obtained in those years by Nail Ibragimov were:

- theory of generalized motions in Riemannian space, including the Killing equation as a special case (1969);
- extension of the Pauli group for the Dirac equations (1969);
- differential–algebraic approach to conservation laws and proof of Noether's converse theorem (1969);
- the discovery of the group-theoretical nature of the Huygens principle in the theory of wave propagation and the solution of the Hadamard problem in spaces with a nontrivial conformal group (1970);
- new conservation laws in hydrodynamics (1973);
- construction of the theory of Lie–Bäcklund transformation groups (1979).

In 1980, N. H. Ibragimov moved to Ufa, where he headed the Laboratory of Mathematical Physics at the Physics and Mathematics Department of the Bashkir Branch of the USSR Academy of Sciences. Ibragimov combined work at the Academy of Sciences with work at Ufa Aviation Institute where he first was a professor of the Department of Higher Mathematics, and then, after 1984, the Head of the Department of Applied Mathematics. During this period, he, his followers, and Ph.D. students developed:

- methods for constructing nonlocal symmetries of the equations of mechanics (1985);
- foundations of the theory of approximate transformation groups and approximate symmetries of equations with a small parameter (1987);
- completed work on the book "*Transformation Groups in Mathematical Physics*" (Moscow: Nauka, 1983), which was distinguished by the USSR State Prize for

Science and Technology (1987) together with L. V. Ovsyannikov's book "*Group Analysis of Differential Equations*" (Moscow: Nauka, 1978).

In 1987, Nail Ibragimov moved to Moscow to work at the M. V. Keldysh Institute of Applied Mathematics by the invitation of A. A. Samarsky, member of the USSR Academy of Sciences. The main direction of his scientific interests at the institute was the symmetry approach to the fundamental solution and the invariance principle in the problems with initial conditions (1992). At the same time, N. H. Ibragimov began to teach the course "Equations of Mathematical Physics" at the Moscow Institute of Physics and Technology. It was within the framework of this course that his course in differential equations, based on the symmetry approach, began to emerge. This activity resulted in two popular brochures "*Primer on group analysis*" (1989) and "*Essay on the group analysis of ordinary differential equations*," Znanie, Moscow (1991).

Since 1976, he lectured intensely all over the world, e.g., at Georgia Tech in the USA, Collége de France, the University of Witwatersrand in South Africa, the University of Catania in Italy, etc. In 1993–1994, N.H. Ibragimov worked as a professor of the Department of Engineering Sciences at Istanbul Technical University (Turkey), in 1994–2000 as professor of the Department of Computational and Applied Mathematics at the University of Witswatersrand (Johannesburg) (until 1997) and the Department of Mathematics at the North-West University (Mmabatho) in South Africa. It is interesting to note that Ibragimov's scientific achievements were highly appreciated in South Africa: he was awarded the highest scientific rating by the National Research Foundation (before him, only one mathematician in the country had such a rating in applied mathematics).

Since 2000, N. H. Ibragimov lived and worked in Sweden: he was a professor at the Department of Mathematics and Natural Sciences at Blekinge Institute of Technology (Karlskrona) and director of the International Center ALGA (Advances in Lie Group Analysis) which he created. The main scientific results obtained by him in recent years are:

- construction of Laplace-type invariants for a parabolic equation (2000);
- solution of the Laplace problem on the invariants of the hyperbolic equation (2004), for which he was awarded the prize, Researcher of the year by Blekinge Research Society, Sweden;
- generalization of Euler's method for solving hyperbolic equations to parabolic equations (2008).

In recent years, Prof. N. H. Ibragimov devoted himself to popularizing the methods of group analysis. Nail H. Ibragimov published about 30 books including 2 graduate textbooks, Elementary Lie group analysis and ordinary differential equations (1999) and A practical course in differential equations and mathematical modelling (2005). The last textbook was also translated into Russian, Swedish, Chinese, and German.

At the research center ALGA, Ibragimov started the journal "Archives of ALGA," which published both new and little-known (or forgotten) results on group analysis of differential equations. Nail Ibragimov was a member of the editorial board of two international journals, Nonlinear Dynamics (since 1987) and Communications in nonlinear sciences and numerical modeling (since 2002), and a member of the editorial board of the Ufa mathematical journal.

N. H. Ibragimov stood at the origins of group analysis and, under his direct leadership, the theory and applications of modern group analysis continued to develop. To a large extent, this was facilitated by the international research conference MOGRAN "Modern Group Analysis" organized by him and held in many universities all over the world.

In 2011, N. H. Ibragimov won a mega-grant of the Ministry of Education and Science of the Russian Federation. He organized research laboratory group analysis of mathematical models in natural and engineering sciences at Ufa State Aviation Technical University, Ufa, Russia (2011–2015).

Since April 2012, he also was the Professor Emeritus in the Department of Mathematics and Natural Sciences, School of Engineering at Blekinge Institute of Technology, Karlskrona, Sweden. A friendly attitude toward people, decency, and scrupulousness both in scientific and in life matters have always distinguished Nail H. Ibragimov. This allowed him to find friends and associates around the world and to spread the idea of invariance in scientific and teaching circles.

Ibragimov's classmate, Edward Kissin (United Kingdom), wrote about him: "He was an amazingly cheerful and positive person. I was always amazed how this guy from a remote village who did not know Russian until 10–12 years old, then joined the army, was able to enroll to the Moscow Institute of Physics and Technology, defend his dissertation in mathematics, and become a professor. Huge willpower and determination!"

Professor Ibragimov passed away on 4 November 2018.

Ufa, Bashkortostan, Russia
April 2021

Rafail K. Gazizov
R. S. Khamitova

Contents

Editors and Contributors

About the Editors

Prof. Albert C. J. Luo has worked at the Southern Illinois University Edwardsville. For over 30 years, his contributions on nonlinear dynamical systems and mechanics lie in (i) the local singularity theory for discontinuous dynamical systems, (ii) dynamical systems synchronization, (iii) analytical solutions of periodic and chaotic motions in nonlinear dynamical systems, (iv) the theory for stochastic and resonant layer in nonlinear Hamiltonian systems, and (v) the full nonlinear theory for a deformable body. Such contributions have been scattered into 25 monographs and over 350 peer-reviewed journal and conference papers. He served editors for the Journal "*Communications in Nonlinear Science and Numerical simulation*" for 14 years, book series on Nonlinear Physical Science (HEP) and Nonlinear Systems and Complexity (Springer). He is the editorial member for two journals (i.e., *IMeCh E Part KJournal of Multibody Dynamics* and *Journal of Vibration and Control*). He also organized over 30 international symposiums and conferences on Dynamics and Control.

Prof. Rafail K. Gazizov has worked at Ufa State Aviation Technical University (Ufa, Russia). For almost 40 years, his contributions on group analysis of differential equations lie in (i) the theory of quasilocal and nonlocal symmetries for equations with Backlund transformations, (ii) the theory of approximate transformation groups, (iii) approximate symmetries of differential equations with a small parameter, and (iv) the symmetry approach for fractional differential equations. His scientific results are also connected with parallel algorithms, supercomputing technologies, and mathematical modeling of oil production processes. Results of his researches have been published in over 200 papers in peer-reviewed journals and 4 books. He is the deputy editor for two journals (Ufa Mathematical Journal and Vestnik of Ufa State Aviation Technical University).

Contributors

A. V. Aksenov Lomonosov Moscow State University, Moscow, Russian Federation

M. S. Bruzón Department of Mathematics, Faculty of Sciences, University of Cádiz, Cádiz, Spain

D. O. Bykov Novosibirsk State University, Novosibirsk, Russia

V. A. Dorodnitsyn Keldysh Institute of Applied Mathematics, Russian Academy of Science, Moscow, Russia

Mikhail M. Dyshaev Chelyabinsk State University, Chelyabinsk, Russia

Vladimir E. Fedorov Chelyabinsk State University, Chelyabinsk, Russia;
South Ural State University (National Research University), Chelyabinsk, Russia

Igor Leite Freire Centro de Matemática, Computação e Cognição, Universidade Federal do ABC, Bairro Bangu, Santo André, SP, Brazil

M. L. Gandarias Department of Mathematics, Faculty of Sciences, University of Cádiz, Cádiz, Spain

V. N. Grebenev Federal Research Center for Information and Computational Technologies, Russian Academy of Sciences, Novosibirsk, Russia

V. F. Kovalev Keldsyh Institute of Applied Mathematics, Moscow, Russia

R. Kozlov Department of Business and Management Science, Norwegian School of Economics, Bergen, Norway

Fazal M. Mahomed DSI-NRF Centre of Excellence in Mathematical and Statistical Sciences School of Computer Science and Applied Mathematics, University of the Witwatersrand, Johannesburg, South Africa;
School of Mathematics and Statistics, The University of New South Wales, Sydney, NSW, Australia

S. B. Medvedev Federal Research Center for Information and Computational Technologies, Russian Academy of Sciences, Novosibirsk, Russia

S. V. Meleshko School of Mathematics, Institute of Science, Suranaree University of Technology, Nakhon Ratchasima, Thailand

V. V. Pukhnachev Lavrentyev Institute of Hydrodynamics, Siberian Branch, Russian Academy of Sciences, Novosibirsk State University, Novosibirsk, Russia

Asghar Qadir Abdus Salam School for Mathematical Sciences, Government College University, Lahore, Pakistan

Júlio César Santos Sampaio Universidade Tecnológica Federal do Paraná, Curitiba, PR, Brazil

A. A. Talyshev Novosibirsk State University, Novosibirsk, Russia

Khristofor V. Yadrikhinskiy M.K. Ammosov North-Eastern Federal University, Yakutsk, Russia

Approximate Symmetry in Nonlinear Physics Problems

V. F. Kovalev

Abstract About 30 years ago, the notion of approximate transformation groups was introduced in modern group analysis by Nail H. Ibragimov with coauthors [1, 2]. Since that moment, the concept of approximate symmetries has been widely used for solving different problems of mathematical physics based on equations with a small parameter. The concept of approximate groups turned out to be in great demand in constructing renormalization-group symmetries used in nonlinear physics since the 90s of the last century. Combining the ideas of an approximate group theory and renormalization-group approach [3, 4] made it possible to construct solutions to a number of boundary value problems in mathematical physics. The chapter presents a review of numerous examples from plasma dynamics and nonlinear optics and acoustics that demonstrate the potentiality of approximate symmetries in constructing approximate analytical solutions.

1 Approximate Transformation Groups and Renormgroup Symmetries

The material presented here is a brief introduction to the theory of approximate transformation groups and symmetries and the method based on renormgroup symmetries. A more detailed discussion of the problem as well as the theory of multi-parameter approximate groups can be found in [1, 2, 4, 5].

V. F. Kovalev (✉)
Keldsyh Institute of Applied Mathematics, Miusskaya sq., 4, Moscow 125047, Russia
e-mail: vfkvvfkv@gmail.com

A. C. J. Luo and R. K. Gazizov (eds.), *Symmetries and Applications of Differential Equations*, Nonlinear Physical Science,
https://doi.org/10.1007/978-981-16-4683-6_1

1.1 Approximate Transformation Groups

1.1.1 Notation and Definitions

In what follows, functions $f(x, \varepsilon)$ of n variables $x = (x^1, \dots, x^n)$ and a parameter ε are considered locally in a neighborhood of $\varepsilon = 0$. These functions are continuous in the x's and ε. This is also true for the derivatives of these functions of such a high order, which is encountered in the subsequent discussion.

If a function $f(x, \varepsilon)$ satisfies the condition

$$\lim_{\varepsilon \to 0} \frac{f(x, \varepsilon)}{\varepsilon^p} = 0,$$

it is written $f(x, \varepsilon) = o(\varepsilon^p)$ and f is said to be *of order less than* ε^p. If

$$f(x, \varepsilon) - g(x, \varepsilon) = o(\varepsilon^p),$$

the functions f and g are said to be *approximately equal* (with an error $o(\varepsilon^p)$) and written

$$f(x, \varepsilon) = g(x, \varepsilon) + o(\varepsilon^p),$$

or, briefly $f \approx g$ when there is no ambiguity.

The approximate equality defines an equivalence relation and we join functions into equivalence classes by letting $f(x, \varepsilon)$ and $g(x, \varepsilon)$ to be members of the same class if and only if $f \approx g$.

Given a function $f(x, \varepsilon)$, let

$$f_0(x) + \varepsilon f_1(x) + \cdots + \varepsilon^p f_p(x)$$

be the approximating polynomial of degree p in ε obtained via the Taylor series expansion of $f(x, \varepsilon)$ in powers of ε about $\varepsilon = 0$. Then any function $g \approx f$ (in particular, the function f itself) has the form

$$g(x, \varepsilon) \approx f_0(x) + \varepsilon f_1(x) + \cdots + \varepsilon^p f_p(x) + o(\varepsilon^p).$$

Consequently the function

$$f_0(x) + \varepsilon f_1(x) + \cdots + \varepsilon^p f_p(x)$$

is called a *canonical representative* of the equivalence class of functions containing f.

Thus, the equivalence class of functions $g(x, \varepsilon) \approx f(x, \varepsilon)$ is determined by the ordered set of $p + 1$ functions

$$f_0(x), \quad f_1(x), \quad \dots, \quad f_p(x).$$

In the theory of approximate transformation groups, one considers ordered sets of smooth vector-functions depending on x's and a group parameter a :

$$f_0(x, a), \quad f_1(x, a), \quad \ldots, \quad f_p(x, a)$$

with coordinates

$$f_0^i(x, a), \; f_1^i(x, a), \ldots, f_p^i(x, a), \quad i = 1, \ldots, n.$$

Let us define the one-parameter family G of *approximate transformations*

$$\bar{x}^i \approx f_0^i(x, a) + \varepsilon f_1^i(x, a) + \cdots + \varepsilon^p f_p^i(x, a), \quad i = 1, \ldots, n, \tag{1}$$

of points $x = (x^1, \ldots, x^n) \in IR^n$ into points $\bar{x} = (\bar{x}^1, \ldots, \bar{x}^n) \in IR^n$ as the class of invertible transformations

$$\bar{x} = f(x, a, \varepsilon) \tag{2}$$

with vector–functions $f = (f^1, \ldots, f^n)$ such that

$$f^i(x, a, \varepsilon) \approx f_0^i(x, a) + \varepsilon f_1^i(x, a) + \cdots + \varepsilon^p f_p^i(x, a).$$

Here a is a real parameter, and the following condition is imposed:

$$f(x, 0, \varepsilon) \approx x.$$

Furthermore, it is assumed that the transformation (2) is defined for any value of a from a small neighborhood of $a = 0$, and that, in this neighborhood, the equation $f(x, a, \varepsilon) \approx x$ yields $a = 0$.

Definition 1.1 The set of transformations (1) is called a one-parameter approximate transformation group if

$$f(f(x, a, \varepsilon), b, \varepsilon) \approx f(x, a + b, \varepsilon)$$

for all transformations (2).

Remark 1.1 Here, unlike the classical Lie group theory, f does not necessarily denote the same function at each occurrence. It can be replaced by any function $g \approx f$ (see the next example).

Example 1.1 Let us take $n = 1$ and consider the functions

$$f(x, a, \varepsilon) = x + a\Big(1 + \varepsilon x + \frac{1}{2}\varepsilon a\Big)$$

and

$$g(x, a, \varepsilon) = x + a(1 + \varepsilon x)\Big(1 + \frac{1}{2}\varepsilon a\Big).$$

They are equal in the first order of precision, namely:

$$g(x, a, \varepsilon) = f(x, a, \varepsilon) + \varepsilon^2 \varphi(x, a), \quad \varphi(x, a) = \frac{1}{2}a^2 x,$$

and satisfy the approximate group property. Indeed,

$$f(g(x, a, \varepsilon), b, \varepsilon) = f(x, a + b, \varepsilon) + \varepsilon^2 \phi(x, a, b, \varepsilon),$$

where

$$\phi(x, a, b, \varepsilon) = \frac{1}{2}a(ax + ab + 2bx + \varepsilon abx).$$

The *generator of an approximate transformation group G* given by (2) is the class of first-order linear differential operators

$$X = \xi^i(x, \varepsilon)\frac{\partial}{\partial x^i} \tag{3}$$

such that

$$\xi^i(x, \varepsilon) \approx \xi^i_0(x) + \varepsilon \xi^i_1(x) + \cdots + \varepsilon^p \xi^i_p(x),$$

where the vector fields $\xi_0, \xi_1, \ldots, \xi_p$ are given by

$$\xi^i_\nu(x) = \frac{\partial f^i_\nu(x, a)}{\partial a}\Big|_{a=0}, \quad \nu = 0, \ldots, p;\ i = 1, \ldots, n.$$

In what follows, an approximate group generator

$$X \approx \left(\xi^i_0(x) + \varepsilon \xi^i_1(x) + \ \ldots\ + \varepsilon^p \xi^i_p(x)\right)\frac{\partial}{\partial x^i}$$

is written simply

$$X = \left(\xi^i_0(x) + \varepsilon \xi^i_1(x) + \cdots + \varepsilon^p \xi^i_p(x)\right)\frac{\partial}{\partial x^i} \cdot \tag{4}$$

In theoretical discussions, approximate equalities are considered with an error $o(\varepsilon^p)$ of an arbitrary order $p \geq 1$. However, in most of the applications, the theory is simplified by letting $p = 1$.

1.1.2 Approximate Lie Equations

Consider one-parameter approximate transformation groups in the first order of precision, i.e., Eqs. (1) of the form

$$\bar{x}^i \approx f_0^i(x,a) + \varepsilon f_1^i(x,a), \quad i = 1, \ldots, n. \tag{5}$$

Let

$$X = X_0 + \varepsilon X_1 \tag{6}$$

be a given approximate operator, where

$$X_0 = \xi_0^i(x)\frac{\partial}{\partial x^i}, \quad X_1 = \xi_1^i(x)\frac{\partial}{\partial x^i}\cdot$$

The corresponding approximate transformation (5) of points x into points $\bar{x} = \bar{x}_0 + \varepsilon \bar{x}_1$ with the coordinates

$$\bar{x}^i = \bar{x}_0^i + \varepsilon \bar{x}_1^i, \tag{7}$$

where

$$\bar{x}^i = f_0^i(x,a), \quad \bar{x}_1^i = f_1^i(x,a),$$

is determined by the following equations:

$$\frac{d\bar{x}_0^i}{da} = \xi_0^i(\bar{x}_0), \quad \bar{x}_0^i\big|_{a=0} = x^i, \quad i = 1, \ldots, n, \tag{8}$$

$$\frac{d\bar{x}_1^i}{da} = \sum_{k=1}^{n} \frac{\partial \xi_0^i(x)}{\partial x^k}\bigg|_{x=\bar{x}_0} \bar{x}_1^k + \xi_1^i(\bar{x}_0), \quad \bar{x}_1^i\big|_{a=0} = 0. \tag{9}$$

The equations (8)–(9) are called the *approximate Lie equations.*

Example 1.2 Let $n = 1$ and let

$$X = (1 + \varepsilon x)\frac{\partial}{\partial x}\cdot$$

Here $\xi_0(x) = 1$, $\xi_1(x) = x$, and Eqs. (8)–(9) are written:

$$\frac{d\bar{x}_0}{da} = 1, \quad \bar{x}_0\big|_{a=0} = x,$$

$$\frac{d\bar{x}_1}{da} = \bar{x}_0, \quad \bar{x}_1\big|_{a=0} = 0.$$

Its solution has the form

$$\bar{x}_0 = x + a, \quad \bar{x}_1 = ax + \frac{a^2}{2}\cdot$$

Hence, the approximate transformation group is given by

$$\bar{x} \approx x + a + \varepsilon\Big(ax + \frac{a^2}{2}\Big).$$

1.2 Approximate Symmetries

In this section, we will carry over the infinitesimal method [6, Vol.3] to approximate symmetries, i.e., approximate transformation groups admitted by differential equations with a small parameter ε. We will consider the approximation in the first order of precision in ε.

1.2.1 Definition of Approximate Symmetries

Definition 1.2 Let G be a one-parameter approximate transformation group

$$\bar{z}^i \approx f(z, a, \varepsilon) \equiv f_0^i(z, a) + \varepsilon f_1^i(z, a), \quad i = 1, \dots, N. \tag{10}$$

An approximate equation

$$F(z, \varepsilon) \equiv F_0(z) + \varepsilon F_1(z) \approx 0 \tag{11}$$

is said to be *approximately invariant* with respect to G, or *admits* G if

$$F(\bar{z}, \varepsilon) \approx (F(f(z, a, \varepsilon), \varepsilon) = o(\varepsilon)$$

whenever $z = (z^1, \dots, z^N)$ satisfies equation (11).

If $z = (x, u, u_{(1)}, \dots, u_{(k)})$, then (11) becomes an approximate differential equation of order k, and G is an *approximate symmetry group* of the differential equation.

1.2.2 Determining Equations. Stable Symmetries

Theorem 1.1 *Equation (11) is approximately invariant under the approximate transformation group (10) with the generator*

$$X = X^0 + \varepsilon X^1 \equiv \xi_0^i(z)\frac{\partial}{\partial z^i} + \varepsilon \xi_1^i(z)\frac{\partial}{\partial z^i}\,, \tag{12}$$

if and only if

$$\big[XF(z,\varepsilon)\big]_{F\approx 0} = o(\varepsilon),$$

or

$$\Big[X^0F_0(z) + \varepsilon\Big(X^1F_0(z) + X^0F_1(z)\Big)\Big]_{(11)} = o(\varepsilon). \tag{13}$$

The proof can be found in [5].

The operator (12) satisfying Eq. (13) is called an *infinitesimal approximate symmetry* of, or an *approximate operator admitted* by Eq. (11). Accordingly, Eq. (13) is termed the *determining equation* for approximate symmetries.

Remark 1.2 The determining equation (13) can be written as follows:

$$X^0F_0(z) = \lambda(z)F_0(z), \tag{14}$$

$$X^1F_0(z) + X^0F_1(z) = \lambda(z)F_1(z). \tag{15}$$

The factor $\lambda(z)$ is determined by Eq. (14) and then substituted in Eq. (15). The latter equation must hold for all solutions of $F_0(z) = 0$.

Comparing Eq. (14) with the determining equation of exact symmetries, we obtain the following statement.

Theorem 1.2 *If Eq. (11) admits an approximate transformation group with the generator $X = X^0 + \varepsilon X^1$, where $X^0 \neq 0$, then the operator*

$$X^0 = \xi_0^i(z)\frac{\partial}{\partial z^i} \tag{16}$$

is an exact symmetry of the equation

$$F_0(z) = 0. \tag{17}$$

Remark 1.3 It is manifest from Eqs. (14), (15) that if X^0 is an *exact* symmetry of Eq. (17) then $X = \varepsilon X^0$ is an *approximate* symmetry of Eq. (11).

Definition 1.3 Equations (17) and (11) are termed an *unperturbed equation* and a *perturbed equation,* respectively. Under the conditions of Theorem 1.2, the operator X^0 is called a *stable symmetry* of the unperturbed equation (17). The corresponding approximate symmetry generator $X = X^0 + \varepsilon X^1$ for the perturbed equation (16) is called a *deformation of the infinitesimal symmetry* X^0 of Eq. (17) caused by the perturbation $\varepsilon F_1(z)$. In particular, if the most general symmetry Lie algebra of Eq. (17) is stable, we say that the perturbed equation (11) *inherits the symmetries of the unperturbed equation.*

1.2.3 Calculation of Approximate Symmetries

Remark 1.2 and Theorem 1.2 provide an infinitesimal method for calculating approximate symmetries (12) for differential equations with a small parameter. Implementation of the method requires the following three steps:

1st step. Calculation of the exact symmetries X^0 of the unperturbed equation (17), e.g., by solving the determining equation

$$X^0 F_0(z)\Big|_{F_0(z)=0} = 0. \tag{18}$$

2nd step. Determination of the *auxiliary function* H by virtue of Eqs. (14), (15), and (11), i.e., by the equation

$$H = \frac{1}{\varepsilon}\left[X^0(F_0(z) + \varepsilon F_1(z))\Big|_{F_0(z)+\varepsilon F_1(z)=0} \right] \tag{19}$$

with known X^0 and $F_1(z)$.

3rd step. Calculation of the operators X^1 by solving the *determining equation for deformations*

$$X^1 F_0(z)\Big|_{F_0(z)=0} + H = 0. \tag{20}$$

Note that equation (20) unlike the determining equation (18) for exact symmetries is *inhomogeneous.*

1.3 Introduction to Renormgroup Symmetries

The Lie transformation group structure discovered by Stueckelberg and Petermann in the early 1950s in calculation results in renormalized quantum field theory and the exact symmetry of solutions related to this structure were used in 1955 by Bogoliubov and Shirkov to develop a regular method for improving approximate solutions of quantum field problems, the renormalization-group (RG) method. This method is based on the use of the infinitesimal form of the exact group property of a solution to improve a perturbative (that is, obtained by means of the perturbation theory) representation of this solution. The improvement of the approximation properties of a solution turns out to be most efficient in the presence of a singularity because the correct structure of the singularity is then recovered. Extending the RG concepts in quantum field theory to boundary value problems of classical mathematical physics led to the development of a regular algorithm for finding symmetries of the RG type by means of the modern theory of transformation groups. It is notable that the algorithm for the construction of renormalization-group (or renormgroup) symmetries proposed in [3, 4, 8] can be applied to problems involving differential and integral equations, as well as linear functionals of the solution. The use of the group property (the symmetry) of a solution underlies both the renormalization-group method in quantum field theory and its analog, the new renormgroup symmetry algorithm in mathematical physics. This section is motivated by a desire to draw attention to this fairly general algorithm based on applying symmetry to an approximate solution for enhancing its approximation power. Searching for renormgroup symmetries in most cases is based on the theory of approximate transformation groups.

1.3.1 The Renormalization-Group Algorithm in Mathematical Physics

We preface the description of the RG algorithm with the following simple argument. It is known that if we treat all the variables (independent or dependent in the standard sense) involved in a differential equation and their derivatives (called differential variables in group analysis) as independent, then the differential equation can be regarded as an algebraic relation for these variables. In the case of one equation, this relation describes a 'surface' in the extended space of all the variables involved in the equation (if there are several equations, then we speak of a manifold), and each solution of the equation defines a 'line' on this surface. The projection onto the $\{x, u\}$ 'plane' defines a family of curves, one of which passes through the 'point' $\{x_0, u_0\}$ corresponding to the boundary condition of the BVP in question.

Transformations of the group G move points on the surface (the manifold) along this surface, and therefore, the equation preserves its form in the transformed variables and each solution of the equation is taken into another solution. A transformation T_a from the group G maps a point on the plane $\{x, u\} \in \mathbb{R}^{n+m}$ into a point $\{\bar{x}, \bar{u}\}$, and the geometrical locus of these points is a continuous curve (a trajectory of the group G) passing through $\{x, u\}$. The locus of images $T_a(\{x, u\})$ is also called the G-orbit of the point $\{x, u\}$. In the general case, the motion along a group trajectory corresponds to the transition from one curve in the family to another, that is, to a 'multiplication' of solutions.

Returning to the renormalization-group point of view, we consider only the group transformations under which points on the curve passing through $\{x_0, u_0\}$ are moved along this curve. This means that the solution of the BVP is the RG orbit of the point $\{x_0, u_0\}$ (of the boundary manifold in the general case) and is an invariant RG manifold (similarly to the invariant charge in quantum field theory [9]). We use the infinitesimal version of this property in our construction of the RG symmetry.

The group property of a solution of a BVP manifests itself as follows: instead of the boundary point $\{x_0, u_0\}$ parameterizing the solution, we can take another point in this curve related to it by an RG transformation. This 'universality' of the solution of a BVP under a change of the way of parameterization is called 'functional self-similarity' [10]. To find RG transformations that map a solution of a BVP into a solution of the same BVP, we use the fact that a physical problem is formulated in terms of differential (integrodifferential) equations whose symmetries can be found by the techniques of group analysis.

RG Symmetry: An Idea of Construction and Its Simple Realization
We now illustrate the characteristic features of the algorithm for constructing an RG symmetry by an example of a BVP for the Hopf equation [11], which is widely used in physics for the description of the initial perturbations at the nonlinear stage of their evolution

$$\partial_t v + v\partial_x v = 0\,, \quad v(0,x) = \epsilon U(x)\,, \tag{21}$$

where U is an invertible function of x and the parameter ϵ defines the 'amplitude' of the initial perturbation 'at the boundary' $t = 0$. For a very small distance $t \ll 1/\epsilon$ from the boundary, the solution of problem (21) given by the perturbation theory is a segment of a power series

$$v = \epsilon U - \epsilon^2 t U \partial_x U + O\left(t^2\right)\,, \tag{22}$$

but this form becomes inapplicable for large t. The RG symmetry allows improving the perturbative result and recovering the correct behavior of the solution in a neighborhood of a singularity (when such a singularity occurs for some values of t).

In constructing an RG symmetry, the algorithm uses the symmetry group of the BVP equations. The boundary data defining a particular solution are involved in RG transformations by extending the space of the variables on which the group acts. In the case of BVP (21), this space involves three independent variables, $x = \{t, x, \epsilon\}$. It is convenient to write differential equation (21) for the function $u = v/\epsilon$ introduced such that the 'amplitude' ϵ is carried over from the boundary condition to the differential equation

$$\partial_t u + \epsilon u \partial_x u = 0\,, \quad u(0,x) = U(x)\,. \tag{23}$$

The general element of the transformation group G for Eq. (23) (for the basic manifold in the general case) can be found by means of the standard Lie techniques (see, e.g., [12]); it is given by a combination of four infinitesimal operators

$$\begin{gathered} X = \sum_i X_i\,, \quad X_1 = \psi^1\left(\partial_t + \epsilon u \partial_x\right)\,, \quad X_2 = \psi^2 \partial_x\,, \\ X_3 = \psi^3\left(x\partial_x + u\partial_u\right)\,, \quad X_4 = \psi^4\left(\epsilon\partial_\epsilon + x\partial_x\right)\,, \end{gathered} \tag{24}$$

where ψ^i $(i = 2, 3, 4)$ are arbitrary functions of ϵ, u and $x - \epsilon u t$ and ψ^1 is an arbitrary function of all the group variables $\{t, x, \epsilon, u\}$. We now use the RG invariance condition for a particular solution of BVP (23) defined by the relation

$$S \equiv u - W(t, x, \epsilon) = 0 \tag{25}$$

with the function W that is unknown at this point; in other words, we check that the RG transformation maps the solution of the BVP into the same solution. In the infinitesimal form, this condition can be written as

$$XS\Big|_{[S]} \equiv \psi^3(W - x\partial_x W) - \psi^2\partial_x W - \psi^4(\epsilon\partial_\epsilon W + x\partial_x W) = 0\,, \qquad (26)$$

where $|_{[S]}$ means that the result of the action of the operator is taken on the manifold defined by the equation $S = 0$ and all its differential consequences. The term containing ψ^1 is absent in (26) because it is proportional to $\partial_t W + \epsilon W\partial_x W$, which vanishes identically on solutions of Eq. (23). Condition (26) holds for all t, and for $t \to 0$ in particular, when W is replaced by the approximate solution

$$W = U - \epsilon t U\partial_x U + O\left(t^2\right) \qquad (27)$$

obtained in the framework of perturbation theory (22). In this limit, Eq. (26) yields a relation for the functions ψ^i $(i = 2, 3, 4)$, which extends in the obvious fashion to $t \neq 0$

$$\psi^2 = -\chi(\psi^3 + \psi^4) + (u/\partial_\chi U)\psi^3\,, \quad \chi = x - \epsilon ut\,, \qquad (28)$$

where the derivative $\partial_\chi U$ must be expressed in terms of χ or u in accordance with the boundary conditions. Using (28) in (24), we arrive at a group of a smaller dimension with the infinitesimal operators

$$\begin{aligned} &R = \sum_i R_i\,, \quad R_1 = \psi^1\left(\partial_t + \epsilon u\partial_x\right)\,, \\ &R_2 = u\psi^3\left[\left(\epsilon t + 1/\partial_\chi U\right)\partial_x + \partial_u\right]\,, \quad R_3 = \epsilon\psi^4\left(tu\partial_x + \partial_\epsilon\right)\,. \end{aligned} \qquad (29)$$

The above procedure reducing (24)–(29) is *the restriction of group* (24) *on a particular solution*, and the set of operators R_i in (29) describes the required RG symmetry. We obtain the solution of the BVP with the use of the corresponding Lie equations for any generator in (29). Without loss of generality, we can take the generator R_3 with $\epsilon\psi^4 = 1$ to obtain the finite RG transformations

$$x' = x + atu\,, \quad \epsilon' = \epsilon + a\,, \quad t' = t\,, \quad u' = u\,, \qquad (30)$$

where a is the group parameter, t and u are invariants, and the transformations of ϵ and x are translations, which in addition depend on t and u for the x variable. For $\epsilon = 0$, in view of (23), the variables x and u are related by $x = H(u)$, where $H(u)$ is the function inverse to $U(x)$. Eliminating a, t, and u from (30) and dropping the dashes in our notation for the variables, we obtain the required solution of BVP (23) in implicit form

$$x - \epsilon tu = H(u)\,. \qquad (31)$$

In effect, this is the improved perturbation theory solution (22), which can be used not only for small $t \ll 1/\epsilon$ provided that (31) defines u uniquely. Depending on $H(u)$, this solution either indicates the correct asymptotic behavior as $t \to \infty$ or gives the correct description of the solution in the neighborhood of finite values $t \to t_{sing}$. One example of the first option is the solution of the BVP for the linear function $U(x) = x$.

This yields the expression $v = \epsilon x(1 + \epsilon t)^{-1}$, which remains finite as $t \to \infty$. For the second option, we can select, for instance, a sine wave $U(x) = -\sin x$ at the boundary. Then solution (31) describes the well-known distortion of the initial profile of a sine wave, transforming it into a saw-tooth shape [13, Ch.6, §1], with a singularity forming at a finite distance $t_{sing} = 1/\epsilon$ from the boundary. We note that for finding solution (31) of the BVP, we use *only* the known symmetry of the solution and the corresponding perturbation theory (PT).

1.3.2 RG Algorithm: Four Steps Scheme

The above example of the construction of RG symmetries illustrates the general algorithm whose detailed description in relation to BVPs for differential equations can be found, e.g., in reviews [3, 8], and whose generalization to nonlocal problems is presented in [11, 14]. We can schematically express the implementation of the RG algorithm as a sequence of four steps (for details, see [4]):

(I) Constructing the *Basic Manifold* $\mathscr{RM}$

The initial issue is to construct the RG symmetry and appropriate transformations that involve the parameters of partial solution. Therefore, the purpose of step **I** is to include all the parameters, both from the equations and from the boundary conditions on which a particular solution depends, in group transformations in one or another way. This purpose is achieved by constructing a special manifold $\mathscr{RM}$ given by a system that consists of s kth-order differential equations and q nonlocal relations

$$F_\sigma(z, u, u_{(1)}, \dots, u_{(k)}) = 0\,, \qquad \sigma = 1, \dots, s\,, \tag{32}$$

$$F_\sigma(z, u, u_{(1)}, \dots, u_{(r)}, J(u)) = 0\,, \qquad \sigma = 1 + s, \dots, q + s\,. \tag{33}$$

The nonlocal variables $J(u)$ here are introduced by integral objects

$$J(u) = \int \mathscr{F}(u(z))\, dz\,. \tag{34}$$

(II) Finding a *Symmetry Group* G Admitted by $\mathscr{RM}$

Step **II** is to calculate the widest admitted group $\mathscr{G}$ for system (32), (33). In application to an $\mathscr{RM}$ defined only by system of differential equations (32), the question is about a local group of transformations in a space of differential functions $\mathscr{A}$, for which, system (32) remains unchanged. This group is defined by the generator of form (24) prolonged on all higher-order derivatives

$$X = \xi^i \frac{\partial}{\partial z^i} + \eta^\alpha \frac{\partial}{\partial u^\alpha} + \zeta_i^\alpha \frac{\partial}{\partial u_i^\alpha} + \zeta_{i_1 i_2}^\alpha \frac{\partial}{\partial u_{i_1 i_2}^\alpha} + \cdots\,, \tag{35}$$

where $\xi^i([z,u]), \eta^\alpha([z,u]) \in \mathscr{A}$ and

$$\zeta_i^\alpha = D_i(\varkappa^\alpha) + \xi^j u_{ij}^\alpha\,, \quad \zeta_{i_1 i_2}^\alpha = D_{i_1} D_{i_2}(\eta^\alpha - \xi^i u_i^\alpha) + \xi^j u_{j i_1 i_2}^\alpha\,.$$

The generalization of the second step of the algorithm to the case where $\mathscr{RM}$ is an integral or integrodifferential manifold is described in [4, 14].

(III) *Restricting the Symmetry Group* G on a Particular Solution of the BVP The group $\mathscr{G}$ found in step **II** and determined by operators (35) is generally wider than the RG of interest, which is related to a particular solution of a boundary value problem. Hence, to obtain the RG symmetry, we need step **III**, *restricting* the group $\mathscr{G}$ on a manifold determined by this particular solution. From the mathematical standpoint, this procedure consists of checking the vanishing conditions for a linear combination of coordinates $\varkappa_j^\alpha$ of a canonical operator equivalent to (35) on some particular boundary value problem solution $U^\alpha(z)$

$$\left\{ \sum_j A^j \varkappa_j^\alpha \equiv \sum_j A^j \left(\eta_j^\alpha - \xi_j^i u_i^\alpha\right) \right\}_{\big| u^\alpha = U^\alpha(z)} = 0\,. \tag{36}$$

The form of the condition set by relation (36) is common for any solution of the boundary value problem, but how the restriction procedure of a group is realized may differ in each partial case. In calculating combination (36) on a particular solution $U^\alpha(z)$, the latter is transformed from a system of differential equations for group invariants to algebraic relations. Note the two consequences of step **III**. First, the restriction procedure results in a set of relations between A^j and thus 'links' the coordinates of various group operators X_j admitted by $\mathscr{RM}$ (32), (33). Second, it (partially or completely) eliminates an arbitrariness that can arise in the values of the coordinates ξ^i and η^α in the case of an infinite group $\mathscr{G}$.

As a rule, the procedure of restricting the group $\mathscr{G}$ reduces its dimension. After performing this procedure, a general element (35) of a new group $\mathscr{RG}$ is represented by a linear combination of new generators R_i with coordinates $\hat{\xi}^i$ and $\hat{\eta}^\alpha$ and arbitrary constants B^j

$$X \Rightarrow R = \sum_j B^j R_j\,, \quad R_j = \hat{\xi}_j^i \frac{\partial}{\partial x^i} + \hat{\eta}_j^\alpha \frac{\partial}{\partial u^\alpha}\,. \tag{37}$$

The set of operators R_j, each containing the required solution of a problem in the invariant manifold, defines a group of transformations $\mathscr{RG}$, which we also call RenormGroup.

(IV) Finding *RG Invariant Solutions* Corresponding to the RG Symmetry The three steps described above completely form the regular algorithm for constructing the RG symmetry, but to finish, a final step is needed. This step **IV** uses the RG symmetry operators to find analytic expressions for new, improved boundary value problem solutions (compared with the input perturbative solution).

From the mathematical standpoint, realizing this step involves the use of *RG-invariance* conditions set by a *joint* system of Eqs. (32) and (33) and the vanishing conditions for a linear combination of the coordinates $\hat{\varkappa}_j^\alpha$ of the canonical operator equivalent to (37)

$$\sum_j R^j \hat{\varkappa}_j^\alpha \equiv \sum_j B^j (\hat{\eta}_j^\alpha - \hat{\xi}_j^i u_i^\alpha) = 0 . \tag{38}$$

Specification of step **IV** concludes the description of the regular algorithm of RG symmetries construction for models with differential and integrodifferential equations.

2 Nonlinear Acoustic Waves in Channels with Variable Cross Sections

In this section, the approximate point symmetry group is studied for the generalized Webster-type equation describing nonlinear acoustic waves in lossy channels with variable cross sections. Approximate analytic solutions to the generalized Webster equation are derived for channels with smoothly varying cross sections and arbitrary initial conditions. A more detailed discussion of the problem can be found in [15].

2.1 Generalized Webster Equation in Nonlinear Acoustics

The generalized Webster-type equation appears in problems on propagation of intense sound in pipes, horns, concentrators, and other waveguides characterized by a varying cross section $S(x)$ [16, 17] and in hemodynamics for describing the nonlinear pulse waves [18]

$$\frac{\partial^2 p}{\partial t^2} - c^2 \frac{\partial^2 p}{\partial x^2} = c^2 \frac{\partial \ln S(x)}{\partial x} \frac{\partial p}{\partial x} + \frac{\varepsilon}{c^2 \rho} \frac{\partial^2 p^2}{\partial t^2} + \frac{b}{\rho} \frac{\partial^3 p}{\partial x \partial t^2} . \tag{39}$$

Here, p is the sound pressure, c is the velocity of sound, x is the coordinate measured along the waveguide axis, ε and b are the nonlinearity and dissipation parameters and ρ is the density of the medium. Equation (39) is applicable to pipes whose characteristic width is small compared to the wavelength. In addition, the cross section is assumed to vary slowly along the x-axis: the area $S(x)$ varies only slightly as x varies by a quantity on the order of the pipe width [19].

In the situation where each of the terms appearing on the right-hand side of the equation is small compared to the terms appearing on the left-hand side, a traveling wave can be considered. In this case, using the method of a slowly varying profile [20], it is possible to reduce the order of nonlinear equation (39). Following the standard procedure [21], we change from the variables x and t involved in Eq. (39) to new independent variables: the 'slow' coordinate $x_1 = \delta x$ (where δ is the small parameter of the problem) and the time $\tau = t - x/c$ in the coordinate system traveling with the velocity of sound. Ignoring small terms on the order of δ^n, where $n \geqslant 2$, we arrive at the evolution equation

$$\frac{\partial p}{\partial x} - ap\frac{\partial p}{\partial \tau} - \nu\frac{\partial^2 p}{\partial \tau^2} + \frac{p}{2}\frac{\partial}{\partial x}(\ln S(x)) = 0\,, \quad p(0,\tau) = P(\tau)\,. \tag{40}$$

Unlike the equations discussed above, Eq. (40) is expressed in dimensionless form. The change from the physical variables involved in Eq. (39) to the more convenient normalized variables appearing in Eq. (40) is performed through the following substitution:

$$x \to \frac{c}{\omega}x\,, \quad \tau \to \frac{\tau}{\omega}\,, \quad p \to p_0 p\,.$$

Here, the normalizing constants ω and p_0 have the meaning of the characteristic frequency and signal amplitude values, respectively. The two parameters involved in Eq. (40) are determined by the following dimensionless combinations of constants:

$$a = \frac{\varepsilon p_0}{c^2\rho}\,, \quad \nu = \frac{b\omega}{2c^2\rho}\,.$$

Their ratio a/ν is called the acoustic Reynolds number [20]. It characterizes the relative contributions of nonlinear and dissipative effects to the distortion of the wave profile. When a/ν is large, nonlinear effects predominate; when this quantity is small, dissipative effects are dominant. Without loss of generality, in Eq. (40), we set $S(0) = 1$.

Although solutions to the GWE simultaneously allowing for the effects of nonlinearity, absorption, and inhomogeneity are important for describing the behavior of sound waves in nonlinear absorbing media, their analytic derivation is a difficult problem even in the case of using approximate methods. The standard practice is either to neglect dissipation, which makes it possible, by changing the independent variable to reduce the initial equation to the Hopf equation [17], or to assume that the nonlinearity is small and to solve the sound wave equation by the method of successive approximations; the latter approach has been used to analyze the second harmonic behavior in a sound channel with a variable cross section [22]. On the other hand, it is evident that, as the cross section of the channel decreases, the second harmonic amplitude grows faster than the fundamental harmonic amplitude, which necessitates analyzing the generalized Webster equation for a finite-amplitude sound wave. This section is devoted to finding an analytic solution to the generalized Webster equation under these conditions using the modern group analysis technique.

We can eliminate the last term from Eq. (40) by changing the variable x and introducing the absorption as a function μ of the coordinate along the channel

$$\zeta = \int \mathrm{d}x/\sqrt{S(x)}\,, \quad p\sqrt{S} = u\,, \quad \mu = \nu\sqrt{S(x(\zeta))}\,, \tag{41}$$

Then, in terms of the new variables, Eq. (40) takes the form

$$\frac{\partial u}{\partial \zeta} - au\frac{\partial u}{\partial \tau} - \mu\frac{\partial^2 u}{\partial \tau^2} = 0\,, \quad u(0,\tau) = P(\tau)\,. \tag{42}$$

Introducing the new variable q related to u by the formula $u = 2(\partial q/\partial \tau)$, we replace Eq. (40) by the modified generalized Webster equation (MGWE)

$$\frac{\partial q}{\partial \zeta} - a\left(\frac{\partial q}{\partial \tau}\right)^2 - \mu\frac{\partial^2 q}{\partial \tau^2} = 0\,, \quad q(0,\tau) = W(\tau)\,. \tag{43}$$

The change to the new variable q in Eq. (42) increases the order of this equation. However, its single integration with respect to τ yields evolutional equation (43) for q. This procedure determines q accurate to the function $C(\zeta)$ (in Eq. (43), this function is omitted), whose choice is fairly arbitrary. For example, for solutions to Eq. (43) that are periodic in τ, this function can be chosen so as to make the period average value of q zero for any values of ζ. At the same time, it is evident that the choice of $C(\zeta)$ does not affect the physical meaning of u.

2.2 *Approximate Symmetry Group and Solutions to the Modified Generalized Webster Equation*

For Eq. (43) in the case of an arbitrary inhomogeneity profile $\mu(\zeta)$, the admitted point transformation group is given by three infinitesimal operators:

$$X_1 = \frac{\partial}{\partial \tau}\,, \quad X_2 = \frac{\partial}{\partial q}\,, \quad X_3 = \zeta\frac{\partial}{\partial \tau} - \frac{\tau}{2a}\frac{\partial}{\partial q}\,. \tag{44}$$

The first two of them represent the translation operators with respect to the variables τ and q, which are evident from the physical point of view; the third operator corresponds to the Galilean transformation group. Transformation group (44) can be extended for the cross section profiles of a specific type

$$M(\frac{\mathrm{d}}{\mathrm{d}\zeta}\ln(\mu(\zeta)))^{-1} = b(\zeta)\,, \quad b(\zeta) = \beta_0 + \beta_1\zeta + \beta_2\zeta^2\,, \quad M = \mathrm{const} \neq 0\,. \tag{45}$$

Condition (45) plays the role of the classifying relation separating the specific profile types for which transformation group (44) is extended by the additional operator X_4

$$X_4 = b\frac{\partial}{\partial\zeta} + \frac{\tau}{2}\big(M + \frac{\mathrm{d}b}{\mathrm{d}\zeta}\big)\frac{\partial}{\partial\tau} + \big(Mq - \frac{(\tau^2 + 2\int \mathrm{d}\zeta\,\mu)}{8a}\frac{\mathrm{d}^2 b}{\mathrm{d}\zeta^2}\big)\frac{\partial}{\partial q}\,. \tag{46}$$

Classifying relation (45) is a first-order differential equation for the function $\mu(\zeta)$; being explicitly integrated, this equation determines a three-parameter (the parameters are β_0, β_1, and β_2) family of curves in the $\{\zeta, \mu\}$ space

$$\ln(\mu/\nu) = d(\zeta) \equiv M\int_{\zeta}^{0}\frac{\mathrm{d}y}{b(y)}\,, \tag{47}$$

where the form of the function $d(\zeta)$ depends on the relative values of the parameters β_i

$$d(\zeta) = \begin{cases} \frac{2M}{\sqrt{4\beta_0\beta_2-\beta_1^2}}\left[\arctan\frac{\beta_1+2\beta_2\zeta}{\sqrt{4\beta_0\beta_2-\beta_1^2}} - \arctan\frac{\beta_1}{\sqrt{4\beta_0\beta_2-\beta_1^2}}\right], & \beta_1^2 < 4\beta_0\beta_2\,, \\ \frac{M}{\sqrt{\beta_0\beta_2}}\frac{z}{z+\sqrt{\beta_0/\beta_2}}\,, & \beta_1^2 = 4\beta_0\beta_2\,, \\ \frac{M}{\sqrt{\beta_1^2-4\beta_0\beta_2}}\ln\frac{(\sqrt{\beta_1^2-4\beta_0\beta_2}-\beta_1-2\beta_2\zeta)(\sqrt{\beta_1^2-4\beta_0\beta_2}+\beta_1)}{(\sqrt{\beta_1^2-4\beta_0\beta_2}+\beta_1+2\beta_2\zeta)(\sqrt{\beta_1^2-4\beta_0\beta_2}-\beta_1)}\,, & \beta_1^2 > 4\beta_0\beta_2\,. \end{cases} \tag{48}$$

The choice of $M = 0$ corresponds to a channel with a constant cross section, $\mathrm{d}\mu(\zeta)/\mathrm{d}\zeta = 0$; then, classifying relation (45) is automatically satisfied for any β_i. In this case, the following three operators appear instead of the operator X_4:

$$X_{41} = \frac{\partial}{\partial\zeta}\,, \quad X_{42} = \zeta\frac{\partial}{\partial\zeta} + \frac{\tau}{2}\frac{\partial}{\partial\tau}\,, \quad X_{43} = \zeta^2\frac{\partial}{\partial\zeta} + \tau\zeta\frac{\partial}{\partial\tau} - \frac{(\tau^2 + 2\zeta\nu)}{4a}\frac{\partial}{\partial q}\,. \tag{49}$$

The first of them, X_{41}, is the translation operator along the ζ axis; the second operator, X_{42}, represents the dilation transformation; and X_{43} corresponds to the projective transformation group. In addition to operators (49), for a channel with a constant cross section, $\mu \equiv \nu$, the MGWE also allows the infinite subgroup operator

$$X_\infty = k(\zeta, \tau)\exp\left(-\frac{aq}{\mu}\right)\frac{\partial}{\partial q}\,, \qquad \frac{\partial k}{\partial\zeta} - \mu\frac{\partial^2 k}{\partial\tau^2} = 0\,. \tag{50}$$

Here, the linear parabolic equation, which is satisfied by the function of two variables $k(\zeta, \tau)$, can be represented in terms of the variables $\{x, \tau\}$:

$$\frac{\partial k}{\partial x} - \nu\frac{\partial^2 k}{\partial\tau^2} = 0\,.$$

The latter fact will be used by us in constructing the approximate point symmetry for the MGWE. We note that the symmetry group given by Eqs. (44), (49), and (50) is well known in the theory of the modified Burgers equation [6], to which the MGWE is reduced in the case under consideration.

Different examples of invariant solutions to the MGWE based on the symmetries (44)–(49) can be found in [15–17, 22, 23]. These invariant solutions to the MGWE have the following drawback: being exact, they can only be obtained for certain specific channel profiles and initial conditions, which are given by classifying relation (45). Here, we present alternative solutions, namely, approximate analytic solutions to nonlinear boundary value problem (43), and these solutions can be constructed for arbitrary initial conditions. An instrument for constructing such solutions is the approximate symmetry group, and the condition for the existence of the latter is the presence of the small parameter related to the relative slowness of variation in the cross-sectional area along the waveguide axis; i.e., the smallness of the derivative $\mathrm{d}(\ln \mu(\zeta))/\mathrm{d}\zeta \equiv \mu_\zeta/\mu \ll 1$ (here, the subscript denotes the derivative with respect to the corresponding argument). In this case, the symmetry of the boundary value problem under study is represented by a series expansion in powers of the small parameter, which allows us to obtain approximate analytic solutions to the problem with an acceptable accuracy.

To construct an approximate analytic solution for a channel with a slowly varying cross section, we use the renormalization-group symmetry algorithm described in Sect. 1.1 for boundary value problem (43). This algorithm allows us to extend the perturbation theory solutions obtained for small values of the nonlinearity parameter a to the region of finite values of this parameter. To take into account the transformation of parameter a, we include the latter in the list of independent variables and represent the infinitesimal operator of this transformation as

$$X_5 = \xi(a)\left(\frac{\partial}{\partial a} - \frac{q}{a}\frac{\partial}{\partial q}\right). \tag{51}$$

The desired renormalization-group symmetry operator is obtained as a linear combination of operator (51) with $\xi(a) = 1$ and infinite subgroup operator (50), which (as we have shown in the previous section), in the zero order in μ_ζ/μ, is allowed by boundary value problem (43)

$$R = \frac{\partial}{\partial a} + \left(k^{(0)}(\zeta, \tau, a)\exp\left(-\frac{aq}{\mu}\right) - \frac{q}{a}\right)\frac{\partial}{\partial q}. \tag{52}$$

Here, the function of three variables $k^{(0)}(\zeta, \tau, a)$ obeys linear parabolic equation (50) with the initial condition $k^{(0)}(0, \tau, a) = W(\tau)/a$ determined by invariance of the solution at $a \to 0$ with respect to the renormalization-group symmetry operator (52). As a result, we arrive at the expression

$$k^{(0)} = \frac{\nu}{a} K_a\,, \quad K(a,x,\tau) = \int\limits_{-\infty}^{\infty} \mathrm{d}\xi \mathrm{e}^{\frac{aW(\xi)}{\nu}} G(x,\tau-\xi)\,, \quad G(x,\tau) = \frac{1}{\sqrt{4\pi\nu x}} \mathrm{e}^{-\frac{\tau^2}{4\nu x}}\,. \tag{53}$$

Here, the subscript marking function K denotes the partial derivative with respect to the corresponding argument, $K_a \equiv \partial K/\partial a$.

Finite transformations of the continuous group are related to the infinitesimal transformation in a one-to-one manner by the Lie equations, i.e., equations of the characteristics for the first-order partial differential equation conjugate to the operator (52), (53). The solution of the Lie equations for operator (52), (53) yields the following approximate analytic solution to initial problem (43):

$$q^{(0)} = \frac{\mu}{a} \ln\left[1 + \frac{\nu}{\mu}(K-1)\right], \tag{54}$$

which is valid in a medium with a slowly varying cross section, $\mu_\zeta/\mu \ll 1$. In fact, the derivation of solution (54) from the solution obtained for a channel with a constant cross section consists of the presence of a factor ν/μ other than unity.

The advantage of the renormalization-group method is the possibility of sequentially refining the resulting analytic approximations. As applied to the problem under study, such a refinement (in the next, i.e., first-order approximation in μ_ζ/μ) is achieved as follows: the function $k^{(0)}(\zeta,\tau,a)$ in generator (52) is replaced by $k^{(1)}(\zeta,\tau,a)$, for which, instead of using the solution to the parabolic equation (50), we use the solution to the inhomogeneous parabolic equation

$$\frac{\partial k^{(1)}}{\partial \zeta} - \mu \frac{\partial^2 k^{(1)}}{\partial \tau^2} = -A\,. \tag{55}$$

Here right-hand side is proportional to the gradient of the channel cross section $\mu_\zeta/\mu \ll 1$ and linearly depends on the function $q^{(0)}$ of the zero-order approximation in this gradient

$$A = ak^{(0)} q^{(0)} \mu_\zeta/\mu^2 = (\mu_\zeta/\mu)(\nu/a)K_a \ln\left[1 + (\nu/\mu)(K-1)\right]\,. \tag{56}$$

Equation (55) is obtained at the stage of calculating the renormalization-group symmetry operator (52) from the so-called *group determining equation*, where the contributions proportional to μ_ζ/μ (which were omitted at the previous step) are calculated using the zero approximation results (53) and (54). The solution to Eq. (55) yields a modified (due to the contribution with the cross section gradient) expression for the function $k^{(0)}(a,x,\tau) \Rightarrow k^{(1)}(a,x,\tau)$

$$k^{(1)} = \frac{\nu}{a} K_a - \int\limits_x^0 dx' \int\limits_{\infty}^{-\infty} d\tau' G(x - x', \tau - \tau') \frac{\nu \mu'_{x'}}{a\mu'} K'_a \ln\left(1 + \frac{\nu}{\mu'}(K' - 1)\right),$$
$$\mu' \equiv \mu(x'), \quad K' \equiv K(a, x', \tau'). \tag{57}$$

The substitution of $k^{(1)}$ instead of $k^{(0)}$ in infinitesimal operator (52) and the subsequent solution of the Lie equations lead to a refined approximation for the desired solution

$$q^{(1)} = \frac{\mu}{a} \ln\left\{1 + \frac{\nu}{\mu}(K - 1) - \frac{\nu}{\mu} \int\limits_x^0 dx' \frac{\mu'_{x'}}{\mu'} \int\limits_{\infty}^{-\infty} d\tau' G(x - x', \tau - \tau') \right. \tag{58}$$
$$\left. \times \left[1 - K' + \left(K' - 1 + \frac{\mu'}{\nu}\right) \ln\left(1 + \frac{\nu}{\mu'}(K' - 1)\right)\right]\right\}.$$

For small values of the nonlinearity parameter a, the first two terms of the expansion of solution (58) in powers of the nonlinearity parameter have the form

$$q^{pt} = \nu K_a^{(0)} + \frac{\nu a}{2}\left[K_{aa}^{(0)} - \frac{\nu}{\mu}(K_a^{(0)})^2 \right. \tag{59}$$
$$\left. - \int\limits_0^x dx' \frac{\nu \mu'_{x'}}{(\mu')^2} \int\limits_{\infty}^{-\infty} d\tau' G(x - x', \tau - \tau')((K')_a^{(0)})^2\right] + O(a^2),$$

where $K_a^{(0)}$ and $K_a^{(0)}$ represent the values of the partial derivatives of function K calculated in the limit a $a \to 0$. By substituting the periodic initial condition $W(\xi) = \cos\xi$ in K and calculating the resulting integrals, it is possible to verify that the expression for q^{pt} agrees well with the result obtained in [22] for a harmonic initial perturbation.

In closing this section, we present the form of the solution to problem (43) for a periodic initial condition $W(\xi) = \cos\xi$: it is given by Eq. (58) with function K determined by the expression

$$K = I_0(a/\nu) + 2\sum_{k=1}^{\infty} I_k(a/\nu) \cos k\tau \, e^{-\nu k^2 x}. \tag{60}$$

The use of Eqs. (58) and (60) allows us to calculate the nonlinear distortion of the spectrum of an acoustic wave propagating in a waveguide with a varying cross section. This opens up better possibilities for diagnostics of acoustic propagation paths as compared to the weakly nonlinear limit [22]. As an example, in Fig. 1, we represent the solutions to the MGWE, $q \equiv q^{(1)}$, versus τ, the solutions being obtained from Eqs. (58) and (60) for different values of the x coordinate along the axis of a channel with an exponentially varying cross-sectional area, $\mu/\nu = \exp(\alpha x)$.

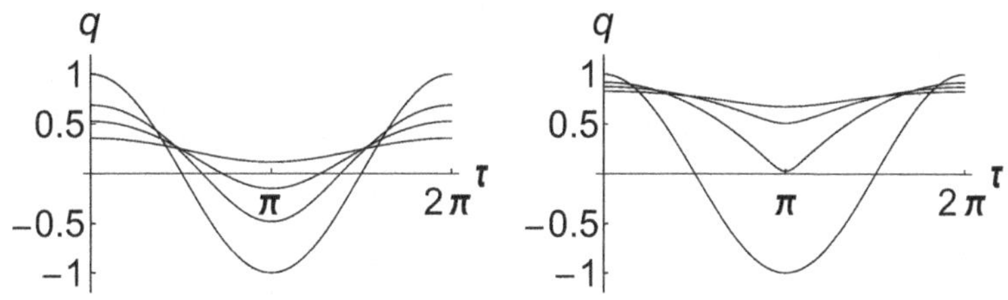

Fig. 1 Curves representing the solution to modified generalized Webster equation (43) versus quantity τ. The curves are obtained from Eqs. (58) and (60) for the periodic initial condition $W(\xi) = \cos\xi$ for different values of the x coordinate along the axis of the channel whose cross-sectional area varies according to the exponential law $\mu(x)/\nu = \exp(\alpha x)$. The left-hand plot shows the curves for $\nu x = 0, 0.5, 1$, and 2 (the increase in νx corresponds to the passage from the upper curves to the lower ones at the axis $\tau = 0$) and for $\alpha/\nu = -0.1$ and $a/\nu = 1$. The right-hand plot shows the curves for $\nu x = 0, 0.2, 0.5$, and 1 (the increase in νx corresponds to the passage from the upper curves to the lower ones at the axis $\tau = 0$) and for $\alpha/\nu = -0.1$ and $a/\nu = 10$

To estimate the accuracy of the approximate analytic solutions obtained by us, we numerically solved initial equation (43). Comparison of the curves plotted with the use of analytic results (58) and (60) with the curves obtained from the numerical solution of initial equation (43) shows good agreement between the numerical and analytic results for the case of a moderate nonlinearity $a/\nu = 1$ with an accuracy of up to fractions of percent. The results of calculating $q \equiv q^{(1)}$ for a greater nonlinearity $a/\nu = 10$ show a difference between the results of numerical and analytic calculations; the difference increases with increasing distance along the pipe, as one can see from the comparison of the curves shown in Fig. 2 at the left. However, the value of the difference is relatively small and, even for $x = 2$, it is of the order of seven percent. We also note that the strongest effect of nonlinearity, which makes the wave profile steeper, manifests itself as early as in the region of x where the difference between numerical and analytic results is small: for example under consideration with $a/\nu = 10$, the corresponding value of the νx coordinate proves to be in the order of 0.08. To illustrate the possibility of increasing the accuracy of analytic calculations with the use of approximate symmetry, in Fig. 2 (right), we present the curves showing the difference between the zero-approximation solution $q^{(0)}$ and the solution q obtained from numerical simulation. One can see that the change from $q^{(0)}$ to $q^{(1)}$ already considerably improves the agreement between the numerical results and the approximate analytic ones. Summarizing the results, we note that the application of renormalization-group symmetries allowed us to determine approximate analytic solutions for arbitrary initial conditions and sufficiently smooth profiles of variation in the waveguide cross section and to demonstrate the way of refining the approximate solutions.

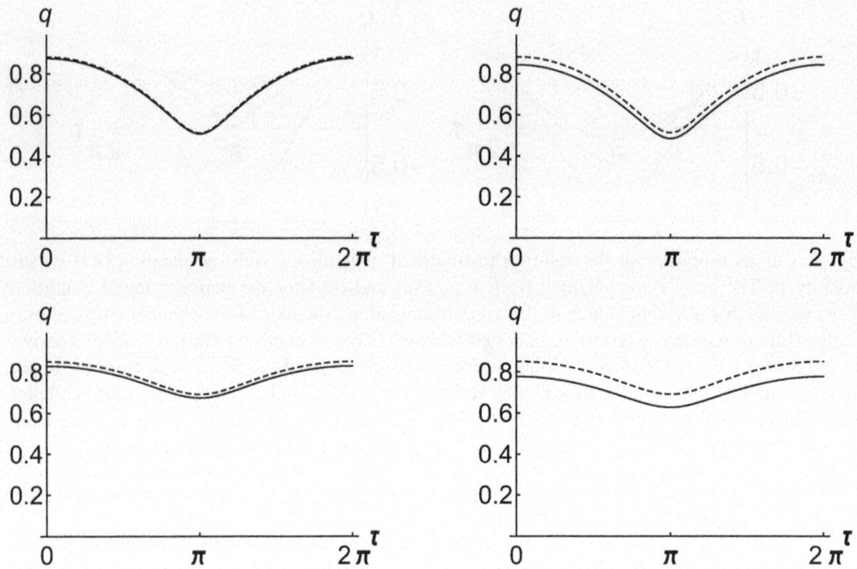

Fig. 2 Comparison between obtained solutions q to modified generalized Webster equation (43) by numerically solving Eq. (43) (dashed curves) and the approximate analytic solutions $q \equiv q^{(1)}$ (solid curves in the left-hand plot) and $q \equiv q^{(0)}$ (solid curves in the right-hand plot) for $\alpha/\nu = -0.1$ and $a/\nu = 10$, and for different values of the coordinate: $\nu x = 0.5$ and 1. The increase in the νx coordinate along the pipe axis corresponds to the passage from the upper plots to the lower ones

3 Acceleration of Ions in a Plasma Channel

In this section, the approximate analytic solution of the Cauchy problem is constructed for a system of kinetic equations of an electron–ion plasma that describe the acceleration of ions and the collisionless heating of electrons caused by the radial ponderomotive force of a laser beam that propagates in the transparent plasma of a gas or other low density target. Under conditions where the Debye radius, r_{De}, of the electrons is considerably smaller than the characteristic localization scale, L, of the laser beam along the radius, $\epsilon = r_{De}/L \ll 1$, this solution is found by a group transformation that is specified by the operator of approximate renormalization-group symmetries over small parameters, ϵ and $\mu = \sqrt{Zm/M}$, of the initial distribution functions of particles. For an axially symmetric geometry of the laser beam, the temporal and spatial dependencies of the distribution functions of particles are obtained and their integral characteristics, such as the density, mean velocity, temperature, and energy spectrum, are found. The formation of a cylindrical density cusp and the localized heating of electrons at the laser channel boundary are analytically described.

The acceleration dynamics of ions in a low density plasma is mainly determined by the force (ponderomotive) action of the laser pulse on plasma electrons. A typical example is the radial acceleration of particles from the laser channel (the traditionally studied rarefied gas plasma [25, 26] or the plasma of new generation low density targets such as aerogels and porous nanocarbon), which are considered in this section. The analytic description of the acceleration of particles by a laser pulse from a plasma channel that is produced via self focusing is not a simple problem even for approximate approaches. That is the reason that the spatiotemporal distributions of laser accelerated particles are studied, as a rule, using kinetic numerical particle in cell (PIC) simulation [27–29]. Some simplification is achieved by using a one-dimensional electrostatic ponderomotive model that describes the plasma expansion dynamics under the action of a radial distribution of pulsed laser radiation [25, 26] along the plasma channel radius. In this case, only the slow dynamics of plasma electrons is taken into account, corresponding to averaging over their fast oscillations in the laser field. Despite the simplicity of such a description, the main results that are obtained in the electrostatic ponderomotive model are also based on numerical PIC simulations, which makes the prediction of the dependencies of the spatiotemporal and spectral characteristics of accelerated particles on arbitrary laser and plasma parameters difficult. Numerical simulations have revealed two distinct effects: (1) The acceleration of ions by a ponderomotive force caused the formation of a cylindrical density cusp at the laser channel boundary [25, 26], where (2) the strong local heating of electrons occurs [26]. Here, we propose an analytic theory using approximate RG symmetries to find the solutions of kinetic equations for the distribution functions of plasma electrons and ions in the electrostatic ponderomotive model and present the analytic description of these effects. The model of radial ponderomotive acceleration of particles from a laser channel considers, along with the self-consistent electric field of plasma in the kinetic equation for electrons that is averaged over fast laser oscillations, the 'outside' electric field that determines the action of the radial ponderomotive force of the laser beam on plasma electrons. The plasma is assumed to be of low density and transparent for the laser beam; the reverse action of plasma fields on the laser beam is neglected.

3.1 The Approximate Renormgroup Symmetry for Solution of the Cauchy Problem for Plasma Equations

We consider the propagation of a cylindrically symmetric laser beam in a transparent plasma. The inhomogeneity of the electric field of laser radiation along the radius results in the displacement of plasma electrons from the region of a strong electric field, producing inhomogeneity of the electron density, which, in turn, causes the redistribution of the ion density and the acceleration of ions to high energies. The dynamics of this process can be described by kinetic equations for the distribution function of plasma particles, in which the influence of a laser's electric field is taken into account in the kinetic equation for electrons in the form of an additional electrostatic term that determines the action of a ponderomotive force averaged over the high laser frequency [30, 31]. Taking the symmetry of the laser beam that propagates along the z axis into account, we will consider kinetic equations for the distribution functions of particles that are integrated over the longitudinal and axial velocity components in the cylindrical coordinate system $\{r, \varphi, z\}$, including their dependencies only on time, t, the radial coordinate, r, and the radial component, $v_r^{e(i)}$, of the velocity of particles of the corresponding type. As a result, we obtain initial kinetic equations for the electron and ion distribution functions of plasma particles,[1] $g(\tau, x, u)$ and $f(\tau, x, w)$,

$$\begin{aligned} \mu \partial_\tau g + \varepsilon u \left(\partial_x g + g/x\right) - \left(p + q - \varepsilon/x\right) \partial_u g = 0\,, \\ \partial_\tau f + \varepsilon w \left(\partial_x f + f/x\right) + \left(p + \varepsilon \Gamma^2/x\right) \partial_w f = 0\,. \end{aligned} \tag{61}$$

Here, m and M are the electron and ion masses with charges $e^e = -e$ and $e^i = Ze$, where Z is the charge number of ions; $\mu = \sqrt{Zm/M} \ll 1$ and $\Gamma^2 = T_{i0}/ZT_{e0}$. In Eqs. (61), dimensionless variables are used: the dimensionless time $\tau = \omega_{Li} t$, where ω_{Li} is the Langmuir ion frequency; the dimensionless coordinate $x = r/L$, where L is the localization scale of the laser beam along the radius; the dimensionless electron velocity $u = v_r^e / V_{Te}$, where $V_{Te} = \sqrt{T_{e0}/m}$; the dimensionless ion velocity $w = v_r^i / c_s$, where $c_s = \sqrt{ZT_{e0}/M}$; the dimensionless electric field $p = \varepsilon (eEL/T_{e0})$, where $\varepsilon = r_{De}/L \ll 1$; the dimensionless distribution functions $f^e = (n_{e0}/V_{Te}) g$ and $f^i = (n_{e0}/(Zc_s)) f$, where $n_{e(i)0}$ is the unperturbed electron (ion) density; $q = \alpha \partial_x \gamma$, where $\alpha = \varepsilon (c^2/V_{Te}^2)$, c is the speed of light, $\gamma = \sqrt{1 + a^2(\tau, x)/2}$, $a^2(\tau, x) = A(\tau) a_0^2 I_0(x)$ is the dimensionless laser radiation intensity, $a_0 = 0.85 \times 10^{-9} \lambda \sqrt{I}$, where λ [μm] is the laser wavelength, I [W/cm^2] is the maximum intensity of the laser pulse, and the function $A(\tau)$ determines the laser pulse shape. The function $I_0(x)$ characterizes the radial distribution of the laser pulse intensity. For example, in [26] the distribution $I_0(x) = \exp(-x^2)$ was analyzed. Kinetic equations (61) should be used together with equations for the self-consistent electric field p `

$$\varepsilon \left(\partial_x p + p/x\right) = \int \mathrm{d} w f - \int \mathrm{d} u g\,, \quad \mu \partial_\tau p = -\mu \int \mathrm{d} w w f + \int \mathrm{d} u u g\,. \tag{62}$$

Equations (61) and (62) are known as the Vlasov–Maxwell system of equations for a collisionless plasma. We are interested in the solution of the Cauchy problem for kinetic equations (61) with initial conditions

$$g|_{\tau=0} = g_0(x, u)\,, \quad f|_{\tau=0} = f_0(x, w)\,, \tag{63}$$

which are determined by the formulation of a particular physical problem. Below, we will analyze the evolution of plasma particles for which the initial distribution functions of particles are assumed to be sufficiently smooth (for example, Maxwellian),

---

[1] For definiteness, we assume that the distributions of particles over the velocities $v_\varphi^{e,i}$ and $v_z^{e,i}$ are Maxwellian, with the electron and ion temperatures T_e and T_i, respectively. In the absence of the distribution over the velocities $v_\varphi^{e,i}$, the contributions that are proportional to $\propto \varepsilon \partial_u g$ and $\varepsilon \partial_w f$ in (61) should be omitted.

with the spatially homogeneous initial electron and ion temperatures T_{e0} and T_{i0} and initial electron and ion densities $n_0^e(x) = \int du g_0$ and $n_0^i(x) = \int dw f_0$ with the characteristic spatial scale, L. We consider the typical situation where L considerably exceeds the Debye electron radius $r_{De} = \sqrt{T_{e0}/(4\pi n_{e0} e^2)}$, i.e., $r_{De}/L \ll 1$. Note that the physical formulation of the problem consists of the specification of the finite initial temperature of particles because a propagating high power laser pulse is always preceded by a long prepulse that heats the plasma prior to the arrival of the main pulse.

Equations of type (61) are usually studied by the method of characteristics (see, for example, [32]). The equations of characteristics for (61) contain the electric field, which is expressed in terms of integrals from the velocity distribution functions via Eqs. (62), which makes the analytic study of these equations difficult and compels us to use numerical methods. Here, the Cauchy problem (61)–(63) is solved using point symmetry groups (exact and approximate) that are admitted by Eqs. (61) and (62) and constructing invariant solutions with their use. The approximate solution of Cauchy problem (61)–(63) is constructed using the possibility of the continuation of this solution written in a small vicinity $\tau \to 0$ as a perturbation power series in τ to a region of considerably longer times $\tau \neq 0$ with the use of a special renormalization-group symmetry. Such a continuation is performed using finite transformations of the group connecting initial distribution functions (63) with the values of these functions at moments $\tau \neq 0$. The required renormalization-group symmetry is found as a subgroup that is admitted by the system (61), (62) of the group of approximate Lie point transformations.

To find an admissible point transformation group, the system of equations (61), (62) should be supplemented with four equalities

$$\partial_w g = 0\,, \quad \partial_u f = 0\,, \quad \partial_u p = 0\,, \quad \partial_w p = 0\,, \tag{64}$$

that have a trivial physical meaning, but are obviously needed for calculating the group of admissible transformations. The Lie point transformation group that is admitted by the system of equations (61), (62) and (64) is defined by an infinitesimal group operator (generator)

$$X = \xi^1 \partial_\tau + \xi^2 \partial_x + \xi^3 \partial_u + \xi^4 \partial_w + \eta^1 \partial_g + \eta^2 \partial_p + \eta^3 \partial_f\,. \tag{65}$$

This operator in the canonical form is

$$Y = \kappa^1 \partial_g + \kappa^2 \partial_p + \kappa^3 \partial_f\,, \tag{66}$$

$$\begin{aligned}
\kappa^1 &= \eta^1 - \xi^1 \partial_\tau g - \xi^2 \partial_x g - \xi^3 \partial_u g\,, \\
\kappa^2 &= \eta^2 - \xi^1 \partial_\tau p - \xi^2 \partial_x p\,, \\
\kappa^3 &= \eta^3 - \xi^1 \partial_\tau f - \xi^2 \partial_x f - \xi^4 \partial_w f\,.
\end{aligned}$$

By acting with group operator (98) on (61), (62), and (64), we arrive at the system of defining equations for the coordinates ξ^i, η^i of operator (65)

$$\begin{aligned} &\mu D_\tau \kappa^1 + \varepsilon u \left(D_x \kappa^1 + \kappa^1/x\right) - (p + q - \varepsilon/x) D_u \kappa^1 - \kappa^2 \partial_u g = 0\,, \\ &D_\tau \kappa^3 + \varepsilon w \left(D_x \kappa^3 + \kappa^3/x\right) + \left(p + \varepsilon \Gamma^2/x\right) D_w \kappa^3 - \kappa^2 \partial_w f = 0\,, \\ &D_w \kappa^1 = 0\,, \quad D_u \kappa^3 = 0\,, \quad D_u \kappa^2 = 0\,, \quad D_w \kappa^2 = 0\,, \end{aligned} \tag{67}$$

$$\begin{aligned} \varepsilon \left(D_x \kappa^2 + \kappa^2/x\right) - \int \mathrm{d}w \kappa^3 + \int \mathrm{d}u \kappa^1 = 0\,, \\ \mu D_\tau \kappa^2 + \mu \int \mathrm{d}w w \kappa^3 - \int \mathrm{d}u u \kappa^1 = 0\,. \end{aligned} \tag{68}$$

Equations (67) and (68) should be solved taking the initial equations and all their differential corollaries into account. Here, D_τ, D_x, D_u, and D_w are the operators of total differentiation over the corresponding variable in the form of a subscript

$$\begin{aligned} D_\tau &= \partial_\tau + g_\tau \partial_g + f_\tau \partial_f + p_\tau \partial_p + g_{\tau\tau} \partial_{g_\tau} + g_{\tau x} \partial_{g_x} + g_{\tau u} \partial_{g_u} \\ &\quad + f_{\tau\tau} \partial_{f_\tau} + f_{\tau x} \partial_{f_x} + f_{\tau w} \partial_{f_w} + p_{\tau\tau} \partial_{p_\tau} + p_{\tau x} \partial_{p_x}\,, \\ D_x &= \partial_x + g_x \partial_g + f_x \partial_f + p_x \partial_p + g_{\tau x} \partial_{g_\tau} + g_{xx} \partial_{g_x} + g_{xu} \partial_{g_u} \\ &\quad + f_{\tau x} \partial_{f_\tau} + f_{xx} \partial_{f_x} + f_{xw} \partial_{f_w} + p_{\tau x} \partial_{p_\tau} + p_{xx} \partial_{p_x}\,, \\ D_u &= \partial_u + g_u \partial_g + g_{\tau u} \partial_{g_\tau} + g_{xu} \partial_{g_x} + g_{uu} \partial_{g_u}\,, \\ D_w &= \partial_w + f_w \partial_f + f_{\tau w} \partial_{f_\tau} + f_{xw} \partial_{f_x} + f_{ww} \partial_{f_w}\,. \end{aligned} \tag{69}$$

For compactness, we use the subscripts to denote partial derivatives g, f, and p over the corresponding variables.

We will find the coordinates ξ^i and η^i from the system of defining questions by using the approach that was developed in [33] (see also [14, Chap. 4]). By applying this approach, we separate the determining equations for ξ^i and η^i into local determining Eqs. (67) that appearing from the invariance condition for (61) and (64) and nonlocal determining Eqs. (68) that follow from the invariance conditions for (62). The solution of the local determining equations gives the so-called intermediate symmetry.

By omitting detailed calculations, we will write, after some simplification, the determining equations that specify the intermediate symmetry

$$\begin{aligned}
&\mu\partial_\tau\eta^1 + \varepsilon u\partial_x\eta^1 - (p+q-\frac{\varepsilon}{x})\partial_u\eta^1 - \frac{\varepsilon u}{x}g\partial_g\eta^1 \\
&\qquad + \frac{\varepsilon ug}{x}(\partial_\tau\xi^1 + \frac{\varepsilon u}{\mu}\partial_x\xi^1 + \frac{\eta^1}{g} + \frac{\xi^3}{u} - \frac{\xi^2}{x}) = 0\,, \\
&\partial_\tau\eta^3 + \varepsilon w\partial_x\eta^3 + (p + \frac{\varepsilon\Gamma^2}{x})\partial_w\eta^3 - \frac{\varepsilon w}{x}f\partial_f\eta^3 \\
&\qquad + \frac{\varepsilon uf}{x}(\partial_\tau\xi^1 + \varepsilon w\partial_x\xi^1 + \frac{\eta^3}{f} + \frac{\xi^4}{w} - \frac{\xi^2}{x}) = 0\,, \\
&\varepsilon u(\partial_\tau\xi^1 + \frac{\varepsilon u}{\mu}\partial_x\xi^1) - \mu\partial_\tau\xi^2 - \varepsilon u\partial_x\xi^2 + \varepsilon\xi^3 = 0\,, \\
&\varepsilon w(\partial_\tau\xi^1 + \varepsilon w\partial_x\xi^1) - \partial_\tau\xi^2 - \varepsilon w\partial_x\xi^2 + \varepsilon\xi^4 = 0\,, \\
&\eta^2 + (p+q-\frac{\varepsilon}{x})(\partial_\tau\xi^1 + \frac{\varepsilon u}{\mu}\partial_x\xi^1 - \partial_u\xi^3) \\
&\qquad + \mu\partial_\tau\xi^3 + \varepsilon u\partial_x\xi^3 + \xi^1\partial_t q + \xi^2\partial_x q + \frac{\varepsilon\xi^2}{x^2} = 0\,, \\
&\eta^2 + (p + \frac{\varepsilon\Gamma^2}{x})(\partial_\tau\xi^1 + \varepsilon w\partial_x\xi^1 - \partial_w\xi^4) - \partial_\tau\xi^4 - \varepsilon w\partial_x\xi^4 - \frac{\varepsilon}{x^2}\Gamma^2\xi^2 = 0\,.
\end{aligned} \tag{70}$$

Here, the coordinates ξ^i and η^i are characterized by the following dependence of variables:

$$\begin{aligned}
&\xi^1 = \xi^1(\tau,x),\ \xi^2 = \xi^2(\tau,x),\ \xi^3 = \xi_0^3(\tau,x) + u\xi_1^3(\tau,x),\ \xi^4 = \xi_0^4(\tau,x) + w\xi_1^4(\tau,x), \\
&\eta^1 = \eta^1(\tau,x,u,g)\,,\quad \eta^2 = \eta^2(\tau,x,p)\,,\quad \eta^3 = \eta^3(\tau,x,w,f)\,.
\end{aligned} \tag{71}$$

For arbitrary values of ε and μ formulas (68), (70) and (71) yield three operators of the exact point symmetry group of Eqs. (61), (62), and (64), which correspond to the time translations and dilations [34]. Invariant solutions with the use of these exact symmetries can be found in [34]. However, a disadvantage of these solutions is that they do not allow us to arbitrarily specify the initial distribution of plasma particles and the spatial distribution of the electric field of the laser beam. This proves to be possible with the use of approximate symmetries. They appear if one consider small parameters ε and μ in (61) and (62) and, as accepted in the theory of approximate transformation groups [1], represent coordinates of the group operator as power series in these parameters

$$\xi^i = \sum_{k,l=0}^{\infty} \varepsilon^k\mu^l\xi^{i(k,l)}, \quad \eta^i = \sum_{k,l=0}^{\infty} \varepsilon^k\mu^l\eta^{i(k,l)}\,. \tag{72}$$

By using the linear dependence of coordinates ξ^3 and ξ^4 on velocities u and w, respectively, substituting (72) into (70) and retaining only contributions linear in ε and μ, we obtain the expressions

$$
\begin{aligned}
\xi^1 &= 1 + 2\varepsilon \int \mathrm{d}\tau' \int \mathrm{d}\tau'' \partial_x(\zeta - q), \quad \xi^2 = \varepsilon \int \mathrm{d}\tau'(\zeta - q), \quad \zeta = \zeta(x), \\
\xi^3 &= \mu(\zeta - q) - \varepsilon\mu\Big[2q \int \mathrm{d}\tau' \int \mathrm{d}\tau'' \partial_x(\zeta - q) + \partial_x \int \mathrm{d}\tau' \int \mathrm{d}\tau''(\zeta - q)\Big] \\
&\quad + \varepsilon u \int \mathrm{d}\tau' \partial_x(\zeta - q), \\
\xi^4 &= \zeta - q - \varepsilon\Big[2q \int \mathrm{d}\tau' \int \mathrm{d}\tau'' \partial_x(\zeta - q) + \partial_x \int \mathrm{d}\tau' \int \mathrm{d}\tau''(\zeta - q)\Big] \\
&\quad + \varepsilon w \int \mathrm{d}\tau' \partial_x(\zeta - q), \\
\eta^2 &= \partial_\tau(\zeta - q) - \varepsilon\Big[2\partial_\tau \Big(q \int \mathrm{d}\tau' \int \mathrm{d}\tau'' \partial_x(\zeta - q)\Big) + \partial_x \int \mathrm{d}\tau'(\zeta - q)\Big] \\
&\quad - 3\varepsilon p \int \mathrm{d}\tau' \partial_x(\zeta - q), \\
\eta^1 &= \varepsilon g C_2 - (\varepsilon g/x) \int \mathrm{d}\tau'(\zeta - q), \quad \eta^3 = \varepsilon f C_2 - (\varepsilon f/x) \int \mathrm{d}\tau'(\zeta - q).
\end{aligned}
\tag{73}
$$

for the coordinates of operator (65). By using the approximate intermediate symmetry (73) in nonlocal determining equations (68) and solving them by using variational differentiation, we can see that these equations are satisfied with the chosen accuracy if the electric field changes slowly enough, $\partial_\tau q \lesssim O(\varepsilon)$.

To construct the solutions of the initial problem (61)–(63), it is necessary to know, rather than all of the operators of the admissible group, only their linear combination that preserves the invariance of the solution of the problem by the perturbation theory in powers of τ, in other words, renormgroup symmetries [4]. Therefore, we should specify the form of the initial distribution functions, f_0 and g_0. We assume that the velocity distributions of particles at the initial instant are Maxwellian

$$
g_0 = \frac{n_0^e(x)}{\sqrt{2\pi}} \exp(-u^2/2), \quad f_0 = \frac{n_0^i(x)}{\sqrt{2\pi}\,\Gamma} \exp(-w^2/2\Gamma^2), \tag{74}
$$

with the initial densities $n_0^e(x)$ and $n_0^i(x)$ and zero mean velocities. These initial distribution functions correspond to the initial electric field distribution that obeys the first equation in (62)

$$
\varepsilon \left(\partial_x p^0(x) + p^0(x)/x\right) = n_0^i(x) - n_0^e(x). \tag{75}
$$

The perturbative expansion of the solutions of the Cauchy problem as a power series in τ gives terms that are proportional to $\propto O(\tau)$ for the electron and ion distribution functions and proportional to $\propto O(\tau^2)$ for the electric field. The invariance conditions for these solutions are specified by coordinates (65) of the group generator (73), which give the form of the function ζ and connect the initial electron and ion densities with the spatial structure of the electric laser field for $\tau = 0$, i.e., with the quantity $q^0(x) = q^0(0, x)$

$$\zeta(x) = -\varepsilon\left(\frac{\partial_x n_0^e}{n_0^e} + \Gamma^2\frac{\partial_x n_0^i}{n_0^i}\right), \quad p^0 = -q^0 - \varepsilon\frac{\partial_x n_0^e}{n_0^e}. \tag{76}$$

For the specified initial distribution of the ion density, n_0^i, Eq. (75) and the second equation in (76) form the system of equations for finding n_0^e. If the laser radiation intensity is not low, $q^0 \gg \varepsilon\partial_x n_0^e/n_0^e$, then can be n_0^e approximately written in the form

$$n_0^e(x) \approx n_0^i + \varepsilon\left(\partial_x q^0 + q^0/x\right). \tag{77}$$

In the next section, we use the obtained symmetry groups for solving problem (61), (62), and (63).

3.2 Finite Group Transformations and Invariant Solutions of Kinetic Equations

The approximate solutions of the initial problem (61)–(63) are expressed in a standard way in terms of invariants of the group (65), (72), which occur from the solutions of Lie equations for the generator of group (65) with coordinates (73) after the exclusion of the group parameter from these solutions. We will write these invariants in the case that corresponds to a constant electric field, $q = q(x)$, independent of time, τ. For the laser pulse of finite duration the analogous result was considered in [34].

We consider a laser pulse with a steep enough leading edge so that plasma ions, in fact, have no time to be displaced during the rise of the pulse intensity to its maximum. Further, we will assume that the laser intensity is constant and will use expressions (72) in which the intensity q is independent of time. Then, group invariants that are generated by the operator (65) with coordinates (72) have the form

$$\begin{aligned}
&I_1 = xf \equiv x'f', \quad I_2 = xg \equiv x'g', \quad I_3 = (p+q)(\zeta - q)^3 \equiv (p'+q')(\zeta' - q')^3, \\
&I_4 = \frac{\tau^2}{2(\zeta - q)^2} - Z(y) \equiv -Z(y'), \quad Z(y) = \int_{y'}^{y} d\xi/(\zeta(\xi) - q(\xi))^3, \quad y = x/\varepsilon, \\
&I_5 = (\zeta - q)u + (\mu/3)(4q - \zeta)(\zeta - q)^2\sqrt{Z(y)/2} - \frac{\mu}{2}\int_{y'}^{y} d\xi\big[(1/3)\sqrt{2/Z(\xi)} \\
&\quad - \sqrt{2Z(\xi)}\zeta(\zeta - q)\partial_\xi(\zeta - q) + \zeta/(\sqrt{2Z(\xi)}(\zeta - q))\big] = (\zeta' - q')u', \\
&I_6 = (\zeta - q)w + (1/3)(4q - \zeta)(\zeta - q)^2\sqrt{Z(y)/2} - \frac{1}{2}\int_{y'}^{y} d\xi\big[(1/3)\sqrt{2/Z(\xi)} \\
&\quad - \sqrt{2Z(\xi)}\zeta(\zeta - q)\partial_\xi(\zeta - q) + \zeta/(\sqrt{2Z(\xi)}(\zeta - q))\big] = (\zeta' - q')w'.
\end{aligned} \tag{78}$$

Here, primed variables correspond to the values of quantities for $\tau \to 0$

$$f' = f^0(x', w'), \; g' = g^0(x', u'), \; p' = p^0(x'), \; q' = q^0(x'). \tag{79}$$

The first three invariants, I_1, I_2, and I_3, define substantially the required approximate analytic solution of the initial problem

$$f = (x'/x) f^0(x', w'), \quad g = (x'/x) g^0(x', u'), \quad p = -q + (p' + q')(\zeta' - q')^3 (\zeta - q)^{-3}, \tag{80}$$

where the primed variables x', u', and w' are expressed in terms of τ, x, u, and w using the invariants I_4, I_5, and I_6. For the initial distributions f_0, g_0, and p^0, which are described by expressions (74) and (76), the solution of (80) is written in the form

$$\begin{aligned} f &= \frac{x' n_0^i(x')}{x\sqrt{2\pi}\Gamma} \exp\Big(- \frac{(\zeta' - q')^2 (w - W)^2}{2(\zeta - q)^2 \Gamma^2} \Big), \quad p = -q - \varepsilon \frac{(\zeta' - q')^3}{(\zeta - q)^3} \frac{\partial_{x'} (n_0^e(x'))}{n_0^e(x')}, \\ g &= \frac{x' n_0^e(x')}{x\sqrt{2\pi}} \exp\Big(- \frac{(\zeta' - q')^2 (u - U)^2}{2(\zeta - q)^2} \Big). \end{aligned} \tag{81}$$

Here,

$$\begin{aligned} W = &-(1/3)(4q - \zeta)(\zeta - q)\sqrt{Z(y)/2} + \frac{1}{2(\zeta - q)} \int_{y'}^{y} \mathrm{d}\xi \big[(1/3)\sqrt{2/Z(\xi)} \\ &- \sqrt{2Z(\xi)}\zeta(\zeta - q)\partial_\xi(\zeta - q) + \zeta/(\sqrt{2Z(\xi)}(\zeta - q)) \big], \quad U = \mu W. \end{aligned} \tag{82}$$

The knowledge of the distribution function (81) of particles allows us to calculate the global characteristics of plasma ions, such as their average velocity, v^i_{av}, and density, n^i_{av}, as well as the ion and electron temperatures, T^i and T^e

$$v^i_{av} = W, \quad n^i_{av} = n^i_0(x') \frac{x'(\zeta' - q')}{x(\zeta - q)}, \quad T^{i(e)} = T_0^{i(e)} \frac{(\zeta' - q')^2}{(\zeta - q)^2}. \tag{83}$$

Aside from expressions (83), that describes the spatial distribution of the average ion velocity and density, the expressions that determine the energy spectrum, N_ϵ, of the accelerated ions are also of interest. This spectrum is specified so that the integral of N_ϵ over all of the admissible ion energies $0 < \epsilon < \infty$ coincides with the total number of plasma ions

$$N_\epsilon = \pi \int_0^\infty \mathrm{d}x x \sqrt{\frac{2}{\epsilon}} \big[f(x, \sqrt{2\epsilon}) + f(x, -\sqrt{2\epsilon}) \big]. \tag{84}$$

By substituting solution (81) into (84), we obtain the spectral density of the ion energy

$$N_\epsilon = \sqrt{\frac{\pi}{\epsilon}}\frac{1}{\Gamma}\int\limits_\infty^0 \mathrm{d}x x' n_0^i(x')\big[\exp\big(-\frac{(\zeta'-q')^2(\sqrt{2\epsilon}-W)^2}{2(\zeta-q)^2\Gamma^2}\big) + \exp\big(-\frac{(\zeta'-q')^2(\sqrt{2\epsilon}+W)^2}{2(\zeta-q)^2\Gamma^2}\big)\big]. \tag{85}$$

At low ion temperatures, $\Gamma^2 \ll 1$, the ion spectrum (85) can be found by using simpler expressions based on asymptotic approaches to the calculations of integrals. As a result, integration in (85) is replaced by summation over contributions from individual 'stationary' points determined from the condition $W(x_k) = \sqrt{2\epsilon}$

$$N_\epsilon \backsimeq \pi\sqrt{\frac{2}{\epsilon}}\sum_k x_k' n_0^i(x_k') \mid \partial_x W_{|x=x_k} \mid^{-1} \big(\frac{\zeta'-q'}{\zeta-q}\big)_{|x=x_k}, \quad \partial_x W_{|x=x_k} \neq 0. \tag{86}$$

According to (86), the spectral density of the ion energy has a singularity at the point $x = x_m$, where $\partial_x W_{|x=x_m} \to 0$ (generally speaking, there can be several such points); it should be replaced by the expression

$$N_\epsilon^m \backsimeq \sqrt{\frac{\pi}{\epsilon\Gamma}}\frac{\Gamma(1/4)}{2^{1/4}}\frac{x_m' n_0^i(x_m')}{\sqrt{\mid \partial_{xx} W_{|x=x_m} \mid}}\big(\frac{\zeta'-q'}{\zeta-q}\big)^{1/2}_{|x=x_m}, \quad \partial_x W_{|x=x_m} = 0, \quad \partial_{xx} W_{|x=x_m} \neq 0. \tag{87}$$

Outside the region $0 < \epsilon < \epsilon_m \equiv W(x_m)^2/2$, the quantity N_ϵ proves to be exponentially small, i.e., ϵ_m determines the upper boundary of the spectrum. The quantity ϵ_m is defined by a pair of equations

$$W(\tau, x_m) - \sqrt{2\epsilon_m} = 0, \quad \partial_x W_{|x=x_m} = 0, \tag{88}$$

which have the explicit form

$$\begin{aligned}
&\big\{\sqrt{2\epsilon_m} - \tau(\zeta-q) + \frac{1}{2(\zeta-q)}\int_{y'}^{y}\mathrm{d}\xi\sqrt{2Z(\xi)}(\zeta-q)\big[3(2\zeta-q)\partial_\xi\zeta \\
&\qquad + (2q-5\zeta)\partial_\xi q\big]\big\}_{|y=y_m} = 0, \quad y_m = \frac{x_m}{\varepsilon}, \\
&\big\{\tau\partial_y(\zeta-q) + \frac{\partial_y(\zeta-q)}{2(\zeta-q)^2}\int_{y'}^{y}\mathrm{d}\xi\sqrt{2Z(\xi)}(\zeta-q)\big[3(2\zeta-q)\partial_\xi\zeta + (2q-5\zeta)\partial_\xi q\big] \\
&\quad + \frac{1}{2(\zeta-q)^4}\big(1+\tau^2\partial_y(\zeta-q)\big)\int_{y'}^{y}\mathrm{d}\xi\frac{(\zeta-q)}{\sqrt{2Z(\xi)}}\big[3(2\zeta-q)\partial_\xi\zeta \\
&\quad - \frac{\tau}{2(\zeta-q)} + (2q-5\zeta)\partial_\xi q\big]\big[3(2\zeta-q)\partial_y\zeta + (2q-5\zeta)\partial_y q\big]\big\}_{|y=y_m} = 0.
\end{aligned} \tag{89}$$

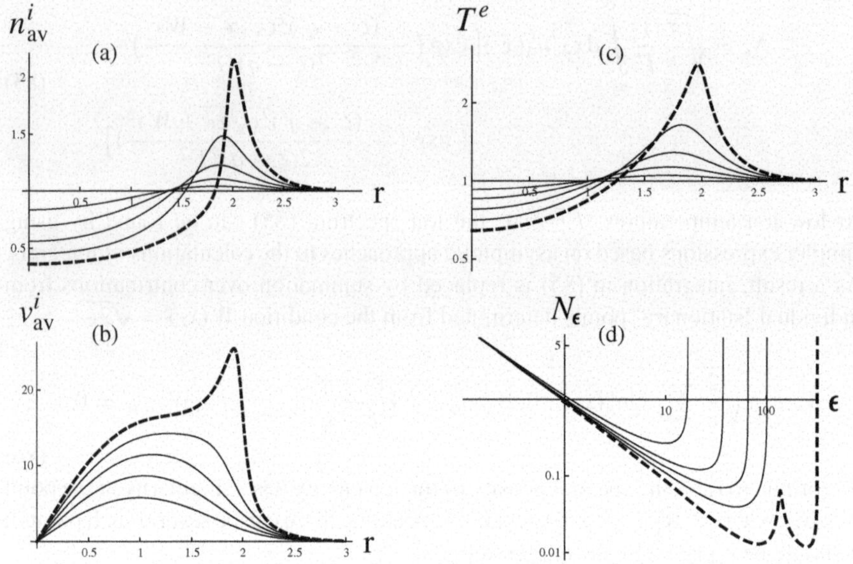

Fig. 3 The spatial distributions of the normalized average density (**a**) and velocity (**b**) of ions, the electron temperature (**c**), and the ion spectrum (**d**) over the dimensionless coordinate r for a stationary laser beam for different $\tau = 2; 3; 4; 5; 6$. The value of τ increases for curves from top to bottom (for $r \to 0$) for (**a**), (**c**), from bottom to top for (**b**), and from left to right for spectral distribution curves. $Z = 2$, $A = 4$, $a_0^2 = 50$, $\alpha = 1$, $\varepsilon = 0.01$, $\mu = \sqrt{1/4000}$, $n^c/n_0^e = 100$ (n^c is the critical density), $n_0^e = 1$; $n_0^i = 0.5$; $\Gamma = 0.001$

Figure 3 shows the spatial distributions of the average ion velocity and density at different instants under the action of stationary laser radiation when $A(\tau) = 1$. One can see that the spatial distributions of ion density and velocity change with time, τ, under the action of the electric field of the laser beam. The acceleration of ions is followed by the formation of an ion density cusp with a minimum at the laser beam center and a maximum at its periphery. Similar behavior is also typical for the electron temperature of the plasma, which is lower at the central part of the laser beam and increases at its periphery. The dashed curves for $\tau = 6$ are formally located outside the region of applicability of our theory based on the approximate symmetry; therefore, they demonstrate only a trend in the change in the ion density and velocity and the electron temperature rather than their accurate quantitative values. Because, according to expressions (83), the product $(\zeta - q)^{-3}(\zeta' - q')^3$ is proportional to $(x/x')\, T^e n^i_{av}$, the increase in the plasma density and electron temperature at the external boundary of the laser beam can lead, with increasing τ, to a difference of the electric field in this plasma region from the laser beam field. For plasma and beam parameters in Fig. 3, the difference of p from $-q$ for $\tau = 6$ and $r \approx 1.95$ is 4%. A greater difference, namely, a decrease in p more than by one-half (compared to $-q$) is observed for $\tau = 7$ and $r \approx 2.155$, although this value of τ lies outside the region of applicability of the theory.

To conclude, in this Section the theory of approximate transformation groups was successfully used to study the spatiotemporal structure of the accelerated plasma particles and analyze the acceleration of ions in an inhomogeneous cylindrical low density plasma caused by laser radiation that propagates along the axis of a cylinder.

4 Approximate Symmetry in Multi-scale Plasma Dynamics

In this section, an application of approximate transformation groups to study dynamics of a system with distinct time scales is discussed. The utilization of the Krylov-Bogoliubov-Mitropolsky method of averaging to find solutions of the Lie equations is considered. Physical illustrations from the plasma kinetic theory demonstrate the potentialities of the suggested approach.

In analyzing various physical systems, we frequently deal with a situation when a complicated dynamics of different systems appears as a superposition of 'fast' and 'slow' motions with incommensurable characteristic scales, for example, slow evolution of 'background' system characteristics accompanied by fast oscillations in the vicinity of a background state. This type of behavior seems typical for various linear and nonlinear problems (numerous examples are found in [35, 36]), e.g., for celestial mechanics in studies of a motion of planets, for mechanics when treating oscillatory regimes of systems with slowly varying parameters, and for various nonlinear problems of multi-component plasma.

The availability of different scales (though the origin of these scales depends upon the particular system of interest) allows to simplify the analysis of the complicated dynamics by treating 'fast' and 'slow' motions separately. These ideas underlie an essence of asymptotic analytical approaches, the method of averaging [37, 38], the method of multiple scales [35], and other asymptotic methods (see, e.g., [35, 36]).

As to relation of modern group analysis to nonlinear dynamics, here we will point to an interpenetration of ideas from both fields: on one hand the use of the Lie group theory in asymptotic methods for integration of nonlinear differential equations gives (in combination with the Hausdorff formula) the theoretical basis for the method of averaging [37] and provides a regular procedure for calculating the asymptotic series in this method [36, 39, 40]. On the other hand, an introduction of multiple-scales approach to modern group analysis [41] enhances the potentiality of approximate transformation groups [1].

In the present section, we demonstarte the Krylov-Bogoliubov-Mitropolsky method (KBM-method) of averaging in group analysis of the system of equations that describes the evolution of plasma particles in multi-component plasma. The linearity of the group determining equations plays the decisive role in separating fast and slow terms in coordinates of a group generator and in successive use of the KBM-method for constructing the asymptotic solutions of the Lie group equations.

4.1 Lie Symmetry Group for Multi-scale Plasma Dynamics

We start with kinetic equations for distribution functions, f^e and f^i

$$\partial_t f^e + v^e \partial_x f^e - (e/m) E \partial_{v^e} f^e = 0 \,, \quad \partial_t f^i + v^i \partial_x f^i + (Ze/M) E \partial_{v^i} f^i = 0 \,, \tag{90}$$

for both species of two-component plasma consisting of electrons and ions with mass m and M and charges $e^e = -e$ and $e^i = Ze$, where Z is a charge number and equations for a self-consistent electric field E

$$\partial_x E = 4\pi\rho \,, \quad \partial_t E = -4\pi j \,, \quad \partial_t \rho + \partial_x j = 0 \,. \tag{91}$$

Here charge ρ and current j densities are related to moments of the distribution functions via nonlocal material relations

$$\rho = e\big[Z \int \mathrm{d}v^i f^i - \int \mathrm{d}v^e f^e\big], \quad j = e\big[Z \int \mathrm{d}v^i v^i f^i - \int \mathrm{d}v^e v^e f^e\big]. \tag{92}$$

Equations (90)–(92) are known as a system of the Vlasov-Maxwell equations for a collisionless plasma. We are interested in the solution of the Cauchy problem for kinetic equations (90) with the initial conditions

$$f^e|_{t=0} = f_0^e(x, v^e) \,, \quad f^i|_{t=0} = f_0^i(x, v^i) \,, \tag{93}$$

which depend on a particular physical problem. In what follows, we consider an evolution of localized plasma bunches and assume sufficiently smooth (e.g., Maxwellian) initial distribution functions with electron T_e and ion T_i temperatures and initial densities of electrons $n^e(x) = \int \mathrm{d}v^e f_0^e$ and ions $n^i(x) = \int \mathrm{d}v^i f_0^i$ with the characteristic scale L. Below we consider a typical situation when L is much greater than the Debye radius of electrons $r_{De} = \sqrt{T_e/(4\pi n_0^e e^2)}$. The difference in mass of plasma particles specifies two different time scales, namely dimensionless time $\omega_{Le} t$ for 'fast' electron motions and $\tau = \mu t$ for 'slow' motions, $\mu = \sqrt{Zm/M} \ll 1$. It is natural that electrons are involved in both fast and slow motions, hence the electron distribution function depends on both t and τ. On the contrary, we consider ions not involved in fast motions. It means that the ion distribution function does not depend upon fast time t, but only upon slow time τ. It also means that averaging upon fast time eliminates the fast component $\tilde{E}$ of the electric field $E = \tilde{E} + \bar{E}$ in the ion kinetic equation that contains only the slow electric field $\bar{E}$. Then introducing dimensionless variables, electron velocity $u = v^e/V_{Te}$, $V_{Te} = \sqrt{T_e/m}$, dimensionless ion velocity $w = v^i/c_s$, $c_s = \sqrt{ZT_e/M}$, dimensionless electric field $p = \varepsilon(eEL/T_e)$, $\varepsilon = r_{De}/L \ll 1$, and dimensionless distribution functions $f^e = (n_0^e/V_{Te})g$, $f^i = (n_0^e/(Zc_s))f$, $n_0^{e(i)} = n^{e(i)}(0)$, we come to the following system of basic equations in dimensionless variables

$$\partial_t g + \mu \partial_\tau g + \varepsilon u \partial_x g - p \partial_u g = 0\,, \quad \partial_\tau f + \varepsilon w \partial_x f + \bar{p} \partial_w f = 0\,, \tag{94}$$

$$\varepsilon \partial_x p = \int \mathrm{d}w f - \int \mathrm{d}u g\,, \quad \partial_t p + \mu \partial_\tau p = -\mu \int \mathrm{d}w w f + \int \mathrm{d}u u g\,. \tag{95}$$

This system of equations should be supplemented by the four additional equalities

$$\partial_w g = 0\,, \quad \partial_u f = 0\,, \quad \partial_u p = 0\,, \quad \partial_w p = 0\,, \tag{96}$$

which are evident from the physical point of view.

The Lie point symmetry group admitted by the system (94) and (95) is defined by a symmetry group generator

$$X = \xi^1 \partial_t + \xi^2 \partial_x + \xi^3 \partial_u + \xi^4 \partial_w + \xi^5 \partial_\tau + \eta^1 \partial_g + \eta^2 \partial_p + \eta^3 \partial_f\,. \tag{97}$$

In the canonical form this generator is written as

$$Y = \kappa^1 \partial_g + \kappa^2 \partial_p + \kappa^3 \partial_f\,, \tag{98}$$

$$\begin{aligned}
\kappa^1 &= \eta^1 - \xi^1 \partial_t g - \xi^2 \partial_x g - \xi^3 \partial_u g - \xi^5 \partial_\tau g\,, \\
\kappa^2 &= \eta^2 - \xi^1 \partial_t p - \xi^2 \partial_x p - \xi^5 \partial_\tau p\,, \\
\kappa^3 &= \eta^3 - \xi^2 \partial_x f - \xi^4 \partial_w f - \xi^5 \partial_\tau f\,.
\end{aligned}$$

When applying group generator (98) to (94), (95) and (96) we get a system of determining equations for coordinates ξ^i, η^i of the generator (97)

$$\begin{aligned}
& D_t \kappa^1 + \mu D_\tau \kappa^1 + \varepsilon u D_x \kappa^1 - p D_u \kappa^1 - \kappa^2 \partial_u g = 0\,, \\
& D_\tau \kappa^3 + \varepsilon w D_x \kappa^3 + \bar{p} D_w \kappa^3 - \bar{\kappa}^2 \partial_w f = 0\,, \\
& D_w \kappa^1 = 0\,, \quad D_u \kappa^3 = 0\,, \quad D_u \kappa^2 = 0\,, \quad D_w \kappa^2 = 0\,,
\end{aligned} \tag{99}$$

$$\begin{aligned}
& \varepsilon D_x \kappa^2 - \int \mathrm{d}w \kappa^3 + \int \mathrm{d}u \kappa^1 = 0\,, \\
& D_t \kappa^2 + \mu D_\tau \kappa^2 + \mu \int \mathrm{d}w w \kappa^3 - \int \mathrm{d}u u \kappa^1 = 0\,,
\end{aligned} \tag{100}$$

which should be solved in view of the basic equations (94)–(96) and all their differential consequences. Here D_t, D_τ, D_x, D_u, and D_w are total differentiations with respect to the variable denoted by lower index

$$\begin{aligned}
D_t &= \partial_t + (\partial_t g)\, \partial_g + (\partial_t p)\, \partial_p\,, \\
D_\tau &= \partial_\tau + (\partial_\tau g)\, \partial_g + (\partial_\tau f)\, \partial_f + (\partial_\tau p)\, \partial_p\,, \\
D_x &= \partial_x + (\partial_x g)\, \partial_g + (\partial_x f)\, \partial_f + (\partial_x p)\, \partial_p\,, \\
D_u &= \partial_u + (\partial_u g)\, \partial_g\,, \quad D_w = \partial_w + (\partial_w f)\, \partial_f\,.
\end{aligned} \tag{101}$$

To find coordinates ξ^i, η^i from the system of determining equations, we use the approach described in [14, Chap. 4]. Following this technique, we separate the determining equations for ξ^i and η^i into local determining equations, (99), which arise from invariance of (94), (96), and nonlocal determining equations, (100), which follow from invariance conditions for (95). Solutions of local determining equations give the so-called intermediate symmetry.

Two distinct moments should be taken into account here: first, in view of multiscale dynamics, we outline in coordinates ξ^1, ξ^2, ξ^3, and η^2, entering local determining equations the fast terms denoted by variables with tilde and slow terms denoted by variables with bar

$$\xi^i = \tilde{\xi}^i + \bar{\xi}^i\,, \quad i = 1, 2, 3\,, \qquad \eta^2 = \tilde{\eta}^2 + \bar{\eta}^2\,. \tag{102}$$

Due to the fact that both local and nonlocal determining equations are linear in ξ and η, we thus can separate terms of different characteristic scales. Then omitting trivial tedious computations, we rewrite fast

$$\begin{gathered}\tilde{\eta}^2 + p\big(\partial_t\tilde{\xi}^1 + \mu\partial_\tau\tilde{\xi}^1 - \partial_u\tilde{\xi}^3\big) + \partial_t\tilde{\xi}^3 + \mu\partial_\tau\tilde{\xi}^3 + \varepsilon up\partial_x\tilde{\xi}^1 + \varepsilon u\partial_x\tilde{\xi}^3 = 0\,, \\ \varepsilon\tilde{\xi}^3 - \partial_t\tilde{\xi}^2 - \mu\partial_\tau\tilde{\xi}^2 + \varepsilon u\big(\partial_t\tilde{\xi}^1 + \mu\partial_\tau\tilde{\xi}^1 - \partial_x\tilde{\xi}^2\big) + \varepsilon^2 u^2\partial_x\tilde{\xi}^1 = 0\,,\end{gathered} \tag{103}$$

and slow local determining equations

$$\begin{gathered}\bar{\eta}^2 + p\left(\partial_\tau\xi^5 + \varepsilon w\partial_x\xi^5 - \partial_w\xi^4\right) - \partial_\tau\xi^4 - \varepsilon w\partial_x\xi^4 = 0\,, \\ \bar{\eta}^2 + p\left(\mu\partial_\tau\bar{\xi}^1 - \partial_u\bar{\xi}^3\right) + \mu\partial_\tau\bar{\xi}^3 + \varepsilon up\partial_x\bar{\xi}^1 + \varepsilon u\partial_x\bar{\xi}^3 = 0\,, \\ \varepsilon\bar{\xi}^3 - \mu\partial_\tau\bar{\xi}^2 + \varepsilon u\left(\mu\partial_\tau\bar{\xi}^1 - \partial_x\bar{\xi}^2\right) + \varepsilon^2 u^2\partial_x\bar{\xi}^1 = 0\,, \\ \varepsilon\xi^4 - \partial_\tau\bar{\xi}^2 + \varepsilon w\left(\partial_\tau\xi^5 - \partial_x\bar{\xi}^2\right) + \varepsilon^2 w^2\partial_x\xi^5 = 0\,.\end{gathered} \tag{104}$$

Here, in (103) and (104) the dependencies of $\tilde{\xi}^i$, $\bar{\xi}^i$, and $\tilde{\eta}^i$, $\bar{\eta}^i$ upon group variables are given by

$$\begin{gathered}\tilde{\xi}^1 = \tilde{\xi}^1(t, \tau, x), \quad \tilde{\xi}^2 = \tilde{\xi}^2(t, \tau, x), \quad \tilde{\xi}^3 = \tilde{\xi}^3(t, \tau, x, u), \quad \tilde{\eta}^2 = \tilde{\eta}^2(t, \tau, x, p), \\ \bar{\xi}^1 = \bar{\xi}^1(\tau, x), \quad \bar{\xi}^2 = \bar{\xi}^2(\tau, x), \quad \bar{\xi}^3 = \bar{\xi}^3(\tau, x, u), \quad \bar{\xi}^4 = \bar{\xi}^4(\tau, x, u), \\ \bar{\xi}^5 = \mu\bar{\xi}^1, \quad \bar{\eta}^2 = \bar{\eta}^2(\tau, x, p), \quad \eta^1 = \eta^1(g), \quad \eta^3 = \eta^3(f).\end{gathered} \tag{105}$$

Second, we shall take an advantage of small parameters in (94), (95) and, as is customary in the approximate group analysis technique [1], express the coordinates of the group generator as power series in ε and μ

$$\xi^i = \sum_{k,l=0}^{\infty} \varepsilon^k\mu^l\xi^{i(k,l)}\,, \quad \eta^i = \sum_{k,l=0}^{\infty} \varepsilon^k\mu^l\eta^{i(k,l)}\,. \tag{106}$$

Collecting terms of the same order, we come to the following infinite set of equations that relate coordinates of different orders for the fast

$$\begin{aligned}
\tilde{\eta}^{2(k,l)} &+ p\big(\partial_t\tilde{\xi}^{1(k,l)} + (1-\delta_{l,0})\partial_\tau\tilde{\xi}^{1(k,l-1)} - \partial_u\tilde{\xi}^{3(k,l)}\big) \\
&+ \partial_t\tilde{\xi}^{3(k,l)} + (1-\delta_{l,0})\partial_\tau\tilde{\xi}^{3(k,l-1)} + up(1-\delta_{k,0})\partial_x\tilde{\xi}^{1(k-1,l)} \\
&+ u(1-\delta_{k,0})\partial_x\tilde{\xi}^{3(k-1,l)} = 0\,, \quad k,l \geqslant 0, \\
-\partial_t\tilde{\xi}^{2(k,l)} &+ (1-\delta_{k,0})\tilde{\xi}^{3(k-1,l)} - (1-\delta_{l,0})\partial_\tau\tilde{\xi}^{2(k,l-1)} \\
&+ u(1-\delta_{k,0})\big(\partial_t\tilde{\xi}^{1(k-1,l)} + (1-\delta_{l,0})\partial_\tau\tilde{\xi}^{1(k-1,l-1)} - \partial_x\tilde{\xi}^{2(k-1,l)}\big) \\
&+ (1-\delta_{k,0})(1-\delta_{k,1})u^2\partial_x\tilde{\xi}^{1(k-2,l)} = 0\,,
\end{aligned} \tag{107}$$

and the slow terms

$$\begin{aligned}
\bar{\eta}^{2(k,l)} &+ p\left(\partial_\tau\xi^{5(k,l)} + (1-\delta_{k,0})w\partial_x\xi^{5(k-1,l)} - \partial_w\xi^{4(k,l)}\right) \\
&- \partial_\tau\xi^{4(k,l)} - (1-\delta_{k,0})w\partial_x\xi^{4(k-1,l)} = 0\,, \\
\bar{\eta}^{2(k,l)} &+ p\left((1-\delta_{l,0})\partial_\tau\bar{\xi}^{1(k,l-1)} - \partial_u\bar{\xi}^{3(k,l)}\right) + (1-\delta_{l,0})\partial_\tau\bar{\xi}^{3(k,l-1)} \\
&+ (1-\delta_{k,0})up\,\partial_x\bar{\xi}^{1(k-1,l)} + (1-\delta_{k,0})u\partial_x\bar{\xi}^{3(k-1,l)} = 0\,, \\
(1-\delta_{k,0})\bar{\xi}^{3(k-1,l)} &+ (1-\delta_{k,0})u\left((1-\delta_{l,0})\partial_\tau\bar{\xi}^{1(k-1,l-1)} - \partial_x\bar{\xi}^{2(k-1,l)}\right) \\
&- (1-\delta_{l,0})\partial_\tau\bar{\xi}^{2(k,l-1)} + (1-\delta_{k,0})(1-\delta_{k,1})u^2\partial_x\bar{\xi}^{1(k-2,l)} = 0\,, \\
(1-\delta_{k,0})\xi^{4(k-1,l)} &- \partial_\tau\bar{\xi}^{2(k,l)} + (1-\delta_{k,0})w\left(\partial_\tau\xi^{5(k-1,l)} - \partial_x\bar{\xi}^{2(k,l)}\right) \\
&+ (1-\delta_{k,0})(1-\delta_{k,1})w^2\partial_x\xi^{5(k-2,l)} = 0\,.
\end{aligned} \tag{108}$$

Using approximate intermediate symmetry, which follows from solutions of equations (107)–(108), in nonlocal determining equations (100), we find a solution of the latter using variational differentiation (see [14, Chap. 4] for details) and obtain the symmetry of the complete system (94)–(96).

In constructing the solution of the b.v.p. (93)–(95), we need not require the entire set of generators, but rather such a combination of group generators that leaves invariant the perturbation theory solution in powers of t and τ, the so-called renormgroup symmetries [4]. Hence, we should specify the initial particle distribution functions $f^0 = f_{|t=0}$, $g^0 = g_{|t=0}$. For concreteness, we assume the initial velocity distribution functions to be maxwellian

$$g^0 = n_0^e(x)\exp(-u^2/2)\,, \quad f^0 = n_0^i(x)\exp(-w^2/2\Gamma^2)\,, \tag{109}$$

with the initial densities $n_0^e(x)$ and $n_0^i(x)$ and the initial zero average velocities. In account of these initial distribution functions, we have the following initial electric field

$$p^0(x) = (1/\varepsilon)\int_0^x \mathrm{d}x\,\left(n_0^i(x) - n_0^e(x)\right)\,. \tag{110}$$

Perturbation expansion of the Cauchy problem solutions in powers of t and τ gives terms $\propto O(t)$ and $\propto O(\tau)$ for the electron distribution function and $\propto O(\tau)$ for the ion distribution function, and $\propto O(t^2)$ and $\propto O(\tau^2)$ for the electric field. Invariance conditions for these solutions specify the coordinates (106) of the group generator (97). Leaving only terms that are linear in ε and μ, we write these coordinates as follows:

$$\begin{aligned} &\xi^1 = 1 + \varepsilon\tau^2\partial_x\xi\,, \quad \xi^2 = \varepsilon\left((\delta/\Omega)\sin\Omega t + \mu\tau\xi\right)\,, \\ &\xi^3 = \delta\cos\Omega t - \varepsilon\mu\tau u\partial_x\xi\,, \quad \xi^4 = \mu\left(\xi - \varepsilon\tau w\partial_x\xi\right)\,, \\ &\eta^2 = \delta\Omega\sin\Omega t - 3\mu\varepsilon\tau p\partial_x\xi\,, \quad \xi^5 = \mu\xi^1\,, \\ &\eta^1 = \eta^3 = 0\,, \quad \Omega^2 = n^i(\tau, x) \equiv \int \mathrm{d}w f\,. \end{aligned} \tag{111}$$

The dependence of functions $\xi(x)$ and $\delta(x)$ upon x is expressed in terms of the initial densities distributions $n_0^{e,i}$ and the initial electric field p^0

$$\xi = -\varepsilon\left((\partial_x n_0^e/n_0^e) + \Gamma^2(\partial_x n_0^i/n_0^i)\right)\,, \; \delta = -p^0 - \varepsilon(\partial_x n_0^e/n_0^e)\,, \; \Gamma = V_{Ti}/c_s\,. \tag{112}$$

For arbitrary parameters ε and μ and arbitrary initial density distributions $n_0^{e,i}$ formulas (111) describe the approximate symmetry. However, in two limiting cases, infinite series (106) terminate and we get the exact symmetry group. The first case is referred to electron plasma with neutralizing homogeneous ion background ($\mu = 0$, $\Omega^2 = n_0^i = const$) [42–44], which gives the generator

$$X = \partial_t + \varepsilon\,(\delta/\Omega)\sin\Omega t\partial_x + \delta\cos\Omega t\partial_u + \delta\Omega\sin\Omega t\partial_p\,. \tag{113}$$

The second case is referred to quasi-neutral approximation for electron-ion plasma with zero current and charge densities $j = \rho = 0$ [45] that is realized for $\delta = 0$ and the initial gaussian densities distribution, $\xi \propto \beta x$

$$X = \left(1 + \beta\varepsilon\tau^2\right)\partial_\tau + \varepsilon\beta\tau x\partial_x + \beta\,(\mu x - \varepsilon\tau u)\,\partial_u + \beta\,(x - \varepsilon\tau w)\,\partial_w - 3\varepsilon\beta\tau p\partial_p\,. \tag{114}$$

The additional term in ξ^3 in (114) that refers to acceleration of electrons is omitted in (111) as it is of the higher order $O(\mu^2)$ as compared to that included in (111).

4.2 Slow and Fast Dynamics of Plasma Particles

To construct group invariant solution for the b.v.p. (93)–(95), we should find solutions of the Lie equations for the group generator (97) with coordinates (106)

$$
\begin{aligned}
&\frac{dt}{da} = 1 + \varepsilon\tau^2\partial_x\xi\,, \quad t_{|a=0} = t'\,,\\
&\frac{dx}{da} = \varepsilon\left((\delta/\Omega)\sin\Omega t + \mu\tau\xi\right)\,, \quad x_{|a=0} = x'\,,\\
&\frac{du}{da} = \delta\cos\Omega t - \varepsilon\mu\tau u\partial_x\xi\,, \quad u_{|a=0} = u'\,,\\
&\frac{dw}{da} = \mu\left(\xi - \varepsilon\tau w\partial_x\xi\right)\,, \quad w_{|a=0} = w'\,,\\
&\frac{d\tau}{da} = \mu\left(1 + \varepsilon\tau^2\partial_x\xi\right)\,, \quad \tau_{|a=0} = \tau'\,,\\
&\frac{dp}{da} = \delta\Omega\sin\Omega t - 3\mu\varepsilon\tau p\partial_x\xi\,, \quad p_{|a=0} = p'\,,\\
&\frac{df}{da} = \frac{dg}{da} = 0\,, \quad g_{|a=0} = g'\,, \quad f_{|a=0} = f'\,.
\end{aligned}
\tag{115}
$$

Solution of the b.v.p. (93)–(95) are expressed as usual in terms of invariants of the group (97), (106) that result from solutions of (115) after excluding the group parameter a. Due to a difference in characteristic time scales, we can separate 'fast' and 'slow' group invariants by applying the averaging procedure to Lie equations.

In fact, at small time $t > 0$, $1/\mu \gg t \gg 1/\Omega$, the'ion'terms that are $\propto \mu$ can be omitted and we come to simplified Lie equations (equations for group invariants f, g, τ, w are omitted here)

$$
\begin{aligned}
&\frac{dt}{da} = 1\,, \quad t_{|a=0} = t'\,; \quad \frac{dx}{da} = \varepsilon\,(\delta/\Omega)\sin\Omega t\,, \quad x_{|a=0} = x'\,;\\
&\frac{du}{da} = \delta\cos\Omega t\,, \quad u_{|a=0} = u'\,; \quad \frac{dp}{da} = \delta\Omega\sin\Omega t\,, \quad p_{|a=0} = p'\,,
\end{aligned}
\tag{116}
$$

the solutions of which define invariants of 'fast' motions at small time $t \ll 1/\mu$:

$$
\begin{aligned}
&J_1 = p + \delta\cos\Omega t \equiv -\varepsilon(\partial_x n_0^e/n_0^e)|_{x=x'}\,,\\
&J_2 = x + (\varepsilon\delta/\Omega^2)\cos\Omega t \equiv x' + (\varepsilon\delta(x')/\Omega^2(x'))\,,\\
&J_3 = u - (\delta/\Omega)\sin\Omega t \equiv u'\,,\\
&J_4 = g \equiv g^0(x', u')\,.
\end{aligned}
\tag{117}
$$

On the contrary, averaging the complete Lie equations on a large time scale $T \gg 1/\mu$, we come to Lie equations defining 'slow' motions (equations for group invariants are again omitted)

$$
\begin{aligned}
&\frac{d\tau}{da} = 1 + \varepsilon\tau^2\partial_{\bar{x}}\bar{\xi}\,, \quad \frac{d\bar{x}}{da} = \varepsilon\tau\bar{\xi}\,, \quad \frac{d\bar{p}}{da} = -3\varepsilon\tau\bar{p}\partial_{\bar{x}}\bar{\xi}\,,\\
&\frac{d\bar{u}}{da} = -\varepsilon\tau\bar{u}\partial_{\bar{x}}\bar{\xi}\,, \quad \frac{dw}{da} = \bar{\xi} - \varepsilon\tau w\partial_{\bar{x}}\bar{\xi}\,,
\end{aligned}
\tag{118}
$$

with the corresponding 'slow' invariants

$$
\begin{aligned}
& I_1 = \bar{f} \equiv f', \quad I_2 = \bar{g} \equiv g', \quad I_3 = \bar{p}\bar{\xi}^3 \equiv p'\xi'^3, \\
& I_4 = \frac{\varepsilon\tau^2}{2\bar{\xi}^2} - \int^{\bar{x}} \mathrm{d}z/\xi^3 \equiv \frac{\varepsilon\tau'^2}{2\xi'^2} - \int^{x'} \mathrm{d}z/\xi^3, \quad I_5 = \bar{\xi}\bar{u} \equiv \xi' u', \\
& I_6 = \bar{\xi}\bar{w} - \frac{1}{\sqrt{2\varepsilon}} \int^{\bar{x}} \mathrm{d}y \left(\int^{y} \mathrm{d}z/\xi^3 \right)^{-1/2} \equiv \xi' w' - \frac{1}{\sqrt{2\varepsilon}} \int^{s} \mathrm{d}y \left(\int^{y} \mathrm{d}z/\xi^3 \right)^{-1/2}.
\end{aligned}
\tag{119}
$$

Here the primed variables are related to values at $\tau \to 0$

$$
\begin{aligned}
& f' = f^0(s, w'), \; g' = g^0(s, u'), \; p' = -\varepsilon\partial_s\varphi, \; s = x' + \frac{\varepsilon\delta(x')}{\Omega^2(x')}, \\
& \varepsilon^2\partial_{xx}\varphi + n_0^i(x) - \mathrm{e}^{\varphi} = 0, \; \partial_x\varphi|_{x=0} = \partial_x\varphi|_{x\to\infty} = 0, \; \varphi|_{x=0} = C < \infty.
\end{aligned}
\tag{120}
$$

In what follows, we use the fast and slow invariants to construct analytical solutions of the Cauchy problem for the kinetic equations (93)–(95).

4.2.1 Slow Dynamics of Particles

Let we consider the slow dynamics of plasma particles under simplifying assumptions, small value of $p_0 < 1$, and low ion temperature, $\gamma \to 0$. Then, following (119)–(120), $s \approx x'$ and ξ coincides with $p' = \bar{p}|_{\tau=0}$, and we come to simplified expressions, which define dynamics of plasma ions

$$
\begin{aligned}
& \bar{f} = f^0(x', w'), \quad \bar{g} = g^0(x', w'), \quad \bar{p} = (\xi'/\bar{\xi})^3 \, p'(x'), \quad \varepsilon\tau^2 = 2\bar{\xi}^2 \int_{x'}^{\bar{x}} \mathrm{d}z/\xi^3, \\
& \bar{w} = (\xi'/\bar{\xi}) \, w' + \frac{1}{\bar{\xi}\sqrt{2\varepsilon}} \int_{x'}^{\bar{x}} \mathrm{d}y \left(\int_{x'}^{y} \mathrm{d}z/\xi^3 \right)^{-1/2}, \quad \bar{u} = (\xi'/\bar{\xi}) \, u'.
\end{aligned}
\tag{121}
$$

For completeness, we also present global characteristics for plasma ions, their average velocity v_{av}^i, density n_{av}^i and temperature T^i

$$
v_{av}^i = \frac{1}{\bar{\xi}\sqrt{2\varepsilon}} \int_{x'}^{\bar{x}} \mathrm{d}y \left(\int_{x'}^{y} \frac{\mathrm{d}z}{\xi^3} \right)^{-1/2}, \; n_{av}^i = n_0^i(x')\frac{\xi'}{\bar{\xi}}, \; T^i = T_0^i \left(\frac{\xi'}{\bar{\xi}} \right)^2. \tag{122}
$$

In exemplification of the results, these formulas are analyzed below for the distinct initial electric field and density distribution. We consider the case when there is very slight difference between p^0 and ξ and the initial electric field practically concises with the initial 'slow' electric field $\bar{p}$ and the amplitude of the 'fast' electric field is small. This is realized, for example, for the Lorentz-type initial density distribution

$$
n_e^0(x) = (1/\pi(1+x^2)), \quad n_i^0(x) = (b/\pi(1+b^2x^2)), \quad |\, b-1 \,| \ll 1. \tag{123}
$$

Substitution of (123) into (110) gives the following formulas for the spatial distribution of the initial electric field and the function ξ

$$p_l^0(x) = \frac{1}{\pi\varepsilon}\left(\arctan bx - \arctan x\right), \quad \xi_l = 2\varepsilon x\Big(\frac{1}{1+x^2} + \frac{\gamma^2 b^2}{1+b^2x^2}\Big). \tag{124}$$

The left panel of Fig. 4 demonstrates the difference between p_l^0 and ξ_l, and the right panel shows the spatial distribution of the 'slow' electric field $\bar{p}$ at different moments of time τ. As for the average ion density, temperature and velocity they are given by the formulas

$$n_{av}^i = \frac{b}{\pi(1+b^2x'^2)}\frac{\xi_l'}{\bar{\xi}_l}, \; T^i = T_0^i\left(\frac{\xi_l'}{\bar{\xi}_l}\right)^2, \; v_{av}^i = \frac{1}{\bar{\xi}_l\sqrt{2\varepsilon}}\int_{x'}^{\bar{x}} \mathrm{d}y\left(\int_{x'}^{y}\mathrm{d}z/\xi_l^3\right)^{-1/2}. \tag{125}$$

and are plotted on the Fig. 5. The opposite situation when the oscillating electric field is of the order of the average electric field that accelerates ions is explained in detail in [46].

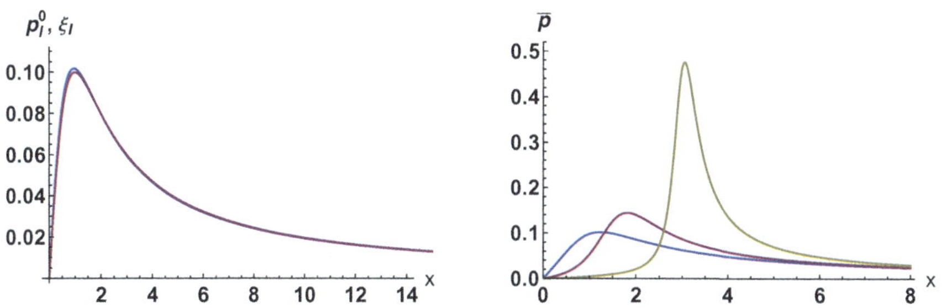

Fig. 4 Plots of p_l^0 (blue line) and ξ_l (red line) at $\tau = 0$ (left), and 'slow' electric field $\bar{p}$ (right) at $\tau = 4$ (blue line), $\tau = 10$ (yellow line), $\tau = 18$ (red line), for $a = 1.0$, $b = 1.0661$, $\varepsilon = 0.1$, $\mu = \sqrt{1/2000}$ and $\gamma = 0.001$

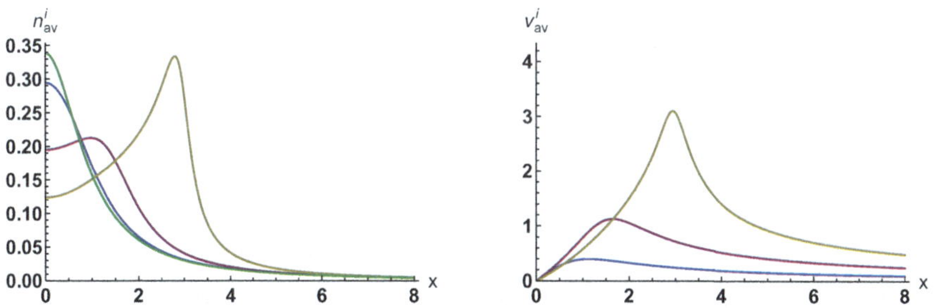

Fig. 5 Density and velocity distributions for ions at $\tau = 0$ (green line), $\tau = 4$ (blue line), $\tau = 10$ (red line), $\tau = 18$ (yellow line), for $b = 1.0661$, $\varepsilon = 0.1$, $\mu = \sqrt{1/2000}$ and $\gamma = 0.001$

4.2.2 Fast Dynamics of Particles

In this subsection, we use slow invariants to restore the complete dynamics of fast particles, electrons. For clarity sake, we consider the case of small values of $\delta\varepsilon \ll 1$, which means that x is identical to $\bar{x}$, and rewrite the Lie equations (115) in a simplified form

$$\frac{dt}{da} = 1 + \varepsilon\tau^2 \partial_x \xi \,, \quad \frac{d(\xi u)}{da} = \delta\xi \cos \Omega t \,, \quad \frac{d(\xi^3 p)}{da} = \delta\Omega\xi^3 \cos \Omega t \,. \tag{126}$$

According to the procedure of averaging [37], we can write the solutions of equations (126) by integrating over fast time t and taking into account the dependence upon slow time by including the dependence upon $\bar{x}$ and τ into δ and ξ. However, the enhanced precision is achieved by direct integration of the Lie equations (126) in account of the slow dependence of τ upon $\bar{x}$ as given by slow motion invariants

$$\begin{aligned} p &= \frac{\xi'^3}{\bar{\xi}^3}\bar{p}' + \frac{1}{\bar{\xi}^3}\int_x^{x'} \mathrm{d}x'' \frac{\bar{\delta}\bar{\Omega}\bar{\xi}^2(x'')}{\varepsilon\mu\tau(x'')} \sin(\bar{\Omega}t(x''))\,, \\ u &= \frac{\xi'}{\bar{\xi}}\bar{u}' + \frac{1}{\bar{\xi}}\int_x^{x'} \mathrm{d}x'' \frac{\bar{\delta}}{\varepsilon\mu\tau(x'')} \cos(\bar{\Omega}t(x''))\,. \end{aligned} \tag{127}$$

Electron distribution function $g = g' \equiv g^0(x', u')$ is the invariant of group transformations. Thus, substituting $x = \bar{x}$ and u' from (127) in g^0 and integrating over the velocity u gives the integral characteristic, the average electron velocity and density

$$n^e_{av} = n^e_0(x')\left(\xi'/\bar{\xi}\right)\,, \quad u^e_{av} = \frac{1}{\xi'}\int_x^{x'} \mathrm{d}x'' \frac{\bar{\delta}}{\varepsilon\mu\tau(x'')} \cos(\bar{\Omega}t(x''))\,. \tag{128}$$

To illustrate these formulas, we employ results of the previous section and consider the Lorentz-type initial densities profiles (123) with the function $\xi = \xi_l$ defined by (124). Substituting ξ_l in (127)–(128), we obtain the formulas for the spatial distribution of the electric field and the average electron velocity that are presented on the figures below for three different time moments. Figure 6 corresponds to moderate values of $\tau = 4$, when the ion density is concentrated mainly in the center of the bunch, thus leading to small-scale spatial oscillations primarily in this region. As the bunch spreads with growth of τ, the small-scale spatial oscillations moves outward as shown in Fig. 7 for $\tau = 10$, The figures also show that the 'complete' electric field p in this case oscillates with the same spatial period as the mean electron velocity and only slightly differs from the average electric field $\bar{p}$ (compare with Fig. 4).

To conclude in the above analysis of a particular physical problem, expansion of a plasma bunch, a new promising tool for analyzing nonlinear multi-scale sys-

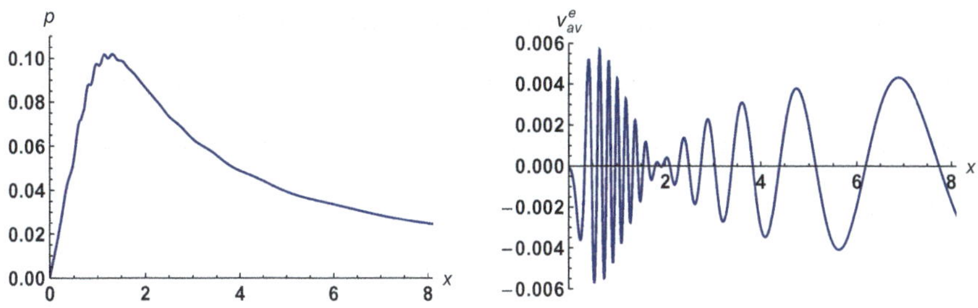

Fig. 6 Electric field and average electron velocity distributions at $\tau = 4$ for $a = 1.0$, $b = 1.0661$, $\varepsilon = 0.1$, $\mu = \sqrt{1/2000}$, and $\gamma = 0.001$

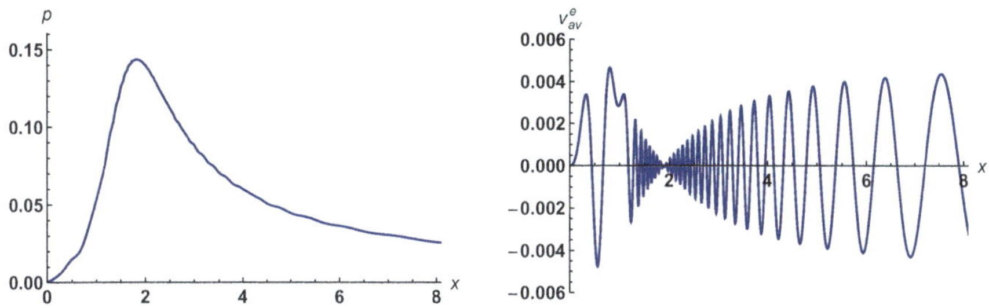

Fig. 7 Electric field and average electron velocity distributions at $\tau = 10$ for $a = 1.0$, $b = 1.0661$, $\varepsilon = 0.1$, $\mu = \sqrt{1/2000}$ and $\gamma = 0.001$

tems was considered. The main idea consists of employing the Krylov-Bogoliubov-Mitropolskii procedure of averaging to construct solutions of Lie equations. The procedure of separating 'fast' and 'slow' terms in coordinates of group generator naturally occurs in determining equations while constructing the symmetry for nonlinear equations that describe multi-scale behavior of any physical system. In our consideration, we use the averaging procedure in combination with a perturbation technique of group analysis [1] that gives approximate symmetries for the analyzed problem and helps to construct approximate RG-invariant solution for arbitrary initial distribution functions of particles.

The use of the averaging procedure in modern group analysis naturally separates invariant manifolds related to slow and fast Lie equations into slow and fast invariant manifolds. This separation is in the root of the theorem of invariant representation [12, §18]: averaging the fast invariant solution that appears as an oscillating curve on fast manifold yields a smooth curve on the slow invariant manifold as shown in the previous section (compare to the method of slow invariant manifold for describing kinetics of dissipative systems [47]). The merits of the approach with different scales that simplifies both the procedure of finding the admitted group and construction of the group invariant solutions point to the quest for future potential applications.

5 Approximate Symmetries in the Theory of Light-Beam Self-focusing

In this section, a method of approximate renormgroup symmetries is applied for analytical solution to the nonlinear Schrödinger equation (NLSE) which describes the electromagnetic beam self-focusing in a plasma with cubic, saturating, and relativistic nonlinearities. Different stationary self-focused waveguide propagation modes with respect to controlling laser-plasma parameters for a Gaussian radial intensity distribution at the plasma boundary are presented that describe self-trapping, self-focusing on the axis, and tubular self-focusing solutions.

The problem of self-focusing of a high power light beam [48–52] plays an important role in nonlinear electrodynamics since early sixties. For example, self-guiding of an intense laser pulse in an underdense plasma for many Rayleigh lengths without significant losses has been demonstrated many times both experimentally and in numerical simulations [53–55]. The need for parametric scaling for the laser beam power trapped in a self-focused channel gave an impulse to a search for effective ways of its analytic description.

The studies of stationary light beam self-focusing via rigorous analytical theories include the inverse scattering method [56] and the classical group analysis [57–60]. Here, a large variety of exact solutions to the NLSE in 1D, 2D, and 3D geometry for cubic and quintic nonlinearity and with additional linear inhomogeneous terms have been obtained. A common disadvantage of rigorous mathematical methods is that they consider some special solution for a specific boundary value problem. However, boundary conditions for these solutions do not correspond to a localized electromagnetic beam at the entry plane. Hence, a natural advance in the theory of self-focusing would be to solve the corresponding Cauchy problem for an intense laser beam that modifies the plasma dielectric permittivity and is characterized by a given intensity distribution at the plasma boundary. Below we draw attention to this problem with the help of the analytical method based on the use of approximate symmetries.

A mathematical model of wave self-focusing is based on the NLSE

$$2ik\partial_z E + \Delta_\perp E + k^2 \frac{\epsilon_{\rm nl}}{\epsilon_0} E = 0, \qquad E(0, \mathbf{r}) = E_0(\mathbf{r}), \tag{129}$$

for the complex electric field amplitude $E(z, \mathbf{r})$ of an electromagnetic wave with the frequency ω slowly varying in the propagation direction z. Equation (129) corresponds to a paraxial (quasioptical) approximation describing the stationary structure of the wave beam. Here, $k = (\omega/c)\sqrt{\epsilon_0}$ is the wave number of the electromagnetic wave, $\Delta_\perp$ is the Laplace operator in the plane $\mathbf{r}$ perpendicular to the beam axis z, $\epsilon_0 = 1 - 4\pi e^2 n_{e0}/(m_0\omega^2)$ is the linear dielectric permittivity of the plasma, and $\epsilon_{\rm nl}$ is the real part of the nonlinear permittivity of the plasma.

The use of quasioptical approximation (129) establishes the applicability conditions for our theory, which are determined by the following constraints on the characteristic longitudinal and transverse scales $\Lambda_{\parallel}$ and $\Lambda_{\perp}$ of the complex amplitude E (also see [62])

$$k\Lambda_{\parallel}, k_p\Lambda_{\parallel} \gg 1, \qquad k\Lambda_{\perp}^2 \approx \Lambda_{\parallel} \max\left\{1; (k\Lambda_{\perp})^2 \frac{\epsilon_{\mathrm{nl}}}{\epsilon_0}\right\}. \tag{130}$$

The inequality in Eq. (130) means that the laser pulse length substantially exceeds the wavelengths of the corresponding electromagnetic and plasma fields. It allows neglecting the contribution of the term with the second derivative of the electric field with respect to the longitudinal coordinate z (along the beam axis) compared with the first term contribution to Eq. (129) when deriving the NLSE from more complicated equations. The approximate equality in Eq. (130) relates the characteristic transverse and longitudinal scales of the electric field caused by diffraction and nonlinearity. Using NLSE (129) involves considering only the electromagnetic wave propagating forward into the nonlinear medium and the absence of backward waves that could arise in the presence of sharp gradients of the dielectric constant of the medium in the longitudinal direction (see, e.g., p. 432 in Sec. 17.12 in [61]). Conditions (130) can be violated if the solution singularity appears where the characteristic longitudinal scale of the complex field amplitude sharply decreases. Hence, the analytic solution that we obtain characterizes the behavior of the beam in some finite spatial domain from the entrance up to the singularity point. This restriction follows from the mathematical model used based on the NLSE (also see the discussion of this question in Sect. 9.2 in [62]).

Using the standard representation for the complex field amplitude $E = A\exp(iks)$ and introducing $w \equiv A^2 = |E|^2$ and the derivative $\mathbf{v} = \{v, 0\} = \nabla_{\perp} s$ of the eikonal s along the radius, we reduce NLSE (129) to the two equations

$$\begin{aligned} &\partial_z v + v\partial_r v - \frac{1}{2}\partial_r\Big(\frac{1}{\sqrt{w}}(\Delta_{\perp}\sqrt{w}) + \rho^2 F\Big) = 0, \\ &\partial_z w + w\partial_r v + v\partial_r w + w\frac{v}{r} = 0\,. \end{aligned} \tag{131}$$

In Eq. (131), we use the following dimensionless variables for the coordinates and complex field amplitude:

$$z \to \sqrt{2\beta}\,\frac{z}{d}, \qquad r \to \frac{r}{d}, \qquad w \to \frac{w}{w_0}, \qquad v \to \frac{v}{\sqrt{2\beta}}, \tag{132}$$

where $\beta = 1/2k^2d^2$, d is the characteristic transverse dimension of the light beam, and w_0 is the maximum value of w at the boundary of the medium. The contributions proportional to $\rho^2 = \omega_{\mathrm{pe}}^2 d^2/c^2$ determine the role of the effects of the relativistic and charge-displacement nonlinearity given by the function $F = (\epsilon_{\mathrm{nl}}/\epsilon_0)(k/k_p)^2$.

Equations (131) should be supplemented by the boundary conditions that determine the structure of the beam at the entrance $z = 0$ of the nonlinear medium. Below, we consider a cylindrically symmetric beam with a plane initial phase front, i.e., with the zero eikonal derivative, $v(0, r) = 0$, and a smooth distribution function for the square of the modulus of the electric field, $w(0, r) \equiv J(r)$. The case with a focused beam with a curved initial phase front, $v(0, r) \neq 0$, was analyzed in a similar way for cubic nonlinearity in [5, 63].

5.1 Approximate Solution of NLSE by the Renormgroup Symmetry Method

We here describe the construction of approximate solutions of NLSE (131) with plasma nonlinearity using the method of renormgroup symmetries [4], described in Sect. 1.3. This method consists in finding symmetries of a special kind under which approximate solutions of (131) constructed by perturbation theory for small distances from the boundary of a nonlinear medium are invariant and then applying these symmetries to extend the approximate solutions to the bulk of the nonlinear medium. Such a procedure is based on the property of a renormgroup symmetry operator to transform a solution of a boundary value problem with given boundary data into a solution of the same boundary value problem. To construct the renormgroup symmetry operator, following the general algorithm [4, 63], we use the Lie–Bäcklund symmetries admitted by the original differential equations (131) and determined by the canonical group operator

$$X = f\,\partial_v + g\,\partial_w. \tag{133}$$

The coordinates f and g of this operator are found by solving the corresponding determining equations expressing the invariance conditions for system (131) with respect to the group with operator (133)

$$\begin{aligned}
&D_z f + v D_r f + f v_1 - \partial_w(B) g - \partial_{w_1}(B) D_r g - \partial_{w_2}(B) D_r^2 g - \partial_{w_3}(B) D_r^3 g = 0,\\
&D_z g + w D_r f + v D_r g + g v_1 + f w_1 + \frac{vg}{r} + \frac{fw}{r} = 0.
\end{aligned} \tag{134}$$

Here,

$$\begin{aligned}
&B = \frac{1}{2} D_r\Big(\frac{D_r(r D_r \sqrt{w})}{r\sqrt{w}}\Big) + \frac{1}{2} D_r(\rho^2 F), \qquad v_s \equiv \frac{\partial^s v}{\partial r^s}, \quad w_s \equiv \frac{\partial^s w}{\partial r^s},\\
&D_r = \partial_r + \sum_{s=0}^{\infty} (v_{s+1}\partial_{v_s} + w_{s+1}\partial_{w_s}),
\end{aligned} \tag{135}$$

and D_z is represented as $D_z = D_z^0 + D_z^1$, where

$$
\begin{aligned}
D_z^0 &= \partial_z - \sum_{s=0}^{\infty} \{ D_r^s(vv_1)\partial_{v_s} + [D_r^{s+1}(wv) + D_r^s(\frac{wv}{r})]\partial_{w_s} \}, \\
D_z^1 &= \sum_{s=0}^{\infty} D_r^s(B)\partial_{v_s}.
\end{aligned} \tag{136}
$$

Because the terms originating from B in the first equation in (131) in the case of a slowly varying electric field amplitude are considered small compared with the first and the second terms, we seek f and g in the form of a series expansion in powers of the 'dimensionless relative amplitude' b of the contributions from B

$$
f = \sum_{i=0}^{\infty} f^i, \qquad g = \sum_{i=0}^{\infty} g^i. \tag{137}
$$

We restrict ourselves to only the first-order corrections

$$
f = f^0 + f^1 + o(b), \qquad g = g^0 + g^1 + o(b), \tag{138}
$$

where $f^0 \propto O(1)$, $g^0 \propto O(1)$, $f^1 \propto O(b)$, and $g^1 \propto O(b)$. We substitute (138) in the determining Eq. (134) and collect the zeroth- and first-order terms, obtaining

$$
\begin{aligned}
&M_0 f^0 = 0, \qquad M_1 g^0 + M_2 f^0 = 0, \\
&M_0 f^1 + D_z^1 f^0 - \partial_w(B) g^0 - \partial_{w_1}(B) D_r g^0 - \partial_{w_2}(B) D_r^2 g^0 - \partial_{w_3}(B) D_r^3 g^0 = 0, \\
&M_1 g^1 + D_z^1 g^0 + M_2 f^1 = 0,
\end{aligned} \tag{139}
$$

where

$$
\begin{aligned}
M_0 &= D_z^0 + vD_r + v_1, \\
M_1 &= D_z^0 + vD_r + v_1 + v/r, \\
M_2 &= wD_r + w_1 + w/r.
\end{aligned} \tag{140}
$$

We now set

$$
f^0 = \frac{1}{2} D_r(v^2), \qquad g^0 = \frac{1}{r} D_r(wvr), \tag{141}
$$

following [63]. This choice obviously satisfies zeroth-order equations (139) and the invariance conditions $f^0 = 0$ and $g^0 = 0$ at the boundary. We can then find f^1 from the first of the first-order equations in (139), which we rewrite as

$$
M_0(f^1 + B) = 0. \tag{142}
$$

The solution of this equation is expressed in terms of invariants of the operator M_0

$$
f^1 = \frac{1}{2} D_r\big(S(\chi) - \rho^2 F - \frac{1}{\sqrt{w}}(\Delta_\perp \sqrt{w})\big), \qquad \chi = r - vz, \tag{143}
$$

where

$$S(\chi) = \rho^2 F(J) + \frac{1}{\chi\sqrt{J(\chi)}}\partial_\chi[\chi\partial_\chi(\sqrt{J(\chi)})], \qquad F(J) = F(J(\chi)). \tag{144}$$

Substituting this result in the second first-order equation in (139) leads to an equation for the function g^1

$$M_1 g^1 + \frac{1}{2r} D_r[wzD_r S(\chi)] = 0. \tag{145}$$

It is easy to show by direct substitution that Eq. (145) can be rewritten as

$$M_0(rg^1) + \frac{1}{2} D_r[wzD_r S(\chi)] = 0.$$

This equation can be integrated similarly to Eq. (142). We then obtain

$$rg^1 = -\frac{1}{2} D_r(rwz\partial_\chi S).$$

Finally, up to the first order in the small parameter b, the Lie–Bäcklund symmetry operators in the canonical form are

$$f = vv_1 + \frac{1}{2} D_r\big(S(\chi) - \rho^2 F - \frac{1}{\sqrt{w}}(\Delta_\perp \sqrt{w})\big), \tag{146}$$

$$g = v\big(w_1 + \frac{w}{r}\big) + wv_1 - \frac{z}{2}\big[w(1 - zv_1)\partial_{\chi\chi} S + \big(w_1 + \frac{w}{r}\big)\partial_\chi S\big]. \tag{147}$$

In view of (131), we can rewrite (146) as

$$f = -\partial_z v + (1 - zv_1)\frac{\partial_\chi S}{2}.$$

Together with Eq. (147), the last equation leads to the two relations

$$v = z\frac{\partial_\chi S}{2}, \tag{148}$$

$$\partial_z v = (1 - zv_1)\frac{\partial_\chi S}{2}, \tag{149}$$

which must be satisfied to preserve the invariance requirement $f = 0$ and $g = 0$.

Keeping in mind the relation between the canonical form of the symmetry operator and the point symmetry group operator [6], we can now write the group symmetry operator

$$
R = \big(1 + \frac{z^2}{2}\partial_{\chi\chi} S\big)\partial_z + \frac{\partial_\chi S}{2}\partial_v + \frac{1}{2}(z\partial_\chi S + vz^2\partial_{\chi\chi} S)\partial_r
- \frac{wz}{2}\big[\big(1 + \frac{vz}{r}\big)\partial_{\chi\chi} S + \frac{1}{r}\,\partial_\chi S\big]\partial_w. \tag{150}
$$

Operator (150) is similar to the one previously obtained in [63] for a collimated beam except that $S(\chi)$ now contains the function F, which describes the laser beam in plasma with arbitrary saturating or relativistic nonlinearities. Operator (150) yields a system of characteristic equations

$$
\frac{dz}{1 + z^2\partial_{\chi\chi} S/2} = \frac{dv}{\partial_\chi S/2} = \frac{d\chi}{-v} = \frac{d\ln(wr)}{-z\partial_{\chi\chi} S/2}. \tag{151}
$$

This system of equations can be easily integrated after Eq. (148) is taken into account. The second and third equations in (151) give

$$
S + (\partial_\chi S)^2\frac{z^2}{4} = S(\mu), \tag{152}
$$

where μ corresponds to the value of χ at the boundary. The third and fourth equations in (151) yield another invariant $rw/\partial_\chi S$, which gives the dependence of w as

$$
w = J(\mu)\frac{\chi}{r}\frac{\partial_{\chi^2} S}{\partial_{\mu^2} S} \tag{153}
$$

in terms of its initial profile $J(\chi)$. Correspondingly, r and χ are related as

$$
r = \chi(1 + z^2\partial_{\chi^2} S). \tag{154}
$$

In summary, the solutions are represented by the equations

$$
v(r,z) = \frac{z}{2}\partial_\chi S, \qquad w(r,z) = J(\mu)\frac{\chi}{r}\frac{\partial_{\chi^2} S}{\partial_{\mu^2} S}, \tag{155}
$$

where χ and μ are defined as functions of z and r by the relations

$$
r = \chi(1 + z^2\partial_{\chi^2} S), \qquad S(\mu) = S(\chi) + \frac{z^2}{4}(\partial_\chi S)^2. \tag{156}
$$

These solutions are discussed in detail in the next sections for particular dependencies of F upon J.

5.2 NLSE Solution for Plasma with Saturating Nonlinearity

In this Section, we consider a plasma whose nonlinearity is due to the ponderomotive force with saturation [64], which corresponds to $F \equiv F^{sat} = 1 - \exp(-\nu w)$. The constant $\nu = w_0/16\pi n_c T_e$ define the saturation strength of nonlinearity in the Schrödinger equation with the nonlinearity $F^{sat} = 1 - \exp(-\nu w^2)$, T_e and n_c are the electron temperature and the critical density of the plasma, respectively. This type of nonlinearity corresponds to the propagation of a laser beam of non-relativistic intensity in a fully ionized plasma. For $\nu w \ll 1$, this function F^{sat} corresponds to a medium with cubic nonlinearity, $\lim_{\nu w \to 0} F^{sat} = F_{cub} = \nu w$.

Using formulas (155), we consider in greater detail the evolution of a beam with the Gaussian initial intensity profile $J(r) = \exp(-r^2)$; for this beam, the function $S^{sat}(\chi)$ is rewritten, after the introduction of a new variable $\eta = \chi^2$, as

$$S^{sat}(p) = \rho^2 \left[1 - \exp\left(-\nu p\right)\right] + \ln(1/p) - 2\,, \quad p = \exp(-\eta)\,. \tag{157}$$

It follows from formulas (155) and (156) that a necessary condition for the existence of rays that deviate toward the beam axis (i.e., rays for which $v < 0$) is given by $\partial_\eta S^{sat} < 0$, where

$$\partial_\eta S^{sat} = 1 - \nu\rho^2 p \mathrm{e}^{-\nu p}\,. \tag{158}$$

The derivative $\partial_\eta S^{sat}$ attains its minimum value, equal to

$$(\partial_\eta S^{sat})^{min} = 1 - \frac{\rho^2}{\mathrm{e}}\,, \tag{159}$$

at the point $p = p_{min} = (1/\nu)$.

Hence, for $\rho^2 < \mathrm{e}$ the derivative of S^{sat}_η is everywhere positive; i.e., all rays deviate away from the axis. As ρ^2 increases, when the opposite condition, $\rho^2 > \mathrm{e}$, is satisfied, there arise domains with $v < 0$, where the rays deviate toward the beam axis. Taking into account the natural constraint $0 \leqslant p \leqslant 1$, we can easily find out that the value $(\partial_\eta S^{sat})^{min}$ is not attained for $\nu < 1$: the function $(\partial_\eta S^{sat})$ monotonically decreases and attains its minimum value $(\partial_\eta S^{sat})^{axis}$ for $p = 1$, i.e., on the beam axis $r = 0$

$$(\partial_\eta S^{sat})^{axis} = 1 - \nu\rho^2 \exp(-\nu)\,. \tag{160}$$

Equality (160) shows that, for $\nu < 1$, the boundary of the domain with $v < 0$ lies below the straight line $\rho^2 = \mathrm{e}$ and is defined by the condition $(\partial_\eta S^{sat})^{axis} = 0$. For $(\partial_\eta S^{sat})^{axis} < 0$, i.e., in a paraxial domain, the rays are directed toward the axis of the beam, and when the beam propagates into the bulk of the plasma, the profile $v(z, r)$ becomes steeper: the derivative $\partial_r v(z, r)$ on the beam axis

$$(\partial_r v(z, r))_{|r \to 0} \equiv \Big[\frac{(z/2)\left(\partial_{\chi\chi} S^{sat}\right)}{1 + (z^2/2)\left(\partial_{\chi\chi} S^{sat}\right)}\Big]_{|r \to 0} = \frac{z\left(1 - \nu\rho^2 \mathrm{e}^{-\nu}\right)}{1 + z^2\left(1 - \nu\rho^2 \mathrm{e}^{-\nu}\right)} \tag{161}$$

increases in absolute value as the coordinate z increases, and, at a distance of z_{axis} from the boundary

$$z_{axis} = \left[\nu\rho^2 e^{-\nu} - 1\right]^{-1/2}, \tag{162}$$

the profile $v(z, r)$ breaks. The limit $\nu \to 0$ corresponds to the well-known result obtained for a medium with cubic nonlinearity [65]; namely, it describes a blow up of intensity on the beam axis when the effect of the nonlinear term turns out to be more significant than the diffraction term.

When $\nu > 1$, the point p_{min} at which the derivative $\partial_\eta S$ attains its minimum lies inside the admissible interval, $p_{min} \in [0, 1]$, and one may expect that the wave front will experience the strongest distortion for the radius defined by the quantity p_{min}. The increase in the distortions of the wave front with increasing coordinate z leads to its breaking away from the beam axis as well. The breaking condition corresponds to the instant, defined by the conditions $\partial_\chi r = \partial_{\chi\chi} r = 0$, when the function $r(\chi)$ ceases to be single valued. As a result, we obtain a pair of relations that define the radial coordinate of the breaking point, $r_{br} = \sqrt{\ln 1/p_{br}}(1 + z_{br}^2(1 - \rho^2 \nu p_{br} e^{-\nu p_{br}}))$ and the corresponding coordinate on the beam axis, z_{br}

$$\begin{aligned} &3(1 - \gamma p_{br}) - 2\ln p_{br}(3\nu p_{br} - 1 - (\nu p_{br})^2) = 0, \\ &z_{br} = \left[\nu p_{br}\rho^2 e^{-\nu p_{br}}(1 + 2(1 - \nu p_{br})\ln p_{br}) - 1\right]^{-1/2}. \end{aligned} \tag{163}$$

To verify the analytical results obtained and to determine their applicability domain, we present the comparison of the numerical and analytical results for two variants of laser-plasma parameters. The most obvious difference and similarity between numerical and analytical results can be demonstrated by the direct comparison of the spatial distributions of intensity and the derivative of the beam eikonal.

As the first example, consider the curves of spatial distribution of the eikonal derivative and the intensity of a beam for the parameters $\nu = 0.1$ and $\rho^2 = 1000$. These curves are shown in Fig. 8. As regards the distribution of intensity, the spatial evolution pattern of a beam in a medium with saturating nonlinearity, shown in Fig. 8a, is similar to the typical pattern of beam evolution in a medium with cubic nonlinearity when the effects of saturation of nonlinearity are inessential. However, in contrast to the variant when there is no saturation of nonlinearity and the intensity on the beam axis tends to infinity, here, for $\nu \neq 0$, the intensity on the beam axis remains constant when approaching the breaking point and behind this point, as it follows from the intensity distribution obtained in the numerical experiment shown in Fig. 8c and in the inset of this figure. This fact suggests that it is necessary to correct the analytical consideration in the neighborhood of the breaking point. A comparison of the curves presented in Fig. 8 shows that the analytical and numerical results are in good agreement up to the point $z \approx 0.75\, z_{axis}$. For larger z, the agreement is only qualitative (the oscillations of the radial intensity distribution, which are observed in the numerical calculations, are not confirmed by analytical calculations); the theory predicts the breaking of the profile $v(z, r)$ for $z = z_{axis}$. There is also a good agreement between the numerical and analytical results for the boundary r_w

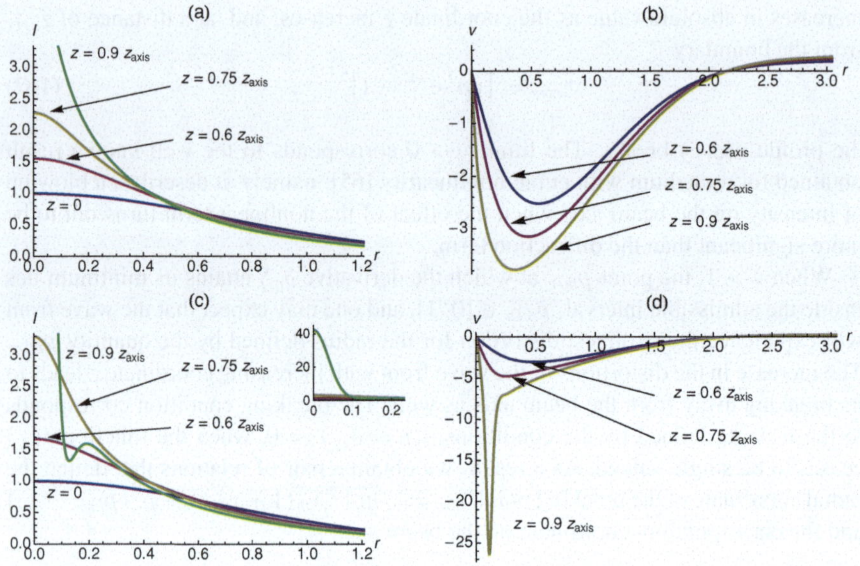

Fig. 8 Spatial distribution of (**a**), (**c**) the intensity $I = w$ and (**b**), (**d**) the eikonal derivative v of a laser beam in a plasma for $\rho^2 = 1000$ and $\nu = 0.1$ at different distances from the plasma boundary: (1) z/z_{axis}= 0.9, (2) 0.75, (3) 0.6, and (4) 0; (**a**), (**b**) analytical results and (**c**), (**d**) results of numerical simulation

of the waveguiding configuration (cf. the points of intersection of the curves of the eikonal derivative with the axis in Fig. 8b, d).

As the second example, we take the case corresponding to $\nu = 8.5$, $\rho^2 = 200$. The analytic curves of the distribution of intensity and eikonal derivative for these parameters are shown in Fig. 9. The maximum beam intensity is formed inside a tubular waveguide at a finite distance from the waveguide axis near the point $r_{br} \approx 0.8983$, and the intensity on the beam axis falls off, although insignificantly. A comparison of Fig. 9a, b with Fig. 9c, d shows that, in the numerical calculation, a ring structure is formed slightly later than that in the analytical calculation, and the formation of the ring itself is accompanied by the emergence of secondary local maxima for larger r. Nevertheless, here we can also speak of a good agreement between the results of analytic and numerical approaches.

5.3 NLSE Solution for Plasma with Relativistic Nonlinearities

In this section, we consider another type of the nonlinearity in NLSE (129) that is determined by the nonlinear refraction of the light beam given by the function $\epsilon_{\rm nl}$

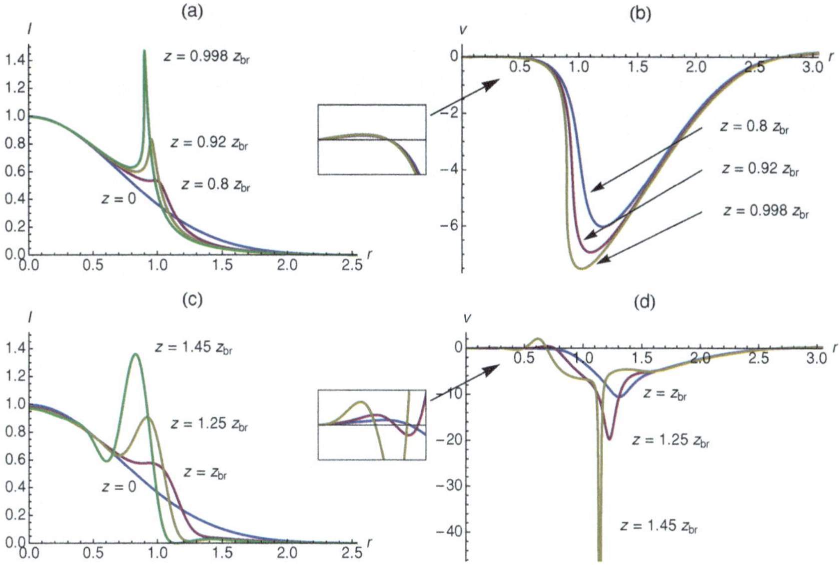

Fig. 9 Spatial distribution of **a**, **c** the intensity $I = w$ and **b**, **d** the eikonal derivative v of a laser beam in a plasma for $\rho^2 = 200$ and $\nu = 8.5$ at different distances from the plasma boundary: (1) z/z_{br}= 0.998, (2) 0.925, (3) 0.8 for **a**, **b**, and (1) z/z_{br}= 1.45, (2) 1.25, (3) 1.0 for **c**, **d**; **a**, **b** analytical results and **c**, **d** results of numerical simulation

$$\epsilon_{\rm nl} = \epsilon_0 \frac{k_p^2}{k^2}\Big(1 - \frac{n_e}{\gamma n_{e0}}\Big), \qquad k_p^2 = \frac{4\pi e^2 n_{e0}}{m_0 c^2}. \tag{164}$$

It is due to two factors: (1) the relativistic nonlinearity of the electron mass, determined by the value of the relativistic factor $\gamma = \sqrt{1 + |E/E_{\rm rel}|^2}$, where $E_{\rm rel}^2 = (\omega c m_0/e)^2$, and (2) the charge-displacement nonlinearity, which determines the nonlinear deformation of the electron density $n_e = n_{e0} N_e(\gamma)$, proportional to $\Delta_\perp \gamma$. Usually, the well-known standard formula

$$N_e = \max\{0, 1 + k_p^{-2} \Delta_\perp \gamma\} \tag{165}$$

is used for N_e, which we follow also taking the standard condition of the nonnegativity of the electron density, $n_e \geq 0$, into account, which allows describing a strong density modulation including the electron cavitation effect [66]. The modification of the piecewise smooth function (165) to obtain a smoothed transition from a vanishingly low electron density, $N_e \to 0$, to a linear dependence on $\Delta_\perp \gamma$ by taking a weak thermal motion of electrons into account was discussed in [67, 68]. As a possible example of such a modification, we can also use the smooth approximation

$$N_e = \frac{1 + k_p^{-2} \Delta_\perp \gamma}{1 - \exp[-\alpha_0 (1 + k_p^{-2} \Delta_\perp \gamma)]}, \tag{166}$$

where the value of the positive parameter $\alpha_0 \gg 1$ determines the transition from the linear dependence of $N_e \propto \Delta_\perp \gamma$ to the exponentially decreasing $N_e \to 0$ with the intensity gradient change. Relying on arguments akin to those used in [67], we believe that there is no need for a concrete identification of the mechanism responsible for α_0 if the results for different $\alpha_0 \gg 1$ differ very little.

Using N_e in $\epsilon_{\rm nl}$ and then in $F \equiv F^{rel} = 1 - N_e(\gamma)/\gamma$, where $\gamma = \sqrt{1 + i_0 w}$, we obtain the formula for the functions S^{rel} and F^{rel}, defining the solution (155), (156)

$$S^{rel}(\chi) = \rho^2 F(J) + \frac{1}{\chi\sqrt{J(\chi)}} \partial_\chi [\chi \partial_\chi (\sqrt{J(\chi)})], \\ F^{rel}(J) = 1 - \frac{N_e\{\gamma[J(\chi)]\}}{\gamma[J(\chi)]}, \tag{167}$$

and accounting for the effects of the relativistic and charge-displacement nonlinearity of the medium. Here, the parameter i_0 can be written as the ratio of the maximum beam intensity $I_0 = (c/4\pi)w_0$ to the characteristic relativistic intensity $I_r = \omega^2 m_0^2 c^3/(4\pi e^2)$, i.e., $i_0 = I_0/I_r$. In the limit $i_0 w \ll 1$, the function F^{rel} corresponds to a medium with cubic nonlinearity, $\lim_{i_0 w \to 0} F^{rel} = F_{\rm cub} = (i_0/2)w$.

Using Eqs. (155), we analyze the evolution of the laser beam with the initial Gaussian profile $J(r) = \exp(-r^2)$ and assume that the effect of electron cavitation does not appear at the boundary of the nonlinear medium. We can then even use in Eq. (167) the simplest expression for N_e, Eq. (165), which is itself a smooth function and does not require smoothing approximation, Eq. (166). The function F^{rel} in Eq. (167) is determined only by the initial distribution of the square of the amplitude of the electric field of the beam at $z = 0$. We rewrite the expression for $S^{rel}(\chi)$ after introducing the variable $p = \exp(-\chi^2)$ in the form

$$S^{rel}(p) = \rho^2 (1 - \frac{1}{\sqrt{1 + i_0 p}}) - \frac{2}{1 + i_0 p} - \frac{\ln p}{(1 + i_0 p)^2}. \tag{168}$$

We recall that the applicability of Eq. (168) is limited only by distributions of electron densities that correspond to the condition that there is no effect of electron cavitation at the boundary of the nonlinear medium $z = 0$, i.e., cavitation does not occur when the inequality

$$\rho \geq \rho_{\rm cav}, \quad \text{where} \quad \rho_{\rm cav}^2 = 2 i_0 (1 + i_0)^{-1/2}, \tag{169}$$

is satisfied. A characteristic feature of the solution given by Eqs. (155)–(156) and Eq. (168) is the possibility of the growth of distortions of the wavefront of the beam with the coordinate z, up to breaking the profile $v(z, r)$. This happens when a single-valued dependence of r on χ is violated, i.e., when two conditions are met, $\partial_{\chi\chi} r = 0$ and $\partial_\chi r = 0$. The first condition defines the radial coordinate of the breaking point $r_{\rm br} = \sqrt{\ln 1/p_{\rm br}} (1 - z_{\rm br}^2 \{p \partial_p S\}_{|p = p_{\rm br}})$ written in terms of $p_{\rm br}$, and the second gives the corresponding coordinate $z_{\rm br}$ along the beam axis. In the variables $\{z, p\}$, the conditions for the appearance of the solution singularity can be written as

$$\left\{p\sqrt{\ln\frac{1}{p}}[3(\partial_p S^{rel} + p\partial_{pp}S^{rel}) + 2\ln p(\partial_p S^{rel} + 3p\,\partial_{pp}S^{rel} + p^2\partial_{ppp}S^{rel})]\right\}_{|p=p_{\rm br}} = 0, \tag{170}$$

$$z_{\rm br}^2 = \{[p\,\partial_p S^{rel} + 2\ln p(p\,\partial_p S^{rel} + p^2\partial_{pp}S^{rel})]^{-1}\}_{|p=p_{\rm br}}. \tag{171}$$

Solutions of the equation for $p_{\rm br}$ in Eq. (171) correspond to either of its two factors vanishing, i.e., either (a) $\ln(1/p_{\rm br}) = 0$, which corresponds to the appearance of the axial singularity $p_{\rm br} = 1$ or (b) the expression in square brackets in Eq. (171) vanishes, which corresponds to the off-axis singularity $p_{\rm br} \neq 0$. We analyze these two cases separately.

a. The appearance of the axial singularity $p_{\rm br} = 1$.

The simplest form of Eqs. (171) is obtained at $p_{\rm br} = 1$, when a singularity appears on the beam axis at the point $\{z_{br}\}_{p=1} \equiv z_{\rm axis}$ defined by the equality

$$z_{\rm axis}^2 = \left[\frac{\rho^2 i_0}{2}\frac{1}{(1+i_0)^{3/2}} + \frac{2i_0 - 1}{(1+i_0)^2}\right]^{-1}. \tag{172}$$

The positivity condition $z_{\rm axis}^2 > 0$ determines the range of parameters ρ^2 and i_0 for which there exists an axial singularity

$$\rho > \rho_{\rm axis}, \quad \text{where} \quad \rho_{\rm axis}^2 = \frac{2}{i_0}(1 - 2i_0)(1+i_0)^{-1/2}, \quad i_0 \leq \sqrt{2} - 1. \tag{173}$$

The minimum beam radius $\rho_{\rm axis}$ depending on i_0 above which an axial singularity appears corresponds to the limit $z_{\rm axis}^2 \to \infty$ in (172). The upper limit of i_0 in Eq. (173) is given by the condition of the absence of the electron cavitation effect at $z = 0$ and corresponds to $\rho_{\rm axis} \geq \rho_{\rm cav}$. For a higher laser beam intensity $i_0 > \sqrt{2} - 1$, the allowable range of the beam radius is cut from the side of small values by the condition $\rho > \rho_{\rm cav}$.

b. The appearance of an off-axis singularity $p_{\rm br} \neq 1$.

The off-axis singularity corresponds to a nonzero value $p_{\rm br} \neq 1$ given by the condition $\partial_{\chi\chi} r = 0$, which in this case becomes

$$[3(\partial_p S^{rel} + p\,\partial_{pp}S^{rel}) + 2\ln p(\partial_p S^{rel} + 3p\,\partial_{pp}S^{rel} + p^2\partial_{ppp}S^{rel})]_{|p=p_{\rm br}} = 0. \tag{174}$$

The boundary of the domain of parameters for which there exists an off-axis singularity is determined by the equation

$$\rho = \rho_{\rm off}, \quad \text{where} \quad \rho_{\rm off}^2 = \frac{8(3 - i_0)}{(i_0 - 2)\sqrt{1 + i_0}}, \tag{175}$$

for $2 < i_0 \leq \sqrt{13} - 1$. For greater beam intensity values $i_0 > \sqrt{13} - 1$, the lower boundary of the beam radius is given, as in case a, by the condition $\rho > \rho_{\rm cav}$. Comparing Eqs. (173) and (175) shows that the boundary of the domain for the solution

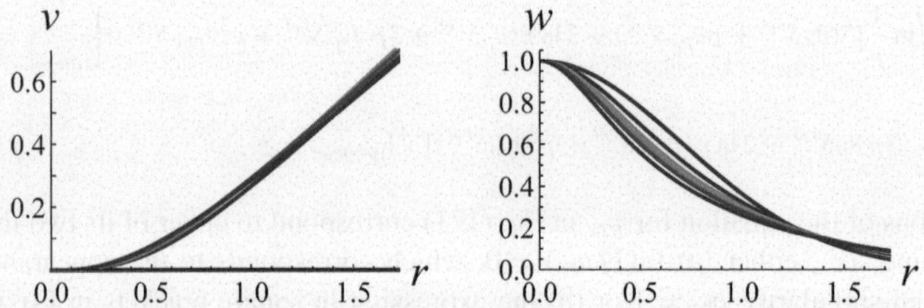

Fig. 10 The spatial distribution of the eikonal derivative v (left image) and the square of the amplitude w of the electric field of the light beam (right image) for different values of the longitudinal coordinate z along the beam axis for $i_0 = 0.1$ and $\rho = 3.99$: the different curves correspond to the transition from $z = 0$ to $z = 0.6; 0.8; 0.9; 1.0; 1.2$. An increase of z is accompanied by a clearly discernible decrease of w in the region of small $r \to 0$

with the off-axis singularity is at greater beam intensities than the domain for the solution with the axial singularity.

Combining the above conditions on the beam-plasma parameters given by Eqs. (169), (173), and (175), we obtain the partitioning of the controlling parameter plane into several domains. A graphical representation on the parameter plane $\{i_0, \rho\}$ in terms of the domain boundaries as well as their detailed analysis can be found in [70]. Knowing the boundaries of the domains in the parameter plane $\{\rho, i_0\}$ makes it much easier to analyze the type of solutions and the conditions for the appearance of the solution singularities. This is an important new finding in the theory of relativistic self-focusing.

To illustrate the results obtained above, we now present plots of the spatial distributions of the eikonal derivative, the square of the amplitude of the light beam electric field, and the electron density obtained above using approximate analytic solution (155) for the three sets of parameters i_0 and ρ.

The first example correspond to those values of i_0 and ρ for which the nonlinearity partially compensates the diffraction spreading of the beam as the z increases but not so much as to lead to the formation of a singularity. Such a regime in which the beam intensity in the near-axis region is preserved at distances exceeding the length at which the beam diffraction divergence occurs in a linear medium is called the self-trapping mode of the wave beam. This mode is illustrated in Fig. 10, where the spatial distribution of the eikonal derivative v and the square of the amplitude w of light beam electric field are shown for $i_0 = 0.1$, $\rho = 3.99$ for different values of the coordinate z along the beam axis.

The next example corresponds to values of the parameters i_0 and ρ for which a solution singularity can arise, developing on the beam axis. The illustration of this focusing on the axis regime is presented for the parameter values $i_0 = 0.2$ and $\rho = 5$. The plots of the spatial distributions of the eikonal derivative v and the square of the amplitude w of the electric field of the light beam for different values of the coordinate z are shown in Fig. 11.

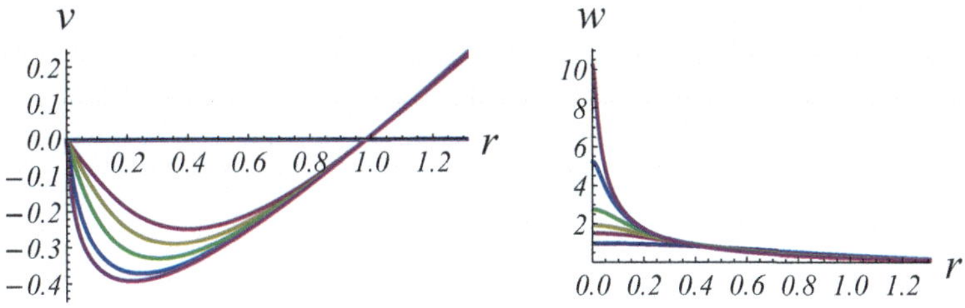

Fig. 11 The spatial distribution of the eikonal derivative v (left image) and the square of the amplitude w of the electric field of the light beam (right image) for different values of the longitudinal coordinate z along the beam axis for $i_0 = 0.2$ and $\rho = 5$: the value of the longitudinal coordinate of the axial singularity for the chosen parameter values $i_0 = 0.2$ and $\rho = 5$, given by condition (172), is equal to $z_{\text{axis}} = 0.82$. The different curves correspond to the transition from $z = 0$ to $z = 0.6z_{\text{axis}}$; $0.7z_{\text{axis}}$; $0.8z_{\text{axis}}$; $0.9z_{\text{axis}}$; $0.95z_{\text{axis}}$ from the top down for curves in the left image and from the bottom up in the right image (for small $r \to 0$)

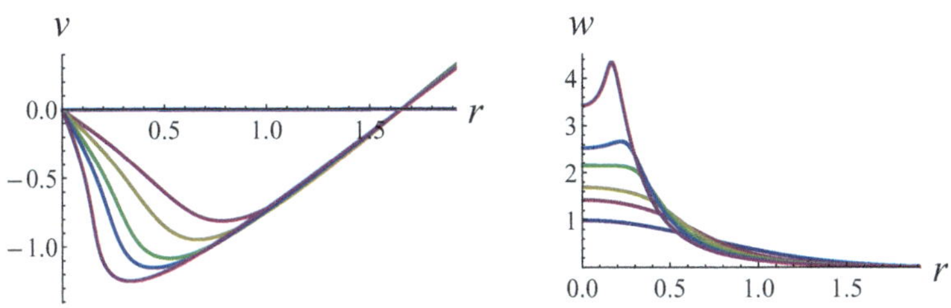

Fig. 12 The spatial distribution of the eikonal derivative v (left image) and the square of the amplitude w of the electric field of the light beam (right image) for different values of the longitudinal coordinate z along the beam axis for $i_0 = 5$ and $\rho = 3$: the values of the longitudinal coordinates of the on- and off-axis singularities for the chosen parameter values are $z_{\text{axis}} \approx 0.564$ and $z_{\text{br}} \approx 0.454$. The different curves correspond to the transition from $z = 0$ to $z = 0.6z_{\text{br}}$; $0.7z_{\text{br}}$; $0.8z_{\text{br}}$; $0.85z_{\text{br}}$; $0.92z_{\text{br}}$ from the top down for curves in the left image and from the bottom up in the right image (for small $r \to 0$)

Finally, the last example corresponds to values of the parameters i_0 and ρ for which there is a possibility to develop both off- and on-axis solution singularities, which for the parameter values $i_0 = 5$, $\rho = 3$ appear at $z_{\text{br}} \approx 0.686$ (off-axis singularity) and at $z_{\text{axis}} \approx 0.749$ (for the on-axis singularity). The plots of the spatial distributions of the eikonal derivative v and the square of the amplitude w of the electric field of the light beam for this example are shown for different values of the coordinate z in Fig. 12.

Because the z_{axis} value exceeds z_{br}, the off-axis feature appears more explicitly in the form of a ring structure shown in Fig. 12, although the square of the amplitude of the electric field on the beam axis also increases.

To conclude, we have demonstrated here the effictiveness of using approximate renormgroup symmetries in constructing analytical solutions to the nonlinear Shrödinger equation which describes self-focusing of a laser beam in a plasma with saturating and relativistic nonlinearities. Detailed discussion of these results in applications can be found in [63, 69, 70].

References

1. Baikov VA, Gazizov RK, Ibragimov NH (1989) Math. USSR-Sbornik. 64:427
2. Baikov VA, Gazizov RK, Ibragimov NH (1991) J Sov Math 55:1450
3. Kovalev VF, Pustovalov VV Shirkov DV (1998) Math Phys 39:1170 hep-th/9706056
4. Kovalev VF, Shirkov DV (2008) Renormalization-group symmetries for solutions of nonlinear boundary value problems, Physics-Uspekhi 51(8):815–830; Prepint arXiv:math-ph/0812.4821
5. Ibragimov NH, Kovalev VF (2009) Approximate and Renormgroup symmetries. In: Luo AC, Ibragimov NH (eds) Nonlinear physical science. Higher Education Press & Springer, ISBN 978-3-642-0027-4, ISBN 978-7-04-026159-3 (2009)
6. Ibragimov NH (1994) CRC handbook of lie group analysis of differential equations
7. Ibragimov NH (ed) (1995) Vol. 1: Symmetries, exact solutions and conservation laws, 1994; vol 2: Applications in engineering and physical sciences, 1995; Vol.3: New trends in theoretical developments and computational methods (1996) CRC handbook of lie group analysis of differential equations. CRC Press, Boca Raton
8. Shirkov DV, Kovalev VF (2009) Phys Rep V 352:219; Prepint arXiv:hep-th/0001210
9. Bogoliubov N, Shirkov D (1959, 1980) Introduction to the theory of quantized fields. Wiley-Interscience, New York
10. Shirkov DV (1982) Sov Phys Dokl 27:197
11. Kovalev VF, Shirkov DV (2006) J Phys A: Math Gen 39:8061
12. Ovsyannikov LV (1978) Gruppovoi Analiz Differentsial'nykh Uravnenii (Group analysis of differential equations). Nauka, Moscow. [Translated into English (New York: Academic Press, 1982)]
13. Rudenko OV, Soluyan SI (1977) Theoretical foundations of nonlinear acoustics consultants bureau. New York
14. Y.N. Grigoriev, N.H. Ibragimov, V.F. Kovalev, S.V. Meleshko, Symmetries of integro-differential equations with applications in mechanics and plasma physics. Series: *Lecture Notes in Physics*, Vol.806, 2010, 330 p
15. *V. F. Kovalev, and O. V. Rudenko* Nonlinear acoustic waves in channels with variable cross sections Acoustical Physics. 2012. V. 58. No. 3. P. 296–303. arXiv:1208.1360v2 [math-ph] 2 Nov 2012
16. *Rudenko O.V.* Nonlinear sawtooth-shaped waves Physics-Uspekhi (Adv.Phys.Sci). 1995. V. 38. P. 965–989
17. *B.O. Enflo, O.V. Rudenko* To the theory of generalizied Burgers' equation Acta Acustica unified with Acustica. 2002. V. 88. P. 155–162
18. Rozanov VV, Rudenko VO, Sysoev NN (2009) Hemodynamics and nonlinear acoustics: General approaches and solutions Acoustical Physics. 55(4–5):602–611
19. Landau LD, Lifshitz EM (1986) Fluid Mechanics. Pergamon Press, Oxford
20. Rudenko OV, Soluyan SI (1977) Theoretical Foundations of Nonlinear Acoustics. Plenum, Consultants Bureau
21. Vinogradova MB, Rudenko OV, Sukhorukov AP (1990) The Theory of Waves. Nauka, Moscow [in Russian]
22. *Yu. R. Lapidus and O. V. Rudenko* Using of nonlinear highest harmonics generation for a profile reconstruction of acoustic waveguides Sov. Phys. Acoust. 1990. V. 36. P. 589–562

23. *O.V. Rudenko, A.B. Shvartsburg* Nonlinear and linear phenomena in narrow pipes Acoustical Physics. 2010. V. 56(4). P. 429–434
24. Rudenko OV, Sukhorukova AK, Sukhorukov AP (1994) Equations of high-frequency nonlinear acoustics for inhomogeneous media Acoustical Physics. 40:264–268
25. Sarkisov GS, Bychenkov VY, Novikov VN, Tikhonchuk VT, Maksimchuk A, Chen S-Y, Wagner R, Mourou G, Umstadter D (1999) Self-focusing, channel formation, and high-energy ion generation in interaction of an intense short laser pulse with a He jet Phys. Rev. E 59:7042
26. Macchi A, Ceccherini F, Cornolti F, Kar S, Borghesi M (2009) Plasma Phys. Control. Fusion 51:024005
27. Kuznetsov AV, Esirkepov TZ, Kamenets FF, Bulanov SV (2001) Plasma Phys Rep 27(3):211
28. Pukhov A, Sheng Z-M, Meyer-ter-Vehn J (1999) Particle acceleration in relativistic laser channels Phys. Plasmas 6:2847
29. Popov KI, Rozmus W, Bychenkov VY, Naseri N, Capjack CE, Brantov AV (2010) Ion Response to Relativistic Electron Bunches in the Blowout Regime of Laser-Plasma Accelerators Phys. Rev. Lett. 105:195002
30. M. A. Miller, Izv. Vyssh. Uchebn. Zaved., Radiofiz. 1, 110 (1958)
31. A. G. Litvak, in Problems of Plasma Theory, Ed. by M. A. Leontovich (Atomizdat, Moscow, 1980), Vol. 10, p. 164 [in Russian]
32. Gurevich AV, Pariĭskaya LV, Pitaevskiĭ LP (1966) Self-similar motion of rarefied plasma Soviet Physics JETP 22(2):449
33. V.F. Kovalev, S. V. Krivenko, and V. V. Pustovalov, Differen. Uravn. 29, 1804 (1993); V. F. Kovalev, S. V. Krivenko, and V. V. Pustovalov, Differ. Uravn. 29, 1971 (1993)
34. Kovalev VF, Bychenkov VY (2015) Radial Acceleration of Ions by a Laser Pulse in a Plasma Channel. J Exp Theor Phys 121(1):1–18. https://doi.org/10.1134/S1063776115060187
35. Nayfeh AH (1973) Perturbation methods. Wiley, New York
36. V. F. Zhuravlev, D. M. Klymov, *Applied methods in the theory of oscillations*, (Moscow, "Nauka", 1988)
37. N. N. Bogoliubov and Yu. A. Mitropolsky, *Asymptotic methods in the theory of nonlinear oscillations*, 4th ed., ("Nauka", Moscow, 1974; English transl. of 2nd ed., Hindustan, Delhi, 1961, and Gordon & Breach, New York, 1962)
38. N. M. Krylov, N. N. Bogolyubov, *Introduction to non-linear mechanics* (Kiev: Izd-vo AN SSSR, 1937 (in Russian); 1947 (in English, partial translation from Russian). Princeton: Princeton Univ. Press.)
39. Hori GI (1966) Theory of general perturbations with unspecified canonical variables. Publ. Astron. Soc. Jap. 18:287–296
40. Yu. A. Mitropolsky, A. K. Lopatin, *A group-theoretic approach in asymptotical methods of nonlinear mechanics*, (Naukova Dumka, Kiev, 1988). (English transl.: *Nonlinear mechanics, groups and symmetry.* Mathematics and its Applications, 319. Kluwer Academic Publishers Group, Dordrecht, 1995.)
41. V. A. Baikov, R. K. Gazizov, N. H. Ibragimov, *Method of multiple scales in approximate group analysis: Boussinesq and Korteweg-de-Vries equations* (Moscow: Nat. Cent. Math. Model., 1991, preprint No.31 (in Russian))
42. Taranov VB (1976) On the symmetry of one-dimensional high-frequency motions of a collisionless plasma. J Tech Phys 46:1271–1277 [in Russian]
43. Grigoriev YN, Meleshko SV (1987) Group analysis of the integro-differential Boltzman equation. Dokl. AS USSR 297(2):323–327
44. Grigoriev YN, Meleshko SV (1995) Group analysis of kinetic equations. Russian J. Numer. Anal. Math. Modelling 10(5):425–447
45. Kovalev VF, Bychenkov VY, Tikhonchuk VT (2002) Particle dynamics during adiabatic expansion of a plasma bunch. Journ. Exper. Theor. Physics 95(2):226–241
46. Kovalev VF (2011) Lie group analysis for multi-scale plasma dynamics. Journal of Nonlinear Mathematical Physics 18(Suppl. 1):163–175. https://doi.org/10.1142/S1402925111001349
47. Gorban AN, Karlin IV (2005) Invariant Manifolds for Physical and Chemical Kinetics, Series: Lecture Notes in Physics, vol 660. Springer, Berlin/Heidelberg

48. Askar'yan GA (1962) Sov. Phys. JETP 15:1567
49. R. Chiao, E. Garmire, and G. Townes, Phys. Rev. Lett. **13**, 479 (1964); **14**, 1056 (1965); **16**, 347 (1966)
50. Talanov VI (1965) JETP Lett 2:138
51. Akhmanov SA, Sukhorukov AP, Khokhlov RV (1966) Sov. Phys. JETP 23:1025
52. Dyshko AL, Lugovoi VN, Prokhorov AM (1967) JETP Lett 6:146
53. Sprangle P, Esarey E, Krall J, Joyce G (1992) Phys Rev Lett 69:2200
54. Bulanov SV, Pegoraro F, Pukhov AM (1995) Phys Rev Lett 74:710
55. Mora P, Antonsen TM Jr (1996) Phys Rev E 53:2068
56. Zakharov VE, Shabat AB (1972) Sov. Phys. JETP 34:62
57. Taranov VB (1986) Preprint KIYaI-80-15, Kiev, 1980. J Plasma Phys 35:141
58. L. Gagnon and P. Winternitz, J. Phys. A: Math. Gen. **21**, 1493 (1988); **22** 469 (1989); aL. Gagnon, B. Grammaticos, A. Ramani, and P. Winternitz, J. Phys. A: Math. Gen. **22**, 499 (1989)
59. L. Gagnon and P. Winternitz, Phys. Lett. A **134**, 276 (1989); J. Phys. A: Math. Gen. **25**, 4425 (1992); Phys. Rev. A **39**, 296 (1989)
60. Izmaylova K, Chupakhin A (2007) Rus. J. Nonlin. Dynamics 3(3):349
61. Boyd RW, Lukishova SG, Shen YR (2009) Self-Focusing: Past and Present. Springer, New York
62. Borovsky AV, Galkin AL, Shiryaev OB, Auguste T (2003) Laser Physics at Relativistic Intensities. Springer, New York
63. Kovalev VF, Bychenkov VY, Tikhonchuk VT (2000) Phys Rev A 61:033809
64. Max CE (1976) Phys Fluids 19:74
65. S. A. Akhmanov, A. P. Sukhorukov, and R. V. Khokhlov, Sov. Phys.—Usp. **10**, 609 (1967)
66. Borisov AB, Borovskiy AV, Shiryaev OB, Korobkin VV, Prokhorov AM, Solem JC, Luk TS, Boyer K, Rhodes CK (1992) Phys Rev A 45:5830
67. Feit MD, Komashko AM, Musher SL, Rubenchik AM, Turitsyn SK (1998) Phys Rev E 57:7122
68. Patil SD, Takale MV, Fulari VJ, Gupta DN, Suk H (2013) Appl Phys B 111:1
69. V. F. Kovalev, K. I. Popov, and V. Yu. Bychenkov, Sov. Phys.—JETP **114**, 25 (2012)
70. Kovalev VF, Bychenkov VY (2019) Phys Rev E 99:043201

One-Dimensional Flows of a Polytropic Gas: Lie Group Classification, Conservation Laws, Invariant and Conservative Difference Schemes

V. A. Dorodnitsyn, R. Kozlov, and S. V. Meleshko

Abstract The chapter considers one-dimensional flows of a polytropic gas in the Lagrangian coordinates in three cases: plain one-dimensional flows, radially symmetric flows, and spherically symmetric flows. The one-dimensional flow of a polytropic gas is described by one second-order partial differential equation in the Lagrangian variables. The Lie group classification of this PDE is performed. Its variational structure allows to construct conservation laws with the help of Noether's theorem. These conservation laws are also recalculated for the gas dynamics variables in the Eulerian and mass Lagrangian coordinates. Additionally, invariant and conservative difference schemes are provided.

1 Introduction

Symmetries of the differential equations of mathematical physics are their fundamental features. They reflect the geometric structure of solutions and physical principles of the considered models. We recall that the Lie group symmetries yield a number of useful properties of differential equations (see [1–6]):

- A group action transforms the complete set of solutions into itself; so it is possible to obtain new solutions from a given one;

V. A. Dorodnitsyn
Keldysh Institute of Applied Mathematics, Russian Academy of Science, Miusskaya Pl. 4, Moscow 125047, Russia
e-mail: Dorodnitsyn@Keldysh.ru

R. Kozlov
Department of Business and Management Science, Norwegian School of Economics, Helleveien 30, 5045 Bergen, Norway
e-mail: Roman.Kozlov@nhh.no

S. V. Meleshko (✉)
School of Mathematics, Institute of Science, Suranaree University of Technology, Nakhon Ratchasima 30000, Thailand
e-mail: sergey@math.sut.ac.th

A. C. J. Luo and R. K. Gazizov (eds.), *Symmetries and Applications of Differential Equations*, Nonlinear Physical Science,
https://doi.org/10.1007/978-981-16-4683-6_2

- There exists a standard procedure to obtain the whole set of invariants of a symmetry group of transformations; it yields the forms of invariant solutions in which they could be found (symmetry reduction of PDEs);
- For ODEs the known symmetry yields the reduction of the order;
- The invariance of ODEs and PDEs is a necessary condition for the application of Noether's theorem to variational problems to obtain conservation laws (first integrals for ODEs).

The symmetry properties of the gas dynamics equations were studied both in the Eulerian coordinates [1, 7] and in the Lagrangian coordinates [8–10]. Extensive group analysis of the one-dimensional gas dynamics equations in the mass Lagrangian coordinates was given in [8–10]. Here, it should be also mentioned that nonlocal conservation laws of the one-dimensional gas dynamics equations in the mass Lagrangian coordinates were found in [11]. The authors of [12, 13] analyzed the Euler–Lagrange equations corresponding to the one-dimensional gas dynamics equations in the mass Lagrangian coordinates: extensions of the known conservation laws were derived. These conservation laws correspond to special forms of the entropy. The group nature of these conservation laws is given in the present chapter.

As mentioned above, besides assisting with the construction of exact solutions, the knowledge of an admitted Lie group allows one to derive conservation laws. Conservation laws provide information on the basic properties of solutions of differential equations. They are also needed in the analyses of stability and global behavior of solutions. Noether's theorem [14] is the tool that relates symmetries and conservation laws. However, an application of Noether's theorem depends on the following condition: the differential equations under consideration need to be presented as the Euler–Lagrange equations with an appropriate Lagrangian, i.e., Noether's theorem requires variational structure. There are also other approaches to find conservation laws, which try to avoid this requirement [15–18].

The application of symmetries to difference and discrete equations is a more recent field of research [19–22]. One of its directions is the discretization of differential equations with the preservation of the Lie point symmetries. It is relevant to the construction of numerical schemes which inherit qualitative properties of the underlying differential equations. This approach was a base for a series of publications [19, 23–27], which are summarized in the book [22]. The method is based on finding finite-difference invariants which correspond to the chosen mesh stencil and using them to construct invariant difference equations and meshes. Recently, this approach was applied to shallow water systems, wave equations, and the Green–Naghdi system [28–30].

The recent paper [31] was devoted to the Lie group classification, conservation laws, and invariant difference schemes of plain one-dimensional flows of a polytropic gas. Here, we extend these results to radially symmetric flows in two-dimensional space and spherically symmetric flows in three-dimensional space. We refer to all such flows as one-dimensional flows. The results of [31] stand as a particular case in this chapter.

There are two distinct ways to model phenomena in gas dynamics (see, e.g., [32–34]). The typical approach uses the Eulerian coordinates, where flow quantities (at each instant of time) are described in fixed points. Alternatively, the Lagrangian description is used: the particles are identified by the positions which they occupy at some initial time. In the Lagrangian description, there are also two ways to analyze the processes occurring in a gas. One of them uses a system of first-order PDEs for the gas dynamics variables. The other approach uses a scalar second-order PDE to which this system can be reduced. The latter way allows one to use variational approach for the analysis of the gas dynamics equations.

The purpose of the chapter is to present an overview of the authors' results concerning the analysis of the gas dynamics equations of a polytropic gas. It is devoted to symmetries, conservation laws, and construction of numerical schemes, which preserve qualitative properties of the gas dynamics equations.

The article is organized as follows. In the forthcoming section, we recall Noether's theorem. Section 3 describes the gas dynamics equations, their reduction to a single second-order PDE, and the Lie point symmetries of this PDE. In Sects. 4 and 5, we consider the general case and the three special cases of the Lie group classification. Invariance and conservative properties of difference schemes are discussed in Sect. 6. Finally, Sect. 7 presents concluding remarks.

2 Symmetries and Noether's Theorem

We briefly remind Noether's theorem [14], which will be used to find conservation laws with the help of symmetries. In the general case, we have several independent variables and dependent variables, which are denoted as $x = (x^1, x^2, \dots, x^n)$ and $u = (u^1, u^2, \dots, u^m)$, respectively. All derivatives of order k are denoted as u_k.

A point symmetry operator has the form

$$X = \xi^i \frac{\partial}{\partial x^i} + \eta^k \frac{\partial}{\partial u^k} + \eta_i^k \frac{\partial}{\partial u_i^k} + \cdots, \tag{1}$$

where we assume that $\xi^i = \xi^i(x, u), \eta^k = \eta^k(x, u)$, and that the operator is prolonged to all derivatives $u^k_{i_1 \dots i_l}$ we need to consider. We denote the considered function as $F(x, u, u_1, \dots, u_k)$. It involves derivatives up to some finite order k.

Noether's theorem is based on the identity [2, 14]

$$XF + FD_i\xi^i = (\eta^k - \xi^i u_i^k)\frac{\delta F}{\delta u^k} + D_i(N^i F), \tag{2}$$

where

$$\frac{\delta}{\delta u^k} = \sum_{s=0}^{\infty}(-1)^s D_{i_1} \dots D_{i_s} \frac{\partial}{\partial u^k_{i_1 i_2 \dots i_s}}, \qquad k = 1, 2, \dots, m, \tag{3}$$

are variational operators, and

$$N^i = \xi^i + \sum_{s=0}^{\infty} D_{i_1} \dots D_{i_s} (\eta^k - \xi^i u_i^k) \frac{\delta}{\delta u^k_{i i_1 i_2 \dots i_s}}, \qquad i = 1, 2, \dots, n. \tag{4}$$

The higher variational operators $\frac{\delta}{\delta u^k_{i_1 i_2 \dots i_s}}$ are obtained from the variational operators (3) by replacing u^k with the corresponding derivatives $u^k_{i_1 i_2 \dots i_s}$.

Theorem 1 (E. Noether) *Let the Lagrangian function $L(x, u, u_1, \dots, u_k)$ satisfy equation*

$$XL + LD_i \xi^i = D_i B^i \tag{5}$$

with a vector $\mathbf{B} = (B^1, B^2, \dots, B^n)$ and a group generator

$$X = \xi^i(x, u)\frac{\partial}{\partial x^i} + \eta^k(x, u)\frac{\partial}{\partial u^k},$$

then the generator X is an admitted symmetry of the system of the Euler–Lagrange equations

$$\frac{\delta L}{\delta u^k} = 0, \qquad k = 1, 2, \dots, m, \tag{6}$$

and the vector

$$(N^1 L - B^1, N^2 L - B^2, \dots, N^n L - B^n) \tag{7}$$

is a conserved vector.

In the case $\mathbf{B} = (B^1, B^2, \dots, B^n) = \mathbf{0}$, we call the symmetry X variational; otherwise, we say that the symmetry X is divergent.

It is well known that variational and divergent symmetries are also symmetries of the Euler–Lagrange equations [1–3]. For the Lie point symmetries, i.e., symmetries with coefficients $\xi^i = \xi^i(x, u)$, $\eta^k = \eta^k(x, u)$, and first-order Lagragians $L = L(x, u, u_1)$, it easily follows from the identities [35]

$$\frac{\delta}{\delta u^j}\left(XL + LD_i\xi^i - D_i B^i\right) = X\Big(\frac{\delta L}{\delta u^j}\Big) + \Big(\frac{\partial \eta^k}{\partial u^j} - \frac{\partial \xi^i}{\partial u^j} u_i^k + \delta_{kj} D_i \xi^i\Big)\frac{\delta L}{\delta u^k},$$
$$j = 1, 2, \dots, m,$$

where δ_{kj} is the Kronecker symbol.

3 Equations of Gas Dynamics for One-Dimensional Flows

We consider three types of gas flows, namely, flows in one-dimensional space, radially symmetric flows in two-dimensional space, and spherically symmetric flows in three-dimensional space. We will refer to these flows as *one-dimensional* flows.

The gas is assumed to be polytropic [34, 36–38]. For a polytropic gas, the pressure p and the density ρ are related as

$$p = S\rho^{\gamma}, \tag{8}$$

where variable S is the function of the entropy $\tilde{S}$

$$S = e^{(\tilde{S}-\tilde{S}_0)/c_v}, \; c_v = \frac{R}{\gamma - 1}.$$

Here, R is the gas constant, c_v is the specific heat capacity at constant volume, and $\tilde{S}_0$ is constant. The adiabatic constant is given as

$$\gamma = 1 + \frac{R}{c_v} > 1.$$

We will also need the equation of state for the polytropic gas, i.e., equation for the specific internal energy

$$\varepsilon = \varepsilon(\rho, p). \tag{9}$$

It has the form

$$\varepsilon = \frac{p}{(\gamma - 1)\rho}. \tag{10}$$

The gas dynamics equations will be given in the Eulerian and Lagrangian coordinates. Eventually, they will be reduced to one scalar PDE of the second order, which will be analyzed for the admitted Lie point symmetries.

3.1 Eulerian Coordinates

In the Eulerian coordinates (t, r), the gas dynamics equation can be written as (see, e.g., [32–34])

$$\rho_t + u\rho_r + \tfrac{\rho}{r^n}(r^n u)_r = 0, \tag{11a}$$

$$u_t + uu_r + \tfrac{1}{\rho}p_r = 0, \tag{11b}$$

$$S_t + uS_r = 0. \tag{11c}$$

Here, we distinguish the case $n = 0$ with coordinate $-\infty < r < \infty$ and velocity u from the cases $n = 1, 2$ with radial distance from the origin $0 < r < \infty$ and the radial velocity u.

We have $n = 0, 1, 2$ for the plain one-dimensional flows, the radially symmetric two-dimensional flows, and the spherically symmetric three-dimensional flows, respectively. Note that for these cases $n = d - 1$, where $d = 1, 2, 3$ is the space dimension.

We also use other representations of Eq. (11c)

$$p_t + up_r + \frac{\gamma p}{r^n}(r^n u)_r = 0 \tag{12}$$

or

$$\varepsilon_t + u\varepsilon_r + \frac{p}{r^n \rho}(r^n u)_r = 0. \tag{13}$$

3.2 Lagrangian Coordinates

As well known [32–34], the mass Lagrangian coordinate s and the Eulerian coordinate r of the particle s are related by the formulas

$$u = \varphi_t, \qquad \rho = \frac{1}{\varphi^n \varphi_s}, \tag{14}$$

where $r = \varphi(t, s)$ defines the motion of a particle s.

In the Eulerian coordinates (t, r), we can introduce the mass Lagrangian coordinate s as a potential by the system

$$s_r = r^n \rho, \qquad s_t = -r^n \rho u, \tag{15}$$

which is equivalent to the 1-form

$$ds = r^n \rho dr - r^n \rho u dt.$$

Here, we rely on the possibility to rewrite the Eq. (11a) as the conservation law

$$(r^n \rho)_t + (r^n \rho u)_r = 0,$$

representing conservation of mass.

In the mass Lagrangian coordinates (t, s), which we will call the Lagrangian coordinates, the time derivative stands for the differentiation along the pathlines. It is called the material derivative. Total derivatives in the Lagrangian coordinates D_t^L and D_s are related to those in the Eulerian coordinates D_t^E and D_r as

$$D_t^L = D_t^E + uD_r, \qquad D_s = \frac{1}{r^n\rho}D_r. \tag{16}$$

We rewrite the gas dynamics equations (11) in the Lagrangian coordinates (t, s) as

$$\rho_t + \rho^2(r^n u)_s = 0, \tag{17a}$$

$$u_t + r^n p_s = 0, \tag{17b}$$

$$S_t = 0. \tag{17c}$$

We remark that here the gas dynamics variables ρ, u, p, and S are functions of the Lagrangian coordinates t and s while in the system (11), they are functions of the Eulerian coordinates t and r.

The Eulerian spatial coordinate $\varphi = r$ is a dependent variable in the mass Lagrangian coordinates. Equations (14) can be rewritten in the form

$$r_t = u, \qquad r_s = \frac{1}{r^n\rho}. \tag{18}$$

It is also possible to use the 1-form

$$dr = \frac{ds}{r^n\rho} + udt.$$

Notice that as for Eq. (11c), we also use other representations of Eq. (17c)

$$p_t + \gamma\rho p(r^n u)_s = 0 \tag{19}$$

or

$$\varepsilon_t + p(r^n u)_s = 0. \tag{20}$$

Equation (17c) can be solved

$$S = S(s), \tag{21}$$

where $S(s)$ is an arbitrary function.

Using these results, it is possible to rewrite the last remaining Eq. (17b), as a partial differential equation of the second order

$$\varphi_{tt} + \varphi^{n(1-\gamma)}\varphi_s^{-\gamma}\left(S' - n\gamma S\frac{\varphi_s}{\varphi} - \gamma S\frac{\varphi_{ss}}{\varphi_s}\right) = 0. \tag{22}$$

This PDE is called the *gas dynamics equation* in the Lagrangian coordinates [34, 36].

PDE (22) has a variational formulation, namely, it is the Euler–Lagrange equation

$$\frac{\delta L}{\delta \varphi} = \frac{\partial L}{\partial \varphi} - D_t^L(\frac{\partial L}{\partial \varphi_t}) - D_s(\frac{\partial L}{\partial \varphi_s}) = 0 \tag{23}$$

for the Lagrangian

$$L = \frac{1}{2}\varphi_t^2 - \frac{S(s)}{\gamma - 1}\varphi^{(1-\gamma)n}\varphi_s^{1-\gamma}. \tag{24}$$

3.3 Conservation Laws

We specify Noether's theorem, given in Sect. 2, for PDE (22). We consider the Lie point symmetries of the form

$$X = \xi^t(t,s,\varphi)\frac{\partial}{\partial t} + \xi^s(t,s,\varphi)\frac{\partial}{\partial s} + \eta^\varphi(t,s,\varphi)\frac{\partial}{\partial \varphi}. \tag{25}$$

Such symmetries of the PDE (22) can be used to compute conservation laws if they are also variational or divergence symmetries of the Lagrangian (24). We require that they satisfy the condition of the elementary action invariance [2]

$$XL + L(D_t^L\xi^t + D_s\xi^s) = D_t^L B_1 + D_s B_2 \tag{26}$$

for some functions $B_1(t,s,\varphi)$ and $B_2(t,s,\varphi)$. If this condition holds with $B_1 = B_2 = 0$, then the symmetry (25) is called variational. We refer to both variational and divergent symmetries as the Noether symmetries.

Given a variational or divergent symmetry, we can find the corresponding conservation law

$$D_t^L[T^t] + D_s[T^s] = 0, \tag{27}$$

where the densities are given by the formulas

$$\begin{aligned} T^t &= \xi^t L + (\eta^\varphi - \xi^t\varphi_t - \xi^s\varphi_s)\tfrac{\partial L}{\partial \varphi_t} - B_1, \\ T^s &= \xi^s L + (\eta^\varphi - \xi^t\varphi_t - \xi^s\varphi_s)\tfrac{\partial L}{\partial \varphi_s} - B_2. \end{aligned} \tag{28}$$

Conservation laws (27) can be rewritten for the Eulerian coordinates as

$$D_t^E[{}^eT^t] + D_r[{}^eT^r] = 0. \tag{29}$$

The relation between two above forms of conservation laws can be proved by direct verification

$$D_t^L T^t + D_s T^s = \varphi_s \left(D_t^E (r^n \rho T^t) + D_r (r^n \rho u T^t + T^s) \right).$$

Therefore, if we have densities T^t and T^s of a conservation law in the Lagrangian coordinates, we can find the corresponding densities in the Eulerian coordinates as

$$^eT^t = r^n \rho T^t, \qquad ^eT^r = r^n \rho u T^t + T^s. \tag{30}$$

3.4 Equivalence Transformations

PDE (22) contains an arbitrary function $S(s)$. Thus, we need the group classification with respect to it. The generators of the equivalence Lie group has the form

$$X^e = \xi^t \frac{\partial}{\partial t} + \xi^s \frac{\partial}{\partial s} + \eta^\varphi \frac{\partial}{\partial \varphi} + \eta^S \frac{\partial}{\partial S}, \tag{31}$$

where generators coefficients $\xi^t, \xi^s, \eta^\varphi and \eta^S depend on (t, s, \varphi, S)$.

Computation gives the generators of the equivalence group. There are five generators

$$X_1^e = \frac{\partial}{\partial t}, \qquad X_2^e = \frac{\partial}{\partial s}, \qquad X_3^e = t\frac{\partial}{\partial t} - 2S\frac{\partial}{\partial S},$$
$$X_4^e = s\frac{\partial}{\partial s} + (1-\gamma)S\frac{\partial}{\partial S}, \qquad X_5^e = \varphi\frac{\partial}{\partial \varphi} + ((n+1)\gamma - n + 1)S\frac{\partial}{\partial S} \tag{32}$$

for the general case. For $n = 0$, there are two additional equivalence transformations given by

$$X_{*,n}^e = \frac{\partial}{\partial \varphi} \qquad \text{and} \qquad X_{**,n}^e = t\frac{\partial}{\partial \varphi}. \tag{33}$$

For the special values of the adiabatic exponent $\gamma_* = \frac{n+3}{n+1}$, we obtain one additional generator

$$X_{*,\gamma}^e = t^2 \frac{\partial}{\partial t} + t\varphi\frac{\partial}{\partial \varphi}. \tag{34}$$

3.5 Group Classification of the Gas Dynamics Equation

The Lie algebra of the admitted transformations is given by the generators

$$X = \sum_{i=1}^{8} k_i Y_i, \tag{35}$$

where

$$Y_1 = \frac{\partial}{\partial t}, \quad Y_2 = \frac{\partial}{\partial s}, \quad Y_3 = \frac{\partial}{\partial \varphi},$$
$$Y_4 = t\frac{\partial}{\partial t}, \quad Y_5 = s\frac{\partial}{\partial s}, \quad Y_6 = \varphi\frac{\partial}{\partial \varphi},$$
$$Y_7 = t\frac{\partial}{\partial \varphi}, \quad Y_8 = t^2\frac{\partial}{\partial t} + t\varphi\frac{\partial}{\partial \varphi}. \tag{36}$$

The coefficients k_i satisfy the system

$$(k_5 s + k_2)S_s = (-2k_4 + (1-\gamma)k_5 + ((n+1)\gamma - n + 1)k_6)S, \tag{37a}$$
$$((n+1)\gamma - n - 3)k_8 = 0, \tag{37b}$$
$$nk_3 = 0, \tag{37c}$$
$$nk_7 = 0. \tag{37d}$$

For the general case, we get two admitted symmetries

$$X_1 = Y_1 = \frac{\partial}{\partial t}, \quad X_2 = ((n+1)\gamma - n + 1)Y_4 + 2Y_6$$
$$= ((n+1)\gamma - n + 1)t\frac{\partial}{\partial t} + 2\varphi\frac{\partial}{\partial \varphi}. \tag{38}$$

For $n = 0$, there are two additional symmetries

$$X_{*,n} = Y_3 = \frac{\partial}{\partial \varphi} \quad \text{and} \quad X_{**,n} = Y_7 = t\frac{\partial}{\partial \varphi}. \tag{39}$$

For the special values $\gamma_* = \frac{n+3}{n+1}$, there is one additional symmetry

$$X_{*,\gamma} = Y_8 = t^2\frac{\partial}{\partial t} + t\varphi\frac{\partial}{\partial \varphi}. \tag{40}$$

The condition (37a) is the classifying equation for function $S(s)$. It can be rewritten as

$$(\alpha_1 s + \alpha_0)S_s = \beta S \tag{41}$$

for some constants α_0, α_1, and β. This classifying equation was studied in [39]. It was shown that one need to consider four cases of the entropy function $S(s)$, the general case and three special cases:

- arbitrary $S(s)$;
- $S(s) = A_0$, $A_0 = \text{const}$;
- $S(s) = A_0 s^q$, $q \neq 0$, $A_0 = \text{const}$;
- $S(s) = A_0 e^{qs}$, $q \neq 0$, $A_0 = \text{const}$.

The same four cases were obtained for plain one-dimensional flows in [31]. Let us note that the equivalence transformations can be used to simplify these cases to $A_0 = 1$.

4 Arbitrary Entropy $S(s)$

Equation (17a) can be rewritten in the form of a conservation law as

$$\left[\frac{1}{\rho}\right]_t - [r^n u]_s = 0.$$

Thus, conservation of mass is included into the equations of the gas dynamics system (17). In the Eulerian coordinates, it has the form

$$[r^n \rho]_t + [r^n \rho u]_r = 0.$$

Equation (17c) gives the conservation of the entropy along pathlines as the conservation law

$$S_t = 0.$$

Let us examine the symmetries of the kernel of admitted Lie algebras (38), (39), and (40) for being variational or divergent symmetries, which provide conservation laws.

4.1 General Case $n \neq 0$, $\gamma \neq \frac{n+3}{n+1}$

In the general cases, the admitted symmetries (38) provide us with one variational symmetry

$$Z_1 = X_1 = \frac{\partial}{\partial t}. \tag{42}$$

It leads to the conservation of energy with densities

$$T_1^t = \frac{\varphi_t^2}{2} + \frac{S}{\gamma - 1}\varphi^{n(1-\gamma)}\varphi_s^{1-\gamma}, \qquad T_1^s = S\varphi^{n(1-\gamma)}\varphi_t\varphi_s^{-\gamma}. \tag{43}$$

For the gas dynamics variables, this conservation law gets rewritten as

$$T_1^t = \frac{u^2}{2} + \frac{S}{\gamma - 1}\rho^{\gamma-1}, \qquad T_1^s = r^n S\rho^\gamma u. \tag{44}$$

In the Eulerian coordinates, it has the densities

$$^{e}T_1^t = r^n\Big(\frac{\rho u^2}{2} + \frac{S}{\gamma-1}\rho^\gamma\Big), \qquad ^{e}T_1^r = r^n\Big(\frac{\rho u^2}{2} + \frac{\gamma S}{\gamma-1}\rho^\gamma\Big)u.$$

4.2 Case $n = 0$, $\gamma \neq \frac{n+3}{n+1}$

We get one more variational symmetry

$$Z_{*,n} = X_{*,n} = \frac{\partial}{\partial\varphi}, \tag{45}$$

and one divergent symmetry

$$Z_{**,n} = X_{**,n} = t\frac{\partial}{\partial\varphi} \quad \text{with} \quad (B_1, B_2) = (\varphi, 0). \tag{46}$$

These symmetries provide conservation laws

$$T^t_{*,n} = \varphi_t, \qquad T^s_{*,n} = S\varphi_s^{-\gamma}; \tag{47}$$

$$T^t_{**,n} = \varphi - \varphi_t t, \qquad T^s_{**,n} = -tS\varphi_s^{-\gamma}, \tag{48}$$

representing the conservation of momentum and the motion of the center of mass, respectively.

In gas dynamics variables, we can rewrite these conservation laws as

$$T^t_{*,n} = u, \qquad T^s_{*,n} = S\rho^\gamma; \tag{49}$$

$$T^t_{**,n} = r - tu, \qquad T^s_{**,n} = -tS\rho^\gamma. \tag{50}$$

Notice that the conserved vector $(T^t_{**,n}, T^s_{**,n})$ contains the function $\varphi \equiv$ r. In the Eulerian coordinates, we get

$$^{e}T^t_{*,n} = \rho u, \qquad ^{e}T^r_{*,n} = \rho u^2 + S\rho^\gamma;$$

$$^{e}T^t_{**,n} = \rho(r - tu), \qquad ^{e}T^r_{**,n} = \rho u(r - tu) - tS\rho^\gamma.$$

4.3 Case $n \neq 0$, $\gamma_* = \frac{n+3}{n+1}$

For $\gamma = \gamma_*$, the symmetries (38) and (40) lead to two variational symmetries: (42) and

$$Z_{*,\gamma} = \frac{1}{2}X_2 = 2t\frac{\partial}{\partial t} + \varphi\frac{\partial}{\partial \varphi} \tag{51}$$

and one divergence symmetry

$$Z_{**,\gamma} = X_{*,\gamma} = t^2\frac{\partial}{\partial t} + t\varphi\frac{\partial}{\partial \varphi} \quad \text{with} \quad (B_1, B_2) = \left(\frac{\varphi^2}{2}, 0\right). \tag{52}$$

In addition to the conservation of energy, given in point Sect. 4.1, there are conservation laws with densities

$$T^t_{*,\gamma} = 2t\Big(\frac{\varphi_t^2}{2} + \frac{S}{\gamma-1}\varphi^{n(1-\gamma)}\varphi_s^{1-\gamma}\Big) - \varphi\varphi_t, \qquad T^s_{*,\gamma} = (2t\varphi_t - \varphi)S\varphi^{n(1-\gamma)}\varphi_s^{-\gamma}; \tag{53}$$

$$T^t_{**,\gamma} = t^2\Big(\frac{\varphi_t^2}{2} + \frac{S}{\gamma-1}\varphi^{n(1-\gamma)}\varphi_s^{1-\gamma}\Big) - t\varphi\varphi_t + \frac{\varphi^2}{2}, \qquad T^s_{**,\gamma} = (t^2\varphi_t - t\varphi)S\varphi^{n(1-\gamma)}\varphi_s^{-\gamma}. \tag{54}$$

We can rewrite these conservation laws for the gas dynamics variables

$$T^t_{*,\gamma} = 2t\Big(\frac{u^2}{2} + \frac{S}{\gamma-1}\rho^{\gamma-1}\Big) - ru, \qquad T^s_{*,\gamma} = r^n(2tu - r)S\rho^\gamma; \tag{55}$$

$$T^t_{**,\gamma} = t^2\Big(\frac{u^2}{2} + \frac{S}{\gamma-1}\rho^{\gamma-1}\Big) - tru + \frac{r^2}{2}, \qquad T^s_{**,\gamma} = r^n(t^2u - tr)S\rho^\gamma \tag{56}$$

as well as in the Eulerian coordinates

$$^eT^t_{*,\gamma} = r^n\big(2t\big(\frac{\rho u^2}{2} + \frac{S}{\gamma-1}\rho^\gamma\big) - r\rho u\big), \qquad ^eT^r_{*,\gamma} = r^n\big(2t\big(\frac{\rho u^2}{2} + \frac{\gamma S}{\gamma-1}\rho^\gamma\big)u - r(\rho u^2 + S\rho^\gamma)\big);$$

$$^eT^t_{**,\gamma} = r^n\big(t^2\big(\frac{\rho u^2}{2} + \frac{S}{\gamma-1}\rho^\gamma\big) - tr\rho u + \frac{r^2\rho}{2}\big), \qquad ^eT^r_{**,\gamma} = r^n\big(t^2\big(\frac{\rho u^2}{2} + \frac{\gamma S}{\gamma-1}\rho^\gamma\big)u - tr(\rho u^2 + S\rho^\gamma) + \frac{r^2\rho u}{2}\big).$$

4.4 *Case $n = 0$, $\gamma_* = 3$*

In this case, the conservation law of the general case get extended by both the conservation laws given in point Sect. 4.2 and by the conservation laws given in point Sect. 4.3.

5 Special Cases of Entropy

Group classification of the PDE (22) gives three special cases of the entropy function. They are examined in this section. These cases inherit the symmetries and conservation laws of the arbitrary entropy $S(s)$, given in the preceding section. We present only additional symmetries and conservation laws.

5.1 *Isentropic Case $S(s) = A_0$*

In the Eulerian coordinates, this case is presented as

$$S(r) = A_0 \qquad \text{or} \qquad S_r = 0. \tag{57}$$

For all cases (the case of general n and γ, the case $n = 0$, and the case of special values $\gamma = \gamma_*$), there are two additional symmetries

$$X_3 = Y_2 = \frac{\partial}{\partial s}, \qquad X_4 = (\gamma - 1)Y_4 - 2Y_5 = (\gamma - 1)t\frac{\partial}{\partial t} - 2s\frac{\partial}{\partial s}. \tag{58}$$

5.1.1 General Case $n \neq 0$, $\gamma \neq \frac{n+3}{n+1}$

In the general case, there are two additional variational symmetries

$$\begin{aligned}
Z_2 &= X_3 = \frac{\partial}{\partial s}, \\
Z_3 &= \frac{\gamma + 1}{2}X_2 + \frac{n + 3 - (n + 1)\gamma}{2}X_4 \\
&= ((n + 3)\gamma - n - 1)t\frac{\partial}{\partial t} + ((n + 1)\gamma - n - 3)s\frac{\partial}{\partial s} + (\gamma + 1)\varphi\frac{\partial}{\partial \varphi}.
\end{aligned} \tag{59}$$

The conservation laws of this case consist of the conservation law given in point Sect. 4.1 (for arbitrary $S(s)$) and the two additional ones, given by densities

$$T_2^t = \varphi_s\varphi_t, \qquad T_2^s = -\frac{\varphi_t^2}{2} + \frac{\gamma S}{\gamma - 1}\varphi^{n(1-\gamma)}\varphi_s^{1-\gamma}; \tag{60}$$

$$T_3^t = ((n+3)\gamma - n - 1)t\Big(\frac{\varphi_t^2}{2} + \frac{S}{\gamma-1}\varphi^{n(1-\gamma)}\varphi_s^{1-\gamma}\Big) + ((n+1)\gamma - n - 3)s\varphi_s\varphi_t$$
$$- (\gamma+1)\varphi\varphi_t,$$
$$T_3^s = ((n+3)\gamma - n - 1)tS\varphi^{n(1-\gamma)}\varphi_t\varphi_s^{-\gamma} + ((n+1)\gamma - n - 3)s$$
$$\left(-\frac{\varphi_t^2}{2} + \frac{\gamma S}{\gamma-1}\varphi^{n(1-\gamma)}\varphi_s^{1-\gamma}\right) - (\gamma+1)S\varphi^{n(1-\gamma)+1}\varphi_s^{-\gamma}. \tag{61}$$

If rewritten for the gas dynamics variables, they take the form

$$T_2^t = \frac{u}{r^n\rho}, \qquad T_2^s = -\frac{u^2}{2} + \frac{\gamma S}{\gamma-1}\rho^{\gamma-1}; \tag{62}$$

$$T_3^t = ((n+3)\gamma - n - 1)t\Big(\frac{u^2}{2} + \frac{S}{\gamma-1}\rho^{\gamma-1}\Big) + ((n+1)\gamma - n - 3)s\frac{u}{r^n\rho} - (\gamma+1)ru,$$
$$T_3^s = ((n+3)\gamma - n - 1)tSr^n\rho^\gamma u + ((n+1)\gamma - n - 3)s\Big(-\frac{u^2}{2} + \frac{\gamma S}{\gamma-1}\rho^{\gamma-1}\Big)$$
$$- (\gamma+1)Sr^{n+1}\rho^\gamma. \tag{63}$$

In the Eulerian coordinates, these conservation laws have densities

$${}^eT_2^t = u, \qquad {}^eT_2^r = \frac{u^2}{2} + \frac{\gamma S}{\gamma-1}\rho^{\gamma-1};$$
$$T_3^t = ((n+3)\gamma - n - 1)tr^n\Big(\frac{\rho u^2}{2} + \frac{S}{\gamma-1}\rho^\gamma\Big) + ((n+1)\gamma - n - 3)su - (\gamma+1)r^{n+1}\rho u,$$
$$T_3^s = ((n+3)\gamma - n - 1)tr^n\Big(\frac{\rho u^2}{2} + \frac{\gamma S}{\gamma-1}\rho^\gamma\Big)u$$
$$+ ((n+1)\gamma - n - 3)s\Big(\frac{u^2}{2} + \frac{\gamma S}{\gamma-1}\rho^{\gamma-1}\Big) - (\gamma+1)r^{n+1}(\rho u^2 + S\rho^\gamma),$$

where $S(s)$, and s is defined by system (15).

5.1.2 Special Cases

For all special cases, namely case $n = 0$, $\gamma \neq \frac{n+3}{n+1}$; case $n \neq 0$, $\gamma_* = \frac{n+3}{n+1}$; and case $n = 0$, $\gamma_* = 3$, we get conservation laws of the arbitrary entropy $S(s)$, which were described in Sect. 4, supplemented by the conservation laws given in point Sect. 5.1.1.

Note that

$$Z_3|_{\gamma=\gamma_*} = 2\frac{n+2}{n+1}Z_{*,\gamma},$$

in other words for $\gamma = \gamma_*$, only the conservation law corresponding to Z_2 in point Sect. 5.1.1 is new.

5.2 Entropy Case $S(s) = A_0 s^q$

In the Eulerian coordinates, this entropy case is described by the differential constraint

$$S_{rr} = \left(\frac{\rho_r}{\rho} + \frac{n}{r}\right) S_r + \frac{q-1}{q}\frac{S_r^2}{S}. \tag{64}$$

For such $S(s)$, there is one additional symmetry

$$X_3 = (\gamma + q - 1)Y_4 - 2Y_5 = (\gamma + q - 1)t\frac{\partial}{\partial t} - 2s\frac{\partial}{\partial s}. \tag{65}$$

5.2.1 General Case $n \neq 0, \gamma \neq \frac{n+3}{n+1}$

For the general case, there is one additional variational symmetry

$$\begin{aligned} Z_2 &= \frac{\gamma + q + 1}{2} X_2 + \frac{n + 3 - (n+1)\gamma}{2} X_3 \\ &= ((n+3)\gamma + 2q - n - 1)t\frac{\partial}{\partial t} + ((n+1)\gamma - n - 3)s\frac{\partial}{\partial s} \\ &\quad + (\gamma + q + 1)\varphi\frac{\partial}{\partial \varphi}. \end{aligned} \tag{66}$$

Thus, in addition to the conservation of energy given in Sect. 4.1, we obtain the conservation law

$$\begin{aligned} T_2^t &= ((n+3)\gamma + 2q - n - 1)t\Big(\frac{\varphi_t^2}{2} + \frac{S}{\gamma - 1}\varphi^{n(1-\gamma)}\varphi_s^{1-\gamma}\Big) \\ &\quad + ((n+1)\gamma - n - 3)s\varphi_s\varphi_t - (\gamma + q + 1)\varphi\varphi_t, \\ T_2^s &= ((n+3)\gamma + 2q - n - 1)tS\varphi^{n(1-\gamma)}\varphi_t\varphi_s^{-\gamma} + ((n+1)\gamma - n - 3)s \\ &\Big(-\frac{\varphi_t^2}{2} + \frac{\gamma S}{\gamma - 1}\varphi^{n(1-\gamma)}\varphi_s^{1-\gamma}\Big) - (\gamma + q + 1)S\varphi^{n(1-\gamma)+1}\varphi_s^{-\gamma}. \end{aligned} \tag{67}$$

For the gas dynamics variables, it takes the form

$$T_2^t = ((n+3)\gamma + 2q - n - 1)t\big(\frac{u^2}{2} + \frac{S}{\gamma - 1}\rho^{\gamma-1}\big) + ((n+1)\gamma - n - 3)s\frac{u}{r^n \rho}$$
$$- (\gamma + q + 1)ru,$$
$$T_2^s = ((n+3)\gamma + 2q - n - 1)tSr^n\rho^\gamma u + ((n+1)\gamma - n - 3)s\big(-\frac{u^2}{2} + \frac{\gamma S}{\gamma - 1}\rho^{\gamma-1}\big)$$
$$- (\gamma + q + 1)Sr^{n+1}\rho^\gamma. \tag{68}$$

To rewrite this conservation laws in the Eulerian coordinates, we use the relation

$$s = qr^n\rho\frac{S}{S_r} \tag{69}$$

to present the Lagrangian coordinate s. This relation allows to write down the densities of the conservation law as follows:

$${}^eT_2^t = ((n+3)\gamma + 2q - n - 1)tr^n\big(\frac{\rho u^2}{2} + \frac{S}{\gamma - 1}\rho^\gamma\big) + ((n+1)\gamma - n - 3)qr^n\rho u\frac{S}{S_r}$$
$$- (\gamma + q + 1)r^{n+1}\rho u,$$
$${}^eT_2^r = ((n+3)\gamma + 2q - n - 1)tr^n\big(\frac{\rho u^2}{2} + \frac{\gamma S}{\gamma - 1}\rho^\gamma\big)u + ((n+1)\gamma - n - 3)qr^n\rho\frac{S}{S_r}$$
$$\big(\frac{u^2}{2} + \frac{\gamma S}{\gamma - 1}\rho^{\gamma-1}\big) - (\gamma + q + 1)r^{n+1}(\rho u^2 + S\rho^\gamma).$$

5.2.2 Case $n = 0$, $\gamma \neq \frac{n+3}{n+1}$

For $n = 0$, the additional Noether symmetries are the same as in the general case. Therefore, we get conservation laws given in points Sects. 4.1, 4.2, and 5.2.1.

5.2.3 Case $n \neq 0$, $\gamma_* = \frac{n+3}{n+1}$

The special case of γ_* splits for values of q. For general q, we get the same Noether symmetries as in the case of arbitrary $S(s)$. Therefore, we obtain the same conservation laws as given in points Sects. 4.1 and 4.3.

For the particular case $q_* = -2\frac{n+2}{n+1}$, there is one additional variational symmetry

$$Z_{*,q} = -\frac{1}{2}X_3 = t\frac{\partial}{\partial t} + s\frac{\partial}{\partial s}. \tag{70}$$

It provides with the following conservation law:

$$T^t_{*,q} = t\Big(\frac{\varphi_t^2}{2} + \frac{S}{\gamma-1}\varphi^{n(1-\gamma)}\varphi_s^{1-\gamma}\Big) + s\varphi_s\varphi_t,$$

$$T^s_{*,q} = tS\varphi^{n(1-\gamma)}\varphi_t\varphi_s^{-\gamma} + s\Big(-\frac{\varphi_t^2}{2} + \frac{\gamma S}{\gamma-1}\varphi^{n(1-\gamma)}\varphi_s^{1-\gamma}\Big). \tag{71}$$

It is also possible to present this conservation laws for the gas dynamics variables

$$T^t_{*,q} = t\Big(\frac{u^2}{2} + \frac{S}{\gamma-1}\rho^{\gamma-1}\Big) + s\frac{u}{r^n\rho},$$

$$T^s_{*,q} = tSr^n\rho^\gamma u + s\Big(-\frac{u^2}{2} + \frac{\gamma S}{\gamma-1}\rho^{\gamma-1}\Big). \tag{72}$$

To rewrite these densities in the Eulerian coordinates, we employ the relation (69) and obtain densities

$${}^eT^t_{*,q} = r^n\Big(t\Big(\frac{\rho u^2}{2} + \frac{S}{\gamma-1}\rho^\gamma\Big) + q\rho u\frac{S}{S_r}\Big),$$

$${}^eT^r_{*,q} = r^n\Big(t\rho u + q\rho\frac{S}{S_r}\Big)\Big(\frac{u^2}{2} + \frac{\gamma S}{\gamma-1}\rho^{\gamma-1}\Big).$$

5.2.4 Case $n = 0$, $\gamma_* = 3$

We get the same conservation laws as described in the previous point. Note that $n = 0$ leads to $q_* = -4$.

5.3 *Entropy Case* $S(s) = A_0e^{qs}$

Let us note that this special case can be given in the Eulerian coordinates by the differential constraint

$$S_r = qr^n\rho S. \tag{73}$$

For all cases of Sect. 4, there is one additional symmetry

$$X_3 = -2Y_2 + qY_4 = qt\frac{\partial}{\partial t} - 2\frac{\partial}{\partial s}. \tag{74}$$

5.3.1 General Case $n \neq 0$, $\gamma \neq \frac{n+3}{n+1}$

For the general case, there is one additional variational symmetry

$$
\begin{aligned}
Z_2 &= \frac{q}{2}X_2 + \frac{n+3-(n+1)\gamma}{2}X_3 \\
&= 2qt\frac{\partial}{\partial t} + ((n+1)\gamma - n - 3)\frac{\partial}{\partial s} + q\varphi\frac{\partial}{\partial \varphi}.
\end{aligned} \tag{75}
$$

The supplementary conservation law has densities

$$
\begin{aligned}
T_2^t &= 2qt\Big(\frac{\varphi_t^2}{2} + \frac{S}{\gamma-1}\varphi^{n(1-\gamma)}\varphi_s^{1-\gamma}\Big) + ((n+1)\gamma - n - 3)\varphi_s\varphi_t - q\varphi\varphi_t, \\
T_2^s &= 2qtS\varphi^{n(1-\gamma)}\varphi_t\varphi_s^{-\gamma} + ((n+1)\gamma - n - 3)\Big(-\frac{\varphi_t^2}{2} + \frac{\gamma S}{\gamma-1}\varphi^{n(1-\gamma)}\varphi_s^{1-\gamma}\Big) \\
&\quad - qS\varphi^{n(1-\gamma)+1}\varphi_s^{-\gamma}.
\end{aligned} \tag{76}
$$

For the gas dynamics variables, we get

$$
\begin{aligned}
T_2^t &= 2qt\Big(\frac{u^2}{2} + \frac{S}{\gamma-1}\rho^{\gamma-1}\Big) + ((n+1)\gamma - n - 3)\frac{u}{r^n\rho} - qru, \\
T_2^s &= 2qtSr^n\rho^\gamma u + ((n+1)\gamma - n - 3)\Big(-\frac{u^2}{2} + \frac{\gamma S}{\gamma-1}\rho^{\gamma-1}\Big) - qSr^{n+1}\rho^\gamma.
\end{aligned} \tag{77}
$$

Finally, we rewrite these densities in the Eulerian coordinates

$$
\begin{aligned}
{}^eT_2^t &= 2qtr^n\Big(\frac{\rho u^2}{2} + \frac{S}{\gamma-1}\rho^\gamma\Big) + ((n+1)\gamma - n - 3)u - qr^{n+1}\rho u, \\
{}^eT_2^r &= 2qtr^n\Big(\frac{\rho u^2}{2} + \frac{\gamma S}{\gamma-1}\rho^\gamma\Big)u + ((n+1)\gamma - n - 3)\Big(\frac{u^2}{2} + \frac{\gamma S}{\gamma-1}\rho^{\gamma-1}\Big) \\
&\quad - qr^{n+1}(\rho u^2 + S\rho^\gamma).
\end{aligned}
$$

5.3.2 Special Cases

For all special cases, we get the same additional conservation law as in the general case of n and γ. We remark that because of

$$
Z_2|_{\gamma=\gamma_*} = qZ_{*,\gamma}
$$

the corresponding conservation law, given in point Sect. 5.3.1, is not new for the special values $\gamma = \gamma_*$.

5.4 Discussion

The complete Lie group classification of the gas dynamics equation in the Lagrangian coordinates (22) allows us to find all conservation laws which can be found using Noether's theorem and admitted symmetries. The group classification has three cases of the entropy for which there exist additional symmetries. In the Eulerian coordinates, these three cases are defined by differential constraints of first or second order. Notice that the overdetermined systems which consist of the gas dynamics equations and one of the considered differential constraints are involutive. The authors of [12, 13] also found conservation laws corresponding to special forms of the entropy. Here, the symmetry nature of these conservation laws is explained.

In contrast to [11], the conservation laws, obtained in this chapter, are local. It should be also noted that these conservation laws are naturally derived: their counterparts in the Lagrangian coordinates were derived directly using Noether's theorem without any additional assumptions. In contrast to the two-dimensional Lagrangian gas dynamics, the special cases of the entropy in the Lagrangian coordinates are given explicitly. In the two-dimensional case [40], the entropy is arbitrary, but the admitted symmetry operators contain functions satisfying quasilinear partial differential equations.

In a conservative form, the one-dimensional gas dynamics equations (11) are [33]

$$[r^n\rho]_t + [r^n\rho u]_r = 0, \tag{78a}$$

$$[r^n\rho u]_t + \left[r^n(\rho u^2 + p)\right]_r = nr^{n-1}p, \tag{78b}$$

$$\left[r^n\left(\rho\varepsilon + \tfrac{\rho u^2}{2}\right)\right]_t + \left[r^n\left(\rho\varepsilon + \tfrac{\rho u^2}{2} + p\right)u\right]_r = 0, \tag{78c}$$

where $[\ldots]_t$ and $[\ldots]_t$ denote total derivatives with respect to time t and the Eulerian coordinate r. One notes that the equation corresponding to the conservation law of momentum is not homogeneous. However, most methods for constructing conservation laws can only construct homogeneous conservation laws.

Consider inhomogeneous conservation laws of the one-dimensional gas dynamics equations

$$D_t\left[f^t\right] + D_r\left[f^r\right] = f, \tag{79}$$

where D_t and D_r are the total derivatives and the functions f^t, f^r, and f depend on (t, r, ρ, u, p). The method which is used to derive such conservation laws consists of obtaining an overdetermined system of partial differential equations for the functions f^t, f^r, and f and finding its general solution. The overdetermined system is derived by substituting the main derivatives ρ_t, u_t, and p_t found from the gas dynamics equations into (79) and splitting it with respect to the parametric derivatives.

Calculations show that the general solution of this system provides the conservation laws

$$[\rho F]_t + [\rho u F]_r = \rho\left(F_t + u\left(F_r - \frac{n}{r}F\right)\right), \tag{80}$$

$$[h\rho u]_t + \left[h(\rho u^2 + p)\right]_r = h_t \rho u + h_r(\rho u^2 + p) - \frac{n}{r} h\rho u^2, \tag{81}$$

$$\begin{aligned} &\left[h(\rho\frac{u^2}{2} + \frac{p}{\gamma - 1})\right]_t + \left[h(\rho\frac{u^2}{2} + \frac{\gamma p}{\gamma - 1})u\right]_r \\ &= h_t(\rho\frac{u^2}{2} + \frac{p}{\gamma - 1}) + (h_r - \frac{n}{r}h)(\rho\frac{u^2}{2} + \frac{\gamma p}{\gamma - 1})u, \end{aligned} \tag{82}$$

where $h(t, r)$ and $F(t, r, p\rho^{-\gamma})$ are arbitrary functions.

Equation (80) becomes a homogeneous conservation law if and only if

$$F(t, r, z) = r^n g(z), \qquad z = \frac{p}{\rho^\gamma},$$

which for $g \equiv 1$ gives Eq. (78a).

Equation (81) can be a homogeneous conservation law only if $n = 0$. Notice that for $h = r^n$, this equation becomes (78b). Equation (82) provides a homogeneous conservation law; only for $h = r^n$, it gives Eq. (78c).

It should be also noted here that if the overdetermined system defined above is extended by the condition $f = 0$, then one obtains all possible homogeneous zero-order conservation laws of the one-dimensional gas dynamics equations.

6 Difference Models Preserving Symmetries

The first problem in discretization of differential equations is the choice of a difference mesh. The peculiarity of our approach is that we add mesh equation(s) into the difference model:

$$F_i(z) = 0, \qquad i = 1, \ldots, I; \tag{83a}$$

$$\Omega_j(z) = 0, \qquad j = 1, \ldots, J. \tag{83b}$$

Here, the first set of equations approximates the underlying differential system and the second set of equations describes the difference mesh; z is a set of difference variables needed for approximation. As it was shown in [19, 22], the invariance of the mesh structure is a necessary condition for the invariance of the difference model. The mesh equations can be presented with the help of difference invariants or, alternatively, one can check the invariance of any chosen mesh by means of a certain criterium (see [19, 22]).

Symmetries of difference schemes allow one to construct difference counterparts of the differential conservation laws. The latter provides the absence of fake sources of energy, momentum, etc. in difference models which play an important role in solutions with high gradients. Moreover, the presence of (local) difference conservation laws gives a possibility to apply the difference counterpart of the Gauss–Ostrogradskii theorem [41] that leads to global conservation properties of the numerical solutions.

For discretization of the gas dynamics system (11a), (11b), (12), which is given in the Eulerian coordinates, the simplest choice seems to be an orthogonal mesh in (t, r) plane. As it will be shown below, this mesh is not invariant with respect to symmetries which we aim to preserve in the difference models. This noninvariance destroys the invariance of difference equations considered on such a mesh. We will choose another coordinate system in which one can preserve mesh geometry and, hence, the invariance of the whole difference model.

6.1 The Gas Dynamics Equations

In Sect. 3, we considered entropy as one of the dependent variables. Since the entropy is conserved along pathlines only for smooth solutions, it is appropriate to choose another form of the gas dynamics equations for numerical modeling.

6.1.1 Eulerian Coordinates

We start with the equations for the gas dynamics variables ρ, u, and p:

$$\rho_t + u\rho_r + \frac{\rho}{r^n}(r^n u)_r = 0, \tag{84a}$$

$$u_t + uu_r + \frac{1}{\rho}p_r = 0, \tag{84b}$$

$$p_t + up_r + \frac{\gamma p}{r^n}(r^n u)_r = 0, \tag{84c}$$

which admits four symmetries for any n and γ

$$X_1 = \frac{\partial}{\partial t}, \qquad X_2 = t\frac{\partial}{\partial t} + r\frac{\partial}{\partial r},$$
$$X_3 = 2t\frac{\partial}{\partial t} + r\frac{\partial}{\partial r} - u\frac{\partial}{\partial u} + 2\rho\frac{\partial}{\partial \rho}, \qquad X_4 = \rho\frac{\partial}{\partial \rho} + p\frac{\partial}{\partial p}. \tag{85}$$

For $n = 0$, there are two additional symmetries

$$X_{*,n} = \frac{\partial}{\partial r} \quad \text{and} \quad X_{**,n} = t\frac{\partial}{\partial r} + \frac{\partial}{\partial u}. \tag{86}$$

For the special values $\gamma_* = \frac{n+3}{n+1}$, there is one additional symmetry

$$X_{*,\gamma} = t^2\frac{\partial}{\partial t} + tr\frac{\partial}{\partial r} + (r - tu)\frac{\partial}{\partial u} - (n+1)t\rho\frac{\partial}{\partial \rho} - (n+3)tp\frac{\partial}{\partial p}. \tag{87}$$

6.1.2 Conservation Laws

System (84) possesses the following conservation laws.

1. General case of n and γ

   In the general case, we get

   - Conservation of mass

$$\left[r^n\rho\right]_t + \left[r^n\rho u\right]_r = 0; \tag{88}$$

   - Conservation of energy

$$\left[r^n\left(\rho\varepsilon + \frac{\rho u^2}{2}\right)\right]_t + \left[r^n\left(\rho\varepsilon + \frac{\rho u^2}{2} + p\right)u\right]_r = 0; \tag{89}$$

   - Conservation law

$$\left[r^n\rho F\left(\frac{p}{\rho^\gamma}\right)\right]_t + \left[r^n\rho u F\left(\frac{p}{\rho^\gamma}\right)\right]_r = 0, \tag{90}$$

     where F is a differentiable function. It holds due to the conservation of mass (88) and conservation of entropy along the pathlines, given by

$$\left(\frac{p}{\rho^\gamma}\right)_t + u\left(\frac{p}{\rho^\gamma}\right)_r = 0.$$

2. Case $n = 0$

   For the particular case $n = 0$ (plain one-dimensional flows), we obtain two additional conservation laws:

   - Momentum

$$[\rho u]_t + \left[\rho u^2 + p\right]_r = 0; \tag{91}$$

   - Motion of the center of mass

$$[\rho(r - tu)]_t + [\rho u(r - tu) - tp]_r = 0. \tag{92}$$

3. Special values of $\gamma_* = \frac{n+3}{n+1}$

   For $\gamma = \gamma_*$, there are two additional conservation laws

$$\begin{aligned}&\left\{r^n\left[2t\left(\rho\varepsilon+\frac{\rho u^2}{2}\right)-r\rho u\right]\right\}_t\\&+\left\{r^n\left[2t\left(\rho\varepsilon+\frac{\rho u^2}{2}+p\right)u-r(\rho u^2+p)\right]\right\}_r=0\end{aligned} \tag{93}$$

and

$$\begin{aligned}&\left\{r^n\left[t^2\left(\rho\varepsilon+\frac{\rho u^2}{2}\right)-tr\rho u+\frac{r^2}{2}\rho\right]\right\}_t\\&+\left\{r^n\left[t^2\left(\rho\varepsilon+\frac{\rho u^2}{2}+p\right)u-tr(\rho u^2+p)+\frac{r^2}{2}\rho u\right]\right\}_r=0.\end{aligned} \tag{94}$$

One can find conservation laws by direct computation or by an appropriate reduction of the three-dimensional conservation laws. Conservation laws of three-dimensional gas dynamics were obtained in [42] (see also [2]) with the help of a variational formulation and Noether's theorem (it requires some assumptions) and by direct computation in [43]. Among the 13 conservation laws of the three-dimensional case, all but one can be integrated over discontinuities [43]. The only conservation law which cannot be integrated over discontinuities gets reduced to (90) in the case of one-dimensional flows. It cannot be integrated over discontinuities because the entropy is not conserved for the discontinuous solutions [37, 44]. In [2, 42], one can find a symmetry interpretation of the conservation laws, i.e., the correspondence between the conservation laws and the Lie point symmetries of the three-dimensional gas dynamics equations.

6.1.3 Lagrangian Coordinates

We rewrite the gas dynamics equations (84) in the Lagrangian coordinates (t, s) as

$$\left(\tfrac{1}{\rho}\right)_t=(r^n u)_s, \tag{95a}$$

$$u_t+r^n p_s=0, \tag{95b}$$

$$\varepsilon_t=-p(r^n u)_s, \tag{95c}$$

$$r_t=u. \tag{95d}$$

Note that in the Lagrangian coordinates, variable r is dependent. It is given by Eq. (95d), which is included in the system of the gas dynamics equations, and the relation

$$r_s=\frac{1}{r^n\rho}. \tag{96}$$

From Eq. (95), it is easy to see that

$$\varepsilon_t = -p\left(\frac{1}{\rho}\right)_t. \tag{97}$$

This relation is important for the balance between the specific internal energy and the specific kinetic energy.

We rewrite symmetries (85) and additional symmetries (86) and (87) in the Lagrangian coordinates. There are four symmetries in the general case

$$\begin{aligned} X_1 = \frac{\partial}{\partial t}, \quad & X_2 = t\frac{\partial}{\partial t} + (n+1)s\frac{\partial}{\partial s} + r\frac{\partial}{\partial r}, \\ & X_3 = 2t\frac{\partial}{\partial t} + (n+3)s\frac{\partial}{\partial s} - u\frac{\partial}{\partial u} + 2\rho\frac{\partial}{\partial \rho} + r\frac{\partial}{\partial r}, \\ & X_4 = s\frac{\partial}{\partial s} + \rho\frac{\partial}{\partial \rho} + p\frac{\partial}{\partial p}; \end{aligned} \tag{98}$$

two additional symmetries for the particular case $n = 0$

$$X_{*,n} = \frac{\partial}{\partial r} \quad \text{and} \quad X_{**,n} = t\frac{\partial}{\partial r} + \frac{\partial}{\partial u}; \tag{99}$$

and one additional symmetry for the special case γ_*

$$X_{*,\gamma} = t^2\frac{\partial}{\partial t} + (r - tu)\frac{\partial}{\partial u} - (n+1)t\rho\frac{\partial}{\partial \rho} - (n+3)tp\frac{\partial}{\partial p} + tr\frac{\partial}{\partial r}. \tag{100}$$

We also include the translation symmetry for the mass Lagrange coordinate, which is given by the generator

$$X_0 = \frac{\partial}{\partial s}. \tag{101}$$

6.1.4 Conservation Laws

Let us rewrite the conservation laws for the Lagrangian coordinates. We obtain

1. General case of n and γ
   There hold

   - Conservation of mass

$$\left[\frac{1}{\rho}\right]_t - [r^n u]_s = 0; \tag{102}$$

   - Conservation of energy

$$[\varepsilon + \frac{1}{2}u^2]_t + [r^n pu]_s = 0; \tag{103}$$

- Conservation of entropy along pathlines

$$[\frac{p}{\rho^\gamma}]_t = 0. \tag{104}$$

2. Case $n = 0$

There are additional

- Conservation of momentum

$$[u]_t + [p]_s = 0; \tag{105}$$

- Motion of the center of mass

$$[r - tu]_t - [tp]_s = 0. \tag{106}$$

3. Special values of $\gamma_* = \frac{n+3}{n+1}$

For $\gamma = \gamma_*$, there are two additional conservation laws

$$\left[2t\left(\varepsilon + \frac{1}{2}u^2\right) - ru\right]_t + [r^n p(2tu - r)]_s = 0 \tag{107}$$

and

$$\left[t^2\left(\varepsilon + \frac{1}{2}u^2\right) - tru + \frac{r^2}{2}\right]_t + [r^n p(t^2 u - tr)]_s = 0. \tag{108}$$

6.2 *The Numerical Schemes*

In this section, we consider numerical schemes and their symmetries. Besides, our goal is to construct schemes that have difference conservation laws analogous to the conservation laws of the underlying differential system. We restrict ourselves by the homogenous conservation laws.

6.2.1 Invariance and Eulerian Coordinates for $n = 0$

For discretization of the gas dynamics system (84), which is given in the Eulerian coordinates, the simplest choice seems to be an orthogonal mesh in (t, r) plane. However, this mesh is not invariant that destroys invariance of difference equations considered on such mesh. Indeed, as it was shown in [19, 22], the necessary condition for a mesh to preserve its orthogonality under a group transformation generated by the operator

$$X = \xi^t \frac{\partial}{\partial t} + \xi^r \frac{\partial}{\partial r} + \cdots \tag{109}$$

is the following:

$$D_{+h}(\xi^t) = -D_{+\tau}(\xi^x), \tag{110}$$

where D_{+h} and $D_{+\tau}$ are the operators of difference differentiation in r and t directions, respectively.

System (84) admits the 6-parameter Lie symmetry group of point transformations that corresponds to the Lie algebra of infinitesimal operators (85) and (86). In the special case $\gamma_* = 3$, there is one more symmetry (87).

It is easy to see that the Galilean transformation given by the operator $X_{**,n}$ does not satisfy the criterion (110). The same is true for $X_{*,\gamma}$. It means that one should look for an invariant moving mesh in the Eulerian coordinates.

To obtain an invariant moving mesh, we chose the following difference stencil with two time layers:

- independent variables:

$$t = t_j, \quad \hat{t} = t_{j+1}; \quad r = r_i^j, \quad r_+ = r_{i+1}^j, \quad \hat{r} = r_i^{j+1}, \quad \hat{r}_+ = r_{i+1}^{j+1};$$

- dependent variables in the nodes of the mesh (the same notation as for r):

$$u, \quad u_+, \quad \hat{u}, \quad \hat{u}_+; \quad \rho, \quad \rho_-, \quad \hat{\rho}, \quad \hat{\rho}_-; \quad p, \quad p_-, \quad \hat{p}, \quad \hat{p}_-.$$

Then, we find the finite-difference invariants for symmetries (85) and (86) as solutions of the system of linear equations

$$X_i I(t, \hat{t}, r, r_+, \hat{r}, \hat{r}_+, \ldots, p, p_-, \hat{p}, \hat{p}_-) = 0 \tag{111}$$

for the considered symmetries. Here, we assume that the operator is prolonged for all variables of the stencil [22]. There are 12 functionally independent invariants

$$\frac{\hat{h}_+}{h_+}, \quad \frac{\tau}{h_+}\sqrt{\frac{p}{\rho}}, \quad \sqrt{\frac{\rho}{p}}\left(\frac{\hat{r}-r}{\tau} - u\right),$$

$$\sqrt{\frac{\rho}{p}}(u_+ - u), \quad \sqrt{\frac{\rho}{p}}(\hat{u} - u), \quad \sqrt{\frac{\rho}{p}}(\hat{u}_+ - \hat{u}),$$

$$\frac{p_+}{p}, \quad \frac{\hat{p}}{p}, \quad \frac{\hat{p}_+}{\hat{p}}, \quad \frac{\hat{\rho}}{\rho}, \quad \frac{\hat{\rho}_+}{\hat{\rho}}, \quad \frac{\rho_+}{\hat{\rho}},$$

where $\tau = \hat{t} - t$, $h_+ = r_+ - r$, and $\hat{h}_+ = \hat{r}_+ - \hat{r}$.

Notice that the only one difference invariant contains the value $\hat{r}$. This invariant suggests, for example, an invariant moving mesh given by

$$\sqrt{\frac{\rho}{p}}\left(\frac{\hat{r}-r}{\tau}-u\right)=0$$

or, equivalently,

$$\frac{\hat{r}-r}{\tau}=u. \tag{112}$$

In the continuous limit, it corresponds to the evolution of the spacial variable r given as

$$\frac{dr}{dt}=u. \tag{113}$$

Thus, we arrive at choosing the mass Lagrangian coordinates with the operator of differentiation with respect to t

$$D_t^L = D_t^E + uD_r.$$

6.2.2 Notations

We introduce the mesh for the mass Lagrangian coordinate s:

$$h^s = s_{i+1} - s_i \qquad \text{and} \qquad h^s_- = s_i - s_{i-1}. \tag{114}$$

Generally, the spacing can be nonuniform. For simplicity, we use a uniform mesh $h^s = h^s_-$.

For time, we consider the mesh with points t_j. Since we consider the schemes with two time layers, we denote the time step as τ. Of course, we can consider nonuniform time meshes with steplengths $\tau_j = t_{j+1} - t_j$.

Now the operators have the form

$$X = \xi^t \frac{\partial}{\partial t} + \xi^s \frac{\partial}{\partial s} + \cdots \tag{115}$$

and the criterium of invariant orthogonality

$$D_{+h_s}(\xi^t) = -D_{+\tau}(\xi^s) \tag{116}$$

holds for all considered symmetries (98), (99), (100), and (101). Here, D_{+h_s} and $D_{+\tau}$ are the operators of difference differentiation in s and t directions, respectively.

We split the dependent variables into kinematic and thermodynamic. The kinematic variables u and r are prescribed to the nodes. For example, for u, we have

$$u = u_i^j, \qquad u_+ = u_{i+1}^j, \qquad \hat{u} = u_i^{j+1}, \qquad \hat{u}_+ = u_{i+1}^{j+1}.$$

The thermodynamic variables ρ and p are taken in the midpoints as

$$\rho_- = \rho_{i-1/2}^j, \qquad \rho = \rho_{i+1/2}^j, \qquad \rho_+ = \rho_{i+3/2}^j,$$
$$\hat{\rho}_- = \rho_{i-1/2}^{j+1}, \qquad \hat{\rho} = \rho_{i+1/2}^{j+1}, \qquad \hat{\rho}_+ = \rho_{i+3/2}^{j+1}.$$

To describe the scheme, we need the time and spatial derivatives

$$u_t = \frac{\hat{u} - u}{\tau}, \qquad u_s = \frac{u_{i+1}^j - u_i^j}{s_{i+1} - s_i} = \frac{u_+ - u}{h^s}. \qquad p_{\bar{s}} = \frac{p_{i+1/2}^j - p_{i-1/2}^j}{\frac{1}{2}(s_{i+1} - s_{i-1})} = \frac{p - p_-}{h^s}$$

and weighted values defined as

$$y^{(\alpha)} = \alpha \hat{y} + (1 - \alpha) y, \qquad 0 \leq \alpha \leq 1.$$

6.2.3 The Samarskii–Popov Scheme

In [45] (see also [32]), the authors introduced a conservative scheme for plain one-dimensional flows ($n = 0$). It was generalized to the other one-dimensional flows ($n = 1, 2$) in [32]. This scheme is a discretization of Eq. (95)

$$\left(\frac{1}{\rho}\right)_t = (Ru^{(0.5)})_s, \tag{117a}$$
$$u_t = -Rp_{\bar{s}}^{(\alpha)}, \tag{117b}$$
$$\varepsilon_t = -p^{(\alpha)}(Ru^{(0.5)})_s, \tag{117c}$$
$$r_t = u^{(0.5)}, \tag{117d}$$

where R is a discretization of r^n chosen as

$$R = \frac{\hat{r}^{n+1} - r^{n+1}}{(n+1)(\hat{r} - r)} = \begin{cases} 1, & n = 0; \\ \dfrac{\hat{r} + r}{2}, & n = 1; \\ \dfrac{\hat{r}^2 + \hat{r}r + r^2}{3}, & n = 2. \end{cases}$$

Scheme (117) has four equations for five variables ρ, u, ε, r, and p. It should be supplemented by a discrete equation of state, a discrete analog of (9). For example, it can be taken in the same form that means

$$\varepsilon_{i+1/2}^j = \varepsilon(\rho_{i+1/2}^j, p_{i+1/2}^j). \tag{118}$$

6.2.4 Properties of the Samarskii–Popov Scheme

For a polytropic gas scheme (117), (118) is invariant with respect to the symmetries (98) and (101) corresponding to the general case. For $n = 0$, it is also invariant to symmetries (99). The scheme is not invariant for the additional symmetry (100), which exists for the special values γ_*.

Let us review important properties of the scheme. It possesses many qualitative properties of the underlying differential equations. For any equation of state $\varepsilon = \varepsilon(\rho, p)$, i.e., not only for the polytropic gas (10), this scheme has the following conservation laws:

- Conservation of mass

$$\left[\frac{1}{\rho}\right]_t - [Ru^{(0.5)}]_s = 0; \tag{119}$$

- Conservation of energy

$$\left[\varepsilon + \frac{u^2 + u_+^2}{4}\right]_t + [Rp_*^{(\alpha)}u^{(0.5)}]_s = 0, \tag{120}$$

where

$$p_*^{(\alpha)} = (p_*)_i^{(\alpha)} = \frac{p_{i-1/2}^{(\alpha)} + p_{i+1/2}^{(\alpha)}}{2}.$$

For $n = 0$, there are two additional conservation laws:

- Conservation of momentum

$$[u]_t + [p^{(\alpha)}]_s = 0; \tag{121}$$

- Motion of the center of mass

$$[r - tu]_t - [t^{(0.5)}p^{(\alpha)}]_s = 0. \tag{122}$$

These conservation laws correspond to (102), (103), (105), and (106). There are no discrete conservation laws corresponding to (107) and (108), which hold for the special values of γ_*.

Remark 1 Modifying the equation of state (10), it is possible to achieve conservation of the conservation laws (107) and (108), which hold for $\gamma_* = \frac{n+3}{n+1}$, under discretization. We refer to [46] for the case $n = 0$ and to [47] for the generalization to $n = 1, 2$.

Scheme (117) consists of four equations for five variables ρ, u, p, ε, and r. We will not impose the discrete equation of state (118). The freedom to choose a discretization of the equation of state will be used to impose one additional conservation law. Let us look for an equation of state which gives us the following difference analog of the additional conservation law (107):

$$\left[2t\left(\varepsilon + \frac{< u^2 >}{2}\right) - < ru >\right]_t + [Rp_*^{(\alpha)}(2t^{(0.5)}u^{(0.5)} - r^{(0.5)})]_s = 0, \tag{123}$$

where we use a special notation for the average value of two function values taken in the neighboring nodes of the same time layer

$$< f(u,r) >= \frac{f(u,r) + f(u_+, r_+)}{2}.$$

It leads to the following specific internal energy equation:

$$\varepsilon^{(0.5)} = \frac{p^{(\alpha)}}{\gamma - 1}\left(\frac{1}{\rho}\right)^{(0.5)} - \frac{\tau^2}{8} < (u_t)^2 > + \frac{1}{2}p^{(\alpha)}\left[r^{(0.5)}R - (r^{n+1})^{(0.5)}\right]_s. \tag{124}$$

We will take it as the *discrete equation of state*, which approximates (10).

In this case, we also get a difference analog of the second additional conservation law (108) as

$$\left[t^2\left(\varepsilon + \frac{< u^2 >}{2}\right) - t < ru > + \frac{< r^2 >}{2} + \frac{\tau^2}{8} < u^2 >\right]_t$$
$$+ [Rp_*^{(\alpha)}((t^2)^{(0.5)}u^{(0.5)} - t^{(0.5)}r^{(0.5)})]_s = 0. \tag{125}$$

Note that it has a correcting term $\frac{\tau^2}{8} < u^2 >$, which disappears in the continuous limit.

Thus, we obtained difference scheme (117) supplemented by discrete state equation (124). In this scheme, the pressure values p and $\hat{p}$ appear only as a weighted value $p^{(\alpha)}$, i.e., α has no longer meaning of a parameter. We can consider this value as the pressure in the midpoint of the cell $(t_{j+1/2}, s_{i+1/2})$, i.e., for $\alpha = 0.5$.

The scheme holds a discrete counterpart of the relation (97), namely

$$\varepsilon_t = -p^{(\alpha)}\left(\frac{1}{\rho}\right)_t. \tag{126}$$

This is an important supplement to the conservation of total energy (120), which provides the balance of the specific internal energy and the specific kinematic energy.

In the case of a polytropic gas, the equations of gas dynamics hold the conservation of entropy (104) along pathlines (for smooth solutions). There is no such property for the scheme (117). However, the scheme holds the relation

$$\frac{\Delta p}{p^{(\alpha)}} = \gamma \frac{\Delta \rho}{\rho^{(\alpha)}}, \qquad \Delta p = \hat{p} - p, \qquad \Delta \rho = \hat{\rho} - \rho \tag{127}$$

that approximates (104) presented with the help of differentials

$$\frac{dp}{p} = \gamma \frac{d\rho}{\rho}.$$

6.3 Invariance of Difference Schemes

In this point, we show how to construct invariant schemes with the help of finite-difference invariants. Scheme (117) can be expressed in terms of invariants for the general case of γ. Its modification described in Remark 1 possesses the additional conservation laws which hold for the special values γ_*. However, it is not invariant with respect to the additional symmetry $X_{*,\gamma}$ which exists for these special values. For the special values γ_*, invariant schemes are constructed. The case $n = 0$ was reported in [31].

6.3.1 General Case $n \neq 0, \gamma \neq \frac{n+3}{n+1}$

We chose an orthogonal mesh in the Lagrangian coordinates and a stencil with the following variables:

- independent variables:

$$t = t_j, \quad \hat{t} = t_{j+1}, \quad s = s_i, \quad s_+ = s_{i+1}, \quad s_- = s_{i-1};$$

- kinematic variables in the nodes:

$$u = u_i^j, \quad u_+ = u_{i+1}^j, \quad \hat{u} = u_i^{j+1}, \quad \hat{u}_+ = u_{i+1}^{j+1}, \quad r, \quad r_+, \quad \hat{r}, \quad \hat{r}_+;$$

- thermodynamic variables in the midpoints:

$$\rho = \rho_i^j, \quad \rho_- = \rho_{i-1}^j, \quad \hat{\rho} = \rho_i^{j+1}, \quad \hat{\rho}_- = \rho_{i-1}^{j+1}, \quad p, \quad p_-, \quad \hat{p}, \quad \hat{p}_-.$$

For these 21 stencil variables, we find $16 = 21 - 5$ invariants of the symmetries (98) and (101):

$$I_1 = \frac{h_-^s}{h^s}, \quad I_2 = \frac{\rho r^{n+1}}{h^s}, \quad I_3 = \frac{\tau}{h^s} r^n \sqrt{\rho p}, \quad I_4 = \frac{\tau u}{r},$$

$$I_5 = \frac{u_+}{u}, \quad I_6 = \frac{\hat{u}}{u}, \quad I_7 = \frac{\hat{u}_+}{\hat{u}}, \quad I_8 = \frac{r_+}{r}, \quad I_9 = \frac{\hat{r}}{r}, \quad I_{10} = \frac{\hat{r}_+}{\hat{r}},$$

$$I_{11} = \frac{\rho_-}{\rho}, \quad I_{12} = \frac{\hat{\rho}}{\rho}, \quad I_{13} = \frac{\hat{\rho}_-}{\hat{\rho}}, \quad I_{14} = \frac{p_-}{p}, \quad I_{15} = \frac{\hat{p}}{p}, \quad I_{16} = \frac{\hat{p}_-}{\hat{p}}.$$

The scheme (117) is invariant with respect to the considered symmetries and can be expressed in terms of the invariants as

$$\frac{1}{I_{12}} - 1 = I_2 I_4 \Big(\frac{(I_9 I_{10})^{n+1} - I_8^{n+1}}{(n+1)(I_9 I_{10} - I_8)} \frac{I_6 I_7 + I_5}{2} - \frac{I_9^{n+1} - 1}{(n+1)(I_9 - 1)} \frac{I_6 + 1}{2}\Big), \tag{128a}$$

$$I_6 - 1 = -\frac{I_3^2}{I_2 I_4} \frac{I_9^{n+1} - 1}{(n+1)(I_9 - 1)} \left(\alpha I_{15}(1 - I_{16}) + (1-\alpha)(1 - I_{14})\right), \tag{128b}$$

$$\begin{aligned} \frac{1}{\gamma - 1}\Big(\frac{I_{15}}{I_{12}} - 1\Big) &= -I_2 I_4 \left(\alpha I_{15} + (1-\alpha)\right) \\ &\times \Big(\frac{(I_9 I_{10})^{n+1} - I_8^{n+1}}{(n+1)(I_9 I_{10} - I_8)} \frac{I_6 I_7 + I_5}{2} - \frac{I_9^{n+1} - 1}{(n+1)(I_9 - 1)} \frac{I_6 + 1}{2}\Big), \end{aligned} \tag{128c}$$

$$I_9 - 1 = \frac{1}{2} I_4 (1 + I_7). \tag{128d}$$

6.3.2 Special Case $n = 0$, $\gamma \neq \frac{n+3}{n+1}$

In the space of 21 stencil variables, there are 14 invariants for 7 symmetries (98), (99), (101):

$$I_1 = \frac{h_-^s}{h^s}, \quad I_2 = \frac{\tau}{h^s}\sqrt{\rho p}, \quad I_3 = \sqrt{\frac{\rho}{p}}\Big(\frac{\hat{r} - r}{\tau} - u\Big), \quad I_4 = \sqrt{\frac{\rho}{p}}\Big(\frac{\hat{r} - r}{\tau} - \hat{u}\Big),$$

$$I_5 = \sqrt{\frac{\rho}{p}}(u_+ - u), \quad I_6 = \sqrt{\frac{\rho}{p}}(\hat{u}_+ - \hat{u}), \quad I_7 = \frac{\rho(r_+ - r)}{h^s}, \quad I_8 = \frac{\hat{\rho}(\hat{r}_+ - \hat{r})}{h^s},$$

$$I_9 = \frac{\rho_-}{\rho}, \quad I_{10} = \frac{\hat{\rho}}{\rho}, \quad I_{11} = \frac{\hat{\rho}_-}{\hat{\rho}}, \quad I_{12} = \frac{p_-}{p}, \quad I_{13} = \frac{\hat{p}}{p}, \quad I_{14} = \frac{\hat{p}_-}{\hat{p}}.$$

One can find the scheme (117) for $n = 0$ approximating the gas dynamics system (95) with the help of these invariants as

$$\frac{1}{I_{10}} - 1 = I_2 \frac{I_5 + I_6}{2}, \tag{129a}$$

$$I_3 - I_4 = -I_2 \left(\alpha \left(I_{13} - I_{13} I_{14}\right) + (1-\alpha)(1 - I_{12})\right), \tag{129b}$$

$$\frac{1}{\gamma - 1}\Big(\frac{I_{13}}{I_{10}} - 1\Big) = -I_2(\alpha I_{13} + (1-\alpha))\frac{I_5 + I_6}{2}, \tag{129c}$$

$$I_3 + I_4 = 0. \tag{129d}$$

6.3.3 Special Case $n \neq 0$, $\gamma_* = \frac{n+3}{n+1}$

We use the same mesh and stencil as for the general case of γ. Due to the additional symmetry (100), we get one invariant less. We obtain the following finite-difference invariants:

$$J_1 = \frac{h_-^s}{h^s}, \quad J_2 = \frac{\rho r^{n+1}}{h_s}, \quad J_3 = \frac{\hat{\rho}\hat{r}^{n+1}}{h^s}, \quad J_4 = \frac{\tau r^n}{h_s}\rho^{\frac{1}{2}-\frac{1}{n+1}}\hat{\rho}^{\frac{1}{n+1}}p^{\frac{1}{2}}, \quad J_5 = \frac{\hat{p}}{p}\left(\frac{\rho}{\hat{\rho}}\right)^{\frac{n+3}{n+1}},$$

$$J_6 = \frac{r + \tau u}{\hat{r}} \quad J_7 = \frac{r_+ + \tau u_+}{\hat{r}_+}, \quad J_8 = \frac{\hat{r} - \tau\hat{u}}{r}, \quad J_9 = \frac{\hat{r}_+ - \tau\hat{u}_+}{r_+},$$

$$J_{10} = \frac{r_+}{r} \quad J_{11} = \frac{\hat{r}_+}{\hat{r}}, \quad J_{12} = \frac{\rho_-}{\rho}, \quad J_{13} = \frac{\hat{\rho}_-}{\hat{\rho}}, \quad J_{14} = \frac{p_-}{p}, \quad J_{15} = \frac{\hat{p}_-}{\hat{p}}.$$

Using these invariants, we suggest an invariant scheme

$$\hat{\rho}(\hat{r}_+^{n+1} - \hat{r}^{n+1}) = \rho(r_+^{n+1} - r^{n+1}), \tag{130a}$$

$$\frac{\hat{u} - u}{\tau} = -\left(\frac{\hat{\rho}}{\rho}\right)^{\frac{2}{n+1}} r^n \, \frac{p - p_-}{h^s}, \tag{130b}$$

$$\frac{\hat{p}}{\hat{\rho}^{\frac{n+3}{n+1}}} = \frac{p}{\rho^{\frac{n+3}{n+1}}}, \tag{130c}$$

$$\frac{\hat{r} - r}{\tau} = u, \tag{130d}$$

which allows explicit computations. It is expressed in terms of the invariants as

$$J_3(J_{11}^{n+1} - 1) = J_2(J_{10}^{n+1} - 1), \tag{131a}$$

$$J_8 - 1 = \frac{J_4^2}{J_2}(1 - J_{14}), \tag{131b}$$

$$J_5 = 1, \tag{131c}$$

$$J_6 = 1. \tag{131d}$$

In addition to the invariance, the scheme (130) possesses conservation of mass, given by Eq. (130a), and conservation of the entropy along pathlines, given by (130c).

We remark that the conservation of mass property can be rewritten as

$$\frac{1}{\tau}\left(\frac{1}{\hat{\rho}} - \frac{1}{\rho}\right) = \frac{R_+u_+ - Ru}{h^s} \quad \text{or} \quad \left[\frac{1}{\rho}\right]_t - [Ru]_s = 0 \tag{132}$$

with

$$h_s = \hat{\rho}\frac{\hat{r}_+^{n+1} - \hat{r}^{n+1}}{n+1} = \rho\frac{r_+^{n+1} - r^{n+1}}{n+1}.$$

6.3.4 Special Case $n = 0$, $\gamma_* = 3$

In comparison to the case $n = 0$, $\gamma \neq 3$, we have one more symmetry, namely (100). Therefore, we get one invariant less. There are 13 invariants:

$$J_1 = \frac{h^s_-}{h^s}, \quad J_2 = \frac{\tau}{h^s}(\rho p \hat{\rho} \hat{p})^{\frac{1}{4}}, \quad J_3 = \sqrt{\frac{\rho}{p}}(\frac{\hat{r} - r}{\tau} - u), \quad J_4 = \sqrt{\frac{\hat{\rho}}{\hat{p}}}(\frac{\hat{r} - r}{\tau} - \hat{u}),$$

$$J_5 = \sqrt{\frac{\rho}{p}}(\frac{h_+}{\tau} + u_+ - u), \quad J_6 = \sqrt{\frac{\hat{\rho}}{\hat{p}}}(-\frac{\hat{h}_+}{\tau} + \hat{u}_+ - \hat{u}), \quad J_7 = \frac{\rho(r_+ - r)}{h^s},$$

$$J_8 = \frac{\hat{\rho}(\hat{r}_+ - \hat{r})}{h^s}, \quad J_9 = \frac{\hat{p}}{p}(\frac{\rho}{\hat{\rho}})^3, \quad J_{10} = \frac{\rho_-}{\rho}, \quad J_{11} = \frac{\hat{\rho}_-}{\hat{\rho}}, \quad J_{12} = \frac{p_-}{p}, \quad J_{13} = \frac{\hat{p}_-}{\hat{p}}.$$

There are many possibilities to approximate the gas dynamics system (95a), (95b), (104), (95d) with the help of these invariants. We propose the following explicit invariant scheme:

$$\hat{\rho}(\hat{r}_+ - \hat{r}) = \rho(r_+ - r), \tag{133a}$$

$$\tfrac{\hat{u} - u}{\tau} = -\left(\tfrac{\hat{\rho}}{\rho}\right)^2 \tfrac{p - p_-}{h^s}, \tag{133b}$$

$$\tfrac{\hat{p}}{\hat{\rho}^3} = \tfrac{p}{\rho^3}, \tag{133c}$$

$$\tfrac{\hat{r} - r}{\tau} = u. \tag{133d}$$

In term of the invariants, this scheme is written as

$$J_7 = J_8, \tag{134a}$$

$$J_4 = J_2 J_9^{-3/4}(1 - J_{12}), \tag{134b}$$

$$J_9 = 1, \tag{134c}$$

$$J_3 = 0. \tag{134d}$$

The scheme conserves the entropy, or S, along the pathlines and possesses conservation of mass (133a). Note that the first equation can be rewritten as

$$\frac{1}{\tau}(\frac{1}{\hat{\rho}} - \frac{1}{\rho}) = \frac{u_+ - u}{h^s} \quad \text{or} \quad \left[\frac{1}{\rho}\right]_t - [u]_s = 0 \tag{135}$$

with

$$h^s = \hat{\rho}(\hat{r}_+ - \hat{r}) = \rho(r_+ - r).$$

We remark that implicit invariant schemes are also possible.

7 Conclusion

In the chapter, we examined one-dimensional flows of a polytropic gas and their Lie point symmetry properties. By the one-dimensional flows we mean plain one-dimensional flows, the gas dynamics flows with radial symmetry, and the gas dynamics flows with spherical symmetry. There was performed the Lie group classification of the gas dynamics equations reduced to a single second-order PDE in the Lagrangian coordinates. The entropy function was a parameter of the classification. Four cases were identified. In the general case, there are conservation laws of mass and energy. For the special cases, there were found additional conservation laws. The conservation laws obtained for the second-order PDE were later rewritten for the gas dynamics variables. They were also transformed from the Lagrangian coordinates to the Eulerian ones.

Difference models were discussed for different cases of n and γ. It is shown that the Samarskii–Popov scheme is invariant for the symmetries of the general case of γ, but not for the additional symmetry of the special case $\gamma_* = \frac{n+3}{n+1}$. This scheme possesses conservation of mass and energy, for $n = 0$ also conservation of momentum and motion of the center of mass. It does not have conservation of the entropy along the pathlines. For the special values γ_*, we suggest invariant schemes, which have conservation of mass and conservation of the entropy along the pathlines.

Acknowledgements The research was supported by Russian Science Foundation Grant no. 18-11-00238 “Hydrodynamics-type equations: symmetries, conservation laws, invariant difference schemes”.

References

1. Ovsiannikov LV (1978) Group analysis of differential equations. Nauka, Moscow. English translation: Ames WF (ed) (1982) Published by Academic, New York
2. Ibragimov NH (1985) Transformation groups applied to mathematical physics. Reidel, Boston
3. Olver PJ (1986) Applications of Lie groups to differential equations. Springer, New York
4. Marsden J, Ratiu T (1994) Introduction to mechanics and symmetry. Springer, New York
5. Ibragimov NH (1999) Elementary Lie group analysis and ordinary differential equations. Wiley, Chichester
6. Bluman GW, Kumei S (1989) Symmetries and differential equations. Applied mathematical sciences, No. 81. Springer, New York
7. Ovsiannikov LV (1994) Program SUBMODELS. Gas dynamics. J Appl Maths Mechs 58(4):30–55
8. Akhatov IS, Gazizov RK, Ibragimov NH (1991) Nonlocal symmetries. Heuristic approach. J Math Sci 55(1):1401–1450
9. Ames WF, Anderson RL, Dorodnitsyn VA, Ferapontov EV, Gazizov RK, Ibragimov NH, Svirshchevskii SR (1994) CRC hand-book of Lie group analysis of differential equations, Volume I: symmetries, exact solutions and conservation laws. CRC Press, Boca Raton
10. Ibragimov NH (ed) (1995) CRC handbook of Lie group analysis of differential equations, vol 2. CRC Press, Boca Raton
11. Sjöberg A, Mahomed FM (2004) Non-local symmetries and conservation laws for one-dimensional gas dynamics equations. Appl Math Comput 150:379–397

12. Webb GM, Zank GP (2009) Scaling symmetries, conservation laws and action principles in one-dimensional gas dynamics. J Phys A: Math Theor 42:475205
13. Webb G (2018) Magnetohydrodynamics and fluid dynamics: action principles and conservation laws, vol 946. Lecture notes in physics. Springer, Heidelberg
14. Noether E (1971) Invariante variationsprobleme, Nachr. d. Königlichen Gesellschaft der Wissenschaften zu Göttingen, Math-phys. Klasse, pp 235–257. English translation: Trans Theory Stat Phys 1(3):186–207. arXiv:physics/0503066 [physics.hist-ph]
15. Seliger RL, Whitham GB (1968) Variational principles in continuum mechanics. Proc R Soc Lond A 305:1–25
16. Shmyglevski YuD (1999) Analytical study of gas dynamics and fluids. Editorial URSS, Moscow (in Russian)
17. Ibragimov NH (2007) A new conservation theorem. J Math Anal Appl 333:311–328
18. Bluman GW, Cheviakov AF, Anco SC (2010) Applications of symmetry methods to partial differential equations. Applied mathematical sciences, vol 168. Springer, New York
19. Dorodnitsyn VA (1991) Transformation groups in a space of net variables. In: VINITI Acad Sci USSR, Itogi Nauki i Techniki, vol 34, pp 149–190, (1989) in Russian. English translation: J Sov Math 55(1490):1991
20. Levi D, Winternitz P (2006) Continuous symmetries of difference equations. J Phys A: Math Gen 39:1–63
21. Winternitz P (2011) Symmetry preserving discretization of differential equations and Lie point symmetries of differential-difference equations. In: Levi D, Olver PJ, Thomova Z, Winternitz P (eds) Symmetries and integrability of difference equations. Cambridge University Press, Cambridge, pp 292–341
22. Dorodnitsyn V (2011) Applications of Lie groups to difference equations. Chapman & Hall/CRC differential and integral equations series
23. Dorodnitsyn VA (1993) Finite-difference analog of the Noether theorem. Dokl Akad Nauk 328:678 (in Russian)
24. Bakirova M, Dorodnitsyn V, Kozlov R (1997) Invariant difference schemes for heat transfer equations with a source. J Phys A: Math Gen 30:8139
25. Dorodnitsyn V, Kozlov R, Winternitz P (2000) Lie group classification of second-order ordinary difference equations. J Math Phys 41:480–504
26. Dorodnitsyn V, Kozlov R (2003) A heat transfer with a source: the complete set of invariant difference schemes. J Nonl Math Phys 10:16–50
27. Dorodnitsyn V, Kozlov R, Winternitz P (2004) Continuous symmetries of Lagrangians and exact solutions of discrete equations. J Math Phys 45:336–359
28. Dorodnitsyn VA, Kaptsov EI (2020) Shallow water equations in Lagrangian coordinates: Symmetries, conservation laws and its preservation in difference models. Commun Nonlinear Sci Numer Simulat 89:105343
29. Cheviakov AF, Dorodnitsyn VA, Kaptsov EI (2020) Invariant conservation law-preserving discretizations of linear and nonlinear wave equations. J Math Phys 61:081504
30. Dorodnitsyn VA, Kaptsov EI, Meleshko SV, Symmetries, conservation laws, invariant solutions and difference schemes of the one-dimensional Green-Naghdi equations. J Nonl Math Phys. To appear
31. Dorodnitsyn VA, Kozlov R, Meleshko SV (2019) One-dimensional gas dynamics equations of a polytropic gas in Lagrangian coordinates: symmetry classification, conservation laws, difference schemes. Commun Nonlinear Sci Numer Simulat 74:201–218
32. Samarskii AA, Popov YuP (1980) Difference methods for solving problems of gas dynamics. Nauka, Moscow (in Russian)
33. Rozhdestvenskii BL, Yanenko NN (1978) Systems of quasilinear equations and their applications to gas dynamics, 2nd ed. Nauka, Moscow. English translation published by Amer Math Soc Providence, RI (1983)
34. Ovsiannikov LV (2003) Lectures on the gas dynamics equations. Institute of computer studies, Moscow-Izhevsk (in Russian)

35. Dorodnitsyn VA, Kozlov R (2011) Lagrangian and Hamiltonian formalism for discrete equations: symmetries and first integrals. Symmetries and integrability of difference equations. London mathematical society lecture notes. Cambridge University Press, Cambridge, pp 7–49
36. Chernyi GG (1988) Gas dynamics. Nauka, Moscow (in Russian)
37. Chorin AJ, Marsden JE (1990) A mathematical introduction to fluid mechanics. Springer, Berlin
38. Toro EF (1997) Riemann solvers and numerical methods for fluid dynamics. Springer, Berlin
39. Andreev VK, Kaptsov OV, Pukhnachov VV, Rodionov AA (1998) Applications of group-theoretic methods in hydrodynamics. Kluwer, Dordrecht
40. Kaptsov EI, Meleshko SV (2019) Conservation laws of the two-dimensional gas dynamics equations. Int J Non-Linear Mech 112:126–132
41. Samarskii AA (2001) The theory of difference schemes. Marcel Deakker, New York
42. Ibragimov NH (1973) Conservation laws in hydrodynamics. Dokl Akad Nauk SSSR 210(6):1307–1309
43. Terent'ev ED, Shmyglevskii YuD (1975) A complete system of equations in divergence form for the dynamics of an ideal gas. Zh Vychisl Mat Mat Fiz 15(6):1535–1544
44. Landau LD, Lifshitz EM (1987) Fluid mechanics, 2nd. ed. Pergamon Press, Oxford
45. Popov YuP, Samarskii AA (1969) Completely conservative difference schemes. Zh Vychisl Mat Mat Fiz 9(4):953–958
46. Korobitsyn VA (1989) Thermodynamically matched difference schemes. Zh Vychisl Mat Mat Fiz 29(2):309–312
47. Kozlov R (2019) Conservative difference schemes for one-dimensional flows of polytropic gas. Commun. Nonlinear Sci Numer Simulat 78:104864

Hamiltonian Structure and Conservation Laws of Three-Dimensional Linear Elasticity Theory

D. O. Bykov, V. N. Grebenev, and S. B. Medvedev

Abstract This is a continuation of the paper [5] wherein the Hamiltonian structure together with the non-canonical singular Poisson bracket and Casimir functionals were established for two-dimensional linear elasticity model. The aim of the present work is the extension of the above-mentioned results to the three-dimension case.

1 Introduction

A three-dimensional, linear-elastic model is widely exploited in studying the small deformations of a medium [1–3]. This model can be considered in various variables referee frames and forms [4]. In the present contribution, the corresponding equations of the model are written in terms of the velocity vector [4]. We establish that this model admits a Hamiltonian structure that presents the extension to the three-dimensional case of the results obtained in [5]. We find the form of Poisson bracket for these equations and demonstrate that this is a singular non-canonical Poisson bracket. Notice that the investigation of Hamiltonian structures of linear equations has a sense since such models are usually derived without using Hamiltonian structures of the original nonlinear models. It is interesting that the existence of a non-canonical form of Poisson bracket enables us to calculate Casimir functionals which are conserved for the Hamiltonian derived. In comparison with the two-dimensional case, we also demonstrate that the three-dimensional Poisson bracket gives more complicated Casimir functionals. We find all zero-order conservation laws and show by the direct calculations that only a unique quadratic conservation law exists which

D. O. Bykov
Novosibirsk State University, Pirogov Str. 1, Novosibirsk 630090, Russia
e-mail: gearbatyan@mail.ru

V. N. Grebenev (✉) · S. B. Medvedev
Federal Research Center for Information and Computational Technologies, Russian Academy of Sciences, Lavrentjev ave. 6, Novosibirsk 630090, Russia

A. C. J. Luo and R. K. Gazizov (eds.), *Symmetries and Applications of Differential Equations*, Nonlinear Physical Science,
https://doi.org/10.1007/978-981-16-4683-6_3

actually coincides with the energy functional. This is similar to the case of the three-dimensional Navier–Stokes equations. Finally, we investigate the linear terms of the density of conservation laws obtained in details using the symmetric form of the equations.

The chapter is structured as follows. In Sect. 2, we present the three-dimensional linear equations of elasticity for the isentropic medium. The Hamiltonian form of this model is derived in Sect. 3. Section 4 is devoted to studying the properties of a Hamiltonian structure admitted by the original model. Casimir functionals associated with this system are found in Sect. 5. In Sect. 6, we find all zero-order conservation laws admitted by this model. A discussion and summary of the results obtained are given in Sect. 7. The appendix contains several well-known formulas from the linear theory of elasticity.

2 System of Equations

A three-dimensional, linear-elastic model for the isentropic medium is considered in the following form [4]

$$\begin{cases} \dfrac{\partial \varepsilon_{ij}}{\partial t} = \dfrac{1}{2}\Big(\dfrac{\partial u_i}{\partial x_j} + \dfrac{\partial u_j}{\partial x_i}\Big), \\ \rho_0 \dfrac{\partial u_i}{\partial t} = \dfrac{\partial \sigma_{ij}}{\partial x_j}, \\ \sigma_{ij} = \lambda \delta_{ij}(\varepsilon_{11} + \varepsilon_{22} + \varepsilon_{33}) + 2\mu\varepsilon_{ij}, \end{cases} \tag{1}$$

where $\rho_0 > 0$ is the density of medium, λ denotes the Lamé parameter, μ is the shear modulus, σ denotes the stress tensor and ε is the tensor of small deformations. The system (1) in the variables $u_1, u_2, u_3, \sigma_{11}, \sigma_{22}, \sigma_{33}, \sigma_{12}, \sigma_{13}, \sigma_{23}$ reads

$$\rho_0 \frac{\partial u_1}{\partial t} = \frac{\partial \sigma_{11}}{\partial x_1} + \frac{\partial \sigma_{12}}{\partial x_2} + \frac{\partial \sigma_{13}}{\partial x_3}, \tag{2}$$

$$\rho_0 \frac{\partial u_2}{\partial t} = \frac{\partial \sigma_{21}}{\partial x_1} + \frac{\partial \sigma_{22}}{\partial x_2} + \frac{\partial \sigma_{23}}{\partial x_3}, \tag{3}$$

$$\rho_0 \frac{\partial u_3}{\partial t} = \frac{\partial \sigma_{31}}{\partial x_1} + \frac{\partial \sigma_{32}}{\partial x_2} + \frac{\partial \sigma_{33}}{\partial x_3}, \tag{4}$$

$$\frac{\partial \sigma_{11}}{\partial t} = (\lambda + 2\mu)\frac{\partial u_1}{\partial x_1} + \lambda \frac{\partial u_2}{\partial x_2} + \lambda \frac{\partial u_3}{\partial x_3}, \tag{5}$$

$$\frac{\partial \sigma_{22}}{\partial t} = \lambda \frac{\partial u_1}{\partial x_1} + (\lambda + 2\mu)\frac{\partial u_2}{\partial x_2} + \lambda \frac{\partial u_3}{\partial x_3}, \tag{6}$$

$$\frac{\partial \sigma_{33}}{\partial t} = \lambda \frac{\partial u_1}{\partial x_1} + \lambda \frac{\partial u_2}{\partial x_2} + (\lambda + 2\mu) \frac{\partial u_3}{\partial x_3}, \tag{7}$$

$$\frac{\partial \sigma_{12}}{\partial t} = \mu (\frac{\partial u_2}{\partial x_1} + \frac{\partial u_1}{\partial x_2}), \tag{8}$$

$$\frac{\partial \sigma_{13}}{\partial t} = \mu (\frac{\partial u_3}{\partial x_1} + \frac{\partial u_1}{\partial x_3}), \tag{9}$$

$$\frac{\partial \sigma_{23}}{\partial t} = \mu (\frac{\partial u_3}{\partial x_2} + \frac{\partial u_2}{\partial x_3}), \tag{10}$$

$$\sigma_{21} = \sigma_{12}, \qquad \sigma_{31} = \sigma_{13}, \qquad \sigma_{32} = \sigma_{23}. \tag{11}$$

For convenience, it is used the following notation of the dependent variables $u_4 = \sigma_{11}$, $u_5 = \sigma_{22}$, $u_6 = \sigma_{33}$, $u_7 = \sigma_{12}$, $u_8 = \sigma_{13}$, $u_9 = \sigma_{23}$ and $\boldsymbol{u} = (u_1, \ldots, u_9)^{\mathrm{T}}$. With this, the system (2)–(11) can be written in the operator form

$$\boldsymbol{u}_t = \hat{L} \boldsymbol{u}, \tag{12}$$

where the matrix operator $\hat{L}$ is

$$\hat{L} = \begin{pmatrix} \Theta & \hat{L}_{12} & \hat{L}_{13} \\ \hat{L}_{21} & \Theta & \Theta \\ \hat{L}_{31} & \Theta & \Theta \end{pmatrix} \tag{13}$$

or explicitly

$$\hat{L}_{12} = \frac{1}{\rho_0} \begin{pmatrix} \partial_{x_1} & 0 & 0 \\ 0 & \partial_{x_2} & 0 \\ 0 & 0 & \partial_{x_3} \end{pmatrix}, \qquad \hat{L}_{13} = \frac{1}{\rho_0} \begin{pmatrix} \partial_{x_2} & \partial_{x_3} & 0 \\ \partial_{x_1} & 0 & \partial_{x_3} \\ 0 & \partial_{x_1} & \partial_{x_2} \end{pmatrix}, \tag{14}$$

$$\hat{L}_{21} = \begin{pmatrix} (\lambda + 2\mu)\, \partial_{x_1} & \lambda \partial_{x_2} & \lambda \partial_{x_3} \\ \lambda \partial_{x_1} & (\lambda + 2\mu)\, \partial_{x_2} & \lambda \partial_{x_3} \\ \lambda \partial_{x_1} & \lambda \partial_{x_2} & (\lambda + 2\mu)\, \partial_{x_3} \end{pmatrix}, \tag{15}$$

$$\hat{L}_{31} = \mu \begin{pmatrix} \partial_{x_2} & \partial_{x_1} & 0 \\ \partial_{x_3} & 0 & \partial_{x_1} \\ 0 & \partial_{x_3} & \partial_{x_2} \end{pmatrix}, \qquad \Theta = \begin{pmatrix} 0 & 0 & 0 \\ 0 & 0 & 0 \\ 0 & 0 & 0 \end{pmatrix}. \tag{16}$$

3 Hamiltonian Structure

The system of Eqs. (2)–(11) is linear, the dissipative-free and has the quadratic conservation law [4], i.e. with the energy functional, see Appendix. Therefore (2)–(11) has to be a Hamiltonian system [6]. The aim is to explicitly present a Hamiltonian structure [7, 8] of the original model, that is, to find the representation of Eq. (12) in a Hamiltonian form:

$$\boldsymbol{u}_t = \hat{J}_{\boldsymbol{u}} \frac{\delta H[\boldsymbol{u}]}{\delta \boldsymbol{u}} = \hat{J}_{\boldsymbol{u}} A_{\boldsymbol{u}} \boldsymbol{u}. \tag{17}$$

Here $\hat{J}_{\boldsymbol{u}}$ is a skew-symmetric operator which defines a local Poisson bracket [7, 8], where the index $\boldsymbol{u}$ in (17) means that the operator acts with respect to the variable $\boldsymbol{u}$. As usually, $\frac{\delta H[\boldsymbol{u}]}{\delta \boldsymbol{u}}$ denotes the variational derivative of the Hamiltonian

$$H[\boldsymbol{u}] = \frac{1}{2} \int \boldsymbol{u}^{\mathrm{T}} A_{\boldsymbol{u}} \boldsymbol{u} \, dx_1 dx_2 dx_3, \tag{18}$$

where $A_{\boldsymbol{u}}$ is a symmetric matrix or an operator. In the two-dimensional case, the operators $\hat{J}_{\boldsymbol{u}}$ and $A_{\boldsymbol{u}}$ have been constructed in [5].

Notice also that all blocks of the operator $\hat{L}$, excepting $\hat{L}_{21}$, can be presented as a differential operator multiplied by a parameter. Also the operator $\hat{L}_{21}$ can be written in a matrix form together with the operator of differentiating:

$$\hat{L}_{21} = \begin{pmatrix} (\lambda+2\mu) & \lambda & \lambda \\ \lambda & (\lambda+2\mu) & \lambda \\ \lambda & \lambda & (\lambda+2\mu) \end{pmatrix} \begin{pmatrix} \partial_{x_1} & 0 & 0 \\ 0 & \partial_{x_2} & 0 \\ 0 & 0 & \partial_{x_3} \end{pmatrix}. \tag{19}$$

Using these properties, we can write $\hat{L}$ in the factor-form, i.e. $\hat{L} = S\hat{I}$, where $\hat{I}$ reads

$$\hat{I} = \begin{pmatrix} 0 & 0 & 0 & \partial_{x_1} & 0 & 0 & \partial_{x_2} & \partial_{x_3} & 0 \\ 0 & 0 & 0 & 0 & \partial_{x_2} & 0 & \partial_{x_1} & 0 & \partial_{x_3} \\ 0 & 0 & 0 & 0 & 0 & \partial_{x_3} & 0 & \partial_{x_1} & \partial_{x_2} \\ \partial_{x_1} & 0 & 0 & 0 & 0 & 0 & 0 & 0 & 0 \\ 0 & \partial_{x_2} & 0 & 0 & 0 & 0 & 0 & 0 & 0 \\ 0 & 0 & \partial_{x_3} & 0 & 0 & 0 & 0 & 0 & 0 \\ \partial_{x_2} & \partial_{x_1} & 0 & 0 & 0 & 0 & 0 & 0 & 0 \\ \partial_{x_3} & 0 & \partial_{x_1} & 0 & 0 & 0 & 0 & 0 & 0 \\ 0 & \partial_{x_3} & \partial_{x_2} & 0 & 0 & 0 & 0 & 0 & 0 \end{pmatrix} \tag{20}$$

$$= \begin{pmatrix} \Theta & \hat{I}_{12} & \hat{I}_{13} \\ \hat{I}_{21} & \Theta & \Theta \\ \hat{I}_{31} & \Theta & \Theta \end{pmatrix}.$$

The symmetric matrix S is of the following block-diagonal form

$$S = \begin{pmatrix} S_{11} & \Theta & \Theta \\ \Theta & S_{22} & \Theta \\ \Theta & \Theta & S_{33} \end{pmatrix}. \tag{21}$$

Here

$$S_{11} = \frac{1}{\rho_0}E, \quad S_{22} = \begin{pmatrix} (\lambda+2\mu) & \lambda & \lambda \\ \lambda & (\lambda+2\mu) & \lambda \\ \lambda & \lambda & (\lambda+2\mu) \end{pmatrix}, \quad S_{33} = \mu E,$$

where E is the identity matrix. With this, the original system takes the form

$$\boldsymbol{u}_t = \left(S\hat{I}S\right)S^{-1}\boldsymbol{u}, \tag{22}$$

for the nonsingular matrix S. Comparing (22) with (17) and taking into account that $\hat{I}$ is a skew-symmetric operator and S is a symmetric matrix, we can see that the operator of Poisson bracket equals a skew-symmetric operator $\hat{J}_{\boldsymbol{u}} = \hat{J} = S\hat{I}S$. Therefore, the Hamiltonian (18) is given by the symmetric matrix $A_{\boldsymbol{u}} = S^{-1}$. Notice that the inverse matrix $S^{-1} = A_{\boldsymbol{u}}$ exists provided that $\det S_{11} = \rho_0^{-3} \neq 0$, $\det S_{22} = 4\mu^2(3\lambda+2\mu) \neq 0$ and $\det S_{33} = \mu^3 \neq 0$. As a result, we get the following conditions[1]

$$\mu \neq 0, \quad \rho_0 \neq 0, \quad 3\lambda + 2\mu \neq 0. \tag{23}$$

Summarising, we explicitly write the operators $A_{\boldsymbol{u}} = S^{-1}$ and $\hat{J}_{\boldsymbol{u}}$ in the following block-matrix forms:

$$A_{\boldsymbol{u}} = S^{-1} = \begin{pmatrix} A_{11} & \Theta & \Theta \\ \Theta & A_{22} & \Theta \\ \Theta & \Theta & A_{33} \end{pmatrix}, \quad \hat{J} = \begin{pmatrix} \Theta & \hat{J}_{12} & \hat{J}_{13} \\ \hat{J}_{21} & \Theta & \Theta \\ \hat{J}_{31} & \Theta & \Theta \end{pmatrix}, \tag{24}$$

where $A_{11} = \rho_0 E$, $A_{33} = \mu^{-1}E$,

$$A_{22} = \begin{pmatrix} \frac{\lambda+\mu}{\mu(3\lambda+2\mu)} & -\frac{\lambda}{2\mu(3\lambda+2\mu)} & -\frac{\lambda}{2\mu(3\lambda+2\mu)} \\ -\frac{\lambda}{2\mu(3\lambda+2\mu)} & \frac{\lambda+\mu}{\mu(3\lambda+2\mu)} & -\frac{\lambda}{2\mu(3\lambda+2\mu)} \\ -\frac{\lambda}{2\mu(3\lambda+2\mu)} & -\frac{\lambda}{2\mu(3\lambda+2\mu)} & \frac{\lambda+\mu}{\mu(3\lambda+2\mu)} \end{pmatrix}, \tag{25}$$

$$\hat{J}_{21} = \hat{J}_{12}^{\mathrm{T}} = \frac{1}{\rho_0}\begin{pmatrix} (\lambda+2\mu)\partial_{x_1} & \lambda\partial_{x_2} & \lambda\partial_{x_3} \\ \lambda\partial_{x_1} & (\lambda+2\mu)\partial_{x_2} & \lambda\partial_{x_3} \\ \lambda\partial_{x_1} & \lambda\partial_{x_2} & (\lambda+2\mu)\partial_{x_3} \end{pmatrix} = \frac{1}{\rho_0}S_{22}\hat{I}_{21}, \tag{26}$$

[1] In the two-dimensional case, the similar condition reads $\lambda + \mu \neq 0$ [5].

$$\hat{J}_{31} = \hat{J}_{13}^{\mathrm{T}} = \frac{\mu}{\rho_0}\begin{pmatrix} \partial_{x_2} & \partial_{x_1} & 0 \\ \partial_{x_3} & 0 & \partial_{x_1} \\ 0 & \partial_{x_3} & \partial_{x_2} \end{pmatrix} = \frac{\mu}{\rho_0}\hat{I}_{31}. \tag{27}$$

4 Linear Transformations

To simplify or change the form of the equations under considerations, we can perform linear transformations of the variables by a nonsingular matrix D

$$\boldsymbol{v} = D\boldsymbol{u}. \tag{28}$$

First of all, we clear up how the Hamiltonian structure is transformed under the change of variables. Using the transformation (28), we get a new form of the system (17)

$$\boldsymbol{v}_t = \hat{J}_v \frac{\delta H[\boldsymbol{v}]}{\delta \boldsymbol{v}} = \hat{J}_v A_v \boldsymbol{v}, \quad H[\boldsymbol{v}] = \frac{1}{2}\int \boldsymbol{v}^{\mathrm{T}} A_v \boldsymbol{v}\, dx_1 dx_2 dx_3, \tag{29}$$

where the elements of Hamiltonian are transformed as follows

$$\hat{J}_v = D\hat{J}_u D^{\mathrm{T}}, \quad A_v = (D^{-1})^{\mathrm{T}} A_u D^{-1}. \tag{30}$$

It is possible to simplify the Hamiltonian structure either by a reduction to the simplest form of the operator of Poisson bracket or by a transformation of the matrix which defines this Hamiltonian. First, we perform this procedure for the operator $\hat{J}_v$ which defines the corresponding Poisson bracket. This form can be achieved by changing the variables

$$\boldsymbol{v} = A_u \boldsymbol{u}, \tag{31}$$

i.e. by setting $D = A_u$. Then the system takes the form

$$\boldsymbol{v}_t = \hat{I}\frac{\delta H[\boldsymbol{v}]}{\delta \boldsymbol{v}} = \hat{I} S \boldsymbol{v}, \tag{32}$$

where

$$S\boldsymbol{v} = \frac{\delta H[\boldsymbol{v}]}{\delta \boldsymbol{v}} = \begin{pmatrix} \frac{\delta H[\boldsymbol{v}]}{\delta v_1} \\ \vdots \\ \frac{\delta H[\boldsymbol{v}]}{\delta v_9} \end{pmatrix}. \tag{33}$$

As a result, the Hamiltonian reads

$$H[\boldsymbol{v}] = \frac{1}{2}\int \boldsymbol{v}^{\mathrm{T}} S \boldsymbol{v}\, dx_1 dx_2 dx_3 \tag{34}$$

and the components of $\boldsymbol{v}$ can be written as follows

$$(v_1, v_2, v_3) = \rho_0(u_1, u_2, u_3), \quad (v_4, v_5, v_6) = (\varepsilon_{11}, \varepsilon_{22}, \varepsilon_{33}), \tag{35}$$

$$(v_7, v_8, v_9) = 2\,(\varepsilon_{12}, \varepsilon_{13}, \varepsilon_{23}). \tag{36}$$

The formula (32) means that now the Poisson bracket is independent of the parameters but the Hamiltonian depends linearly on the Lamé parameters λ and μ.

4.1 Positive Definiteness

The existence of Hamiltonian for the system (2)–(11) does not guarantee the stability of solutions to this system. In the norm given by (18), this property takes place provided that this quadratic integral is positive definite. For this, it is necessary and sufficient to show that A_u (or S in (34)) obeys the same property, that is, A_u is a positive definite matrix. This is occur under the following conditions

$$\mu > 0, \quad \rho_0 > 0, \quad 3\lambda + 2\mu > 0. \tag{37}$$

With this, the conditions (23) follow from positiveness of the quantities in (37).

4.2 Simplification of the Hamiltonian

The Poisson bracket has been reduced above to the simplest form by changing the variables. The transformation of the Hamiltonian can be preformed by reducing to the principal axes that presents the simplest form of positive quadratic forms. Namely, since S is a positive definite matrix then the operator $S^{1/2}$ is defined and we can perform the change of variables

$$\boldsymbol{w} = S^{1/2}\boldsymbol{v}, \quad S^{1/2} = \mathrm{diag}\left(\rho_0^{-1/2}E, S_{22}^{1/2}, \mu^{1/2}E\right), \tag{38}$$

where

$$S_{22}^{1/2} = \frac{\sqrt{3\lambda + 2\mu}}{3}\begin{pmatrix} 1 & 1 & 1 \\ 1 & 1 & 1 \\ 1 & 1 & 1 \end{pmatrix} + \frac{\sqrt{2\mu}}{3}\begin{pmatrix} 2 & -1 & -1 \\ -1 & 2 & -1 \\ -1 & -1 & 2 \end{pmatrix}. \tag{39}$$

Then the system (32) takes the form

$$\boldsymbol{w}_t = \hat{J}_w \frac{\delta H[\boldsymbol{w}]}{\delta \boldsymbol{w}} = \hat{J}_w \boldsymbol{w}, \tag{40}$$

with the Hamiltonian (after reducing to the principal axes)

$$H[w] = \frac{1}{2} \int w^{\mathrm{T}} w dx_1 dx_2 dx_3 \tag{41}$$

and the operator of Poisson bracket

$$\hat{J}_w = S^{1/2} \hat{I} S^{1/2}. \tag{42}$$

The system (40) is relevant for numerical modelling by using finite-difference methods which conserve the Hamiltonian (41). In particular, the Crank–Nicolson scheme can be applied for the numerical integrating:

$$\frac{\psi^{j+1} - \psi^j}{\tau} = K \frac{\psi^{j+1} + \psi^j}{2}, \tag{43}$$

where ψ^j is a finite-dimensional approximation of the function w, K is a skew-symmetric approximation of the operator $\hat{J}_w$. It is well-known that the Crank–Nicolson scheme conserves the quadratic integral

$$H = \frac{1}{2}(\psi^j, \psi^j)$$

for all moments of discrete time j. Other finite-difference approximations which conserve the Hamiltonian structure can be found in [9].

4.3 Symmetric Form

If we multiply from left the system $\boldsymbol{u}_t = \hat{L}\boldsymbol{u}$ on the matrix $A_{\boldsymbol{u}}$, then we get

$$A_{\boldsymbol{u}} \boldsymbol{u}_t = \hat{I} \boldsymbol{u}, \tag{44}$$

which can be presented as a linear symmetric hyperbolic system in the sense of Friedrichs [4]

$$A_{\boldsymbol{u}} \frac{\partial \boldsymbol{u}}{\partial t} + B_1 \frac{\partial \boldsymbol{u}}{\partial x_1} + B_2 \frac{\partial \boldsymbol{u}}{\partial x_2} + B_3 \frac{\partial \boldsymbol{u}}{\partial x_3} = 0. \tag{45}$$

Here $A_{\boldsymbol{u}}$ is a symmetric positive definite matrix but B_1, B_2 and B_3 are symmetric matrixes in general. These matrixes can be easily calculated by exploited the form of (20) and using the decomposition

$$\hat{I} = -B_1 \partial_{x_1} - B_2 \partial_{x_2} - B_3 \partial_{x_3}. \tag{46}$$

The forms of these matrixes are omitted in view of their large size.

Let us consider the Hamiltonian structure (45) and prove the following general assertion: any linear hyperbolic system of the form

$$A\frac{\partial \boldsymbol{u}}{\partial t}+B_1\frac{\partial \boldsymbol{u}}{\partial x_1}+B_2\frac{\partial \boldsymbol{u}}{\partial x_2}+B_3\frac{\partial \boldsymbol{u}}{\partial x_3}+C\boldsymbol{u}=0, \tag{47}$$

where A is a symmetric positive definite matrix, B_1, B_2 and B_3 are symmetric matrixes and C is a skew-symmetric matrix, admits the Hamiltonian structure (17) with the operator

$$\hat{J}=-A^{-1}B_1A^{-1}\frac{\partial}{\partial x_1}-A^{-1}B_2A^{-1}\frac{\partial}{\partial x_2}-A^{-1}B_3A^{-1}\frac{\partial}{\partial x_3}-A^{-1}CA^{-1}, \tag{48}$$

and the Hamiltonian equals

$$H[\boldsymbol{u}]=\frac{1}{2}\int \boldsymbol{u}^{\mathrm{T}}A\boldsymbol{u}\,dx_1dx_2dx_3. \tag{49}$$

The proof is based on algebraic manipulations. Namely, multiplying equation (47) on A^{-1}, we rewrite the system in the form

$$\frac{\partial \boldsymbol{u}}{\partial t}=-\big(A^{-1}B_1A^{-1}\frac{\partial}{\partial x_1}+A^{-1}B_2A^{-1}\frac{\partial}{\partial x_2}+A^{-1}B_3A^{-1}\frac{\partial}{\partial x_3}+A^{-1}CA^{-1}\big)A\boldsymbol{u}, \tag{50}$$

where the terms in bracket present a differential operator which acts on $A\boldsymbol{u}$. Actually, this is the desired skew-symmetric operator $\hat{J}$ which is independent of the variables $\boldsymbol{u}$. Therefore, this operator defines a Poisson bracket. Further, it follows immediately from the equality

$$\frac{\delta H[\boldsymbol{u}]}{\delta \boldsymbol{u}}=A\boldsymbol{u}, \tag{51}$$

that (50) is a Hamiltonian system.

5 Casimir Functionals

Except for the Hamiltonian admitted, the system (32) may conserve different functionals due to the singularity of $\hat{I}$. Notice that both the operators $\hat{\boldsymbol{M}}=\hat{\boldsymbol{M}}_i(\partial_{x_1},\partial_{x_2},\partial_{x_3})$ and some vector-functions $\boldsymbol{N}(x_1,x_2,x_3)$ may vanish the operator $\hat{I}$.

In order to find these functions, we consider the following operator equation

$$\hat{I}\hat{M} = \begin{pmatrix} 0 & 0 & 0 & \partial_{x_1} & 0 & 0 & \partial_{x_2} & \partial_{x_3} & 0 \\ 0 & 0 & 0 & 0 & \partial_{x_2} & 0 & \partial_{x_1} & 0 & \partial_{x_3} \\ 0 & 0 & 0 & 0 & 0 & \partial_{x_3} & 0 & \partial_{x_1} & \partial_{x_2} \\ \partial_{x_1} & 0 & 0 & 0 & 0 & 0 & 0 & 0 & 0 \\ 0 & \partial_{x_2} & 0 & 0 & 0 & 0 & 0 & 0 & 0 \\ 0 & 0 & \partial_{x_3} & 0 & 0 & 0 & 0 & 0 & 0 \\ \partial_{x_2} & \partial_{x_1} & 0 & 0 & 0 & 0 & 0 & 0 & 0 \\ \partial_{x_3} & 0 & \partial_{x_1} & 0 & 0 & 0 & 0 & 0 & 0 \\ 0 & \partial_{x_3} & \partial_{x_2} & 0 & 0 & 0 & 0 & 0 & 0 \end{pmatrix} \begin{pmatrix} \hat{m}_1 \\ \hat{m}_2 \\ \hat{m}_3 \\ \hat{m}_4 \\ \hat{m}_5 \\ \hat{m}_6 \\ \hat{m}_7 \\ \hat{m}_8 \\ \hat{m}_9 \end{pmatrix} = 0, \quad (52)$$

where $\hat{m}_l = \hat{m}_l(\partial_{x_1}, \partial_{x_2}, \partial_{x_3})$ are differential operators with constant coefficients. To resolve this equation, we apply the Fourier transformation to (52). Then instead of the operator Eq. (52), we get the algebraic system

$$IM = \begin{pmatrix} 0 & 0 & 0 & ik_1 & 0 & 0 & ik_2 & ik_3 & 0 \\ 0 & 0 & 0 & 0 & ik_2 & 0 & ik_1 & 0 & ik_3 \\ 0 & 0 & 0 & 0 & 0 & ik_3 & 0 & ik_1 & ik_2 \\ ik_1 & 0 & 0 & 0 & 0 & 0 & 0 & 0 & 0 \\ 0 & ik_2 & 0 & 0 & 0 & 0 & 0 & 0 & 0 \\ 0 & 0 & ik_3 & 0 & 0 & 0 & 0 & 0 & 0 \\ ik_2 & ik_1 & 0 & 0 & 0 & 0 & 0 & 0 & 0 \\ ik_3 & 0 & ik_1 & 0 & 0 & 0 & 0 & 0 & 0 \\ 0 & ik_3 & ik_2 & 0 & 0 & 0 & 0 & 0 & 0 \end{pmatrix} \begin{pmatrix} m_1 \\ m_2 \\ m_3 \\ m_4 \\ m_5 \\ m_6 \\ m_7 \\ m_8 \\ m_9 \end{pmatrix} = 0, \quad (53)$$

where k_n is the wave number of the differential operator ∂_{x_n}, $n = 1, 2, 3$ and $m_l = m_l(ik_1, ik_2, ik_3)$ denotes the symbol of the operator $\hat{m}_l$. The rank of the matrix I equals 6 for nonzero wave numbers k_n. Therefore, the system (53) admits only three independent solutions. Further, it follows from the fourth, fifth and sixth equations of (53) that $m_1 = m_2 = m_3 = 0$. By the quantities m_7, m_8, m_9 we can find m_4, m_5, m_6. Hence, the general solution of (53) looks like a three-parametric vector-function

$$\mathbf{M} = \left(0, 0, 0, -\frac{k_2 m_7 + k_3 m_8}{k_1}, -\frac{k_1 m_7 + k_3 m_9}{k_2}, -\frac{k_1 m_8 + k_2 m_9}{k_3}, m_7, m_8, m_9\right)^{\mathrm{T}}. \quad (54)$$

Recall that we look for a solution of (52) in the class of differential operators. It means that each component of $\mathbf{M}$ is a polynomial of k_n, $n = 1, 2, 3$ because in the opposite case the vector $\mathbf{M}$ will contain the divisors k_n that leads to arising the integral operator. In view of this argument, we also assume that m_7, m_8 and m_9 are polynomials of k_n.

We start with the consideration of the fourth component of (54). It will be a polynomial if we have (0) $k_2 m_7 + k_3 m_8 = 0$; (1) m_7 and m_8 has the common divisor k_1. The same holds for the fifth and sixth components of (54). Therefore, we get in total the eight different variants which are denoted by (h_1, h_2, h_3) where h_i equals 0 or 1.

Let us consider the variant (1, 1, 1) and assume that the property (1) holds for the fourth component of (54). Then we get that $m_7 = k_1 p_7, m_8 = k_1 p_8$ where p_7 and p_8 are polynomials of k_n. With this, the fifth component of (54) reads

$$-\frac{k_1^2 p_7 + k_3 m_9}{k_2}. \tag{55}$$

The property (1) also holds for the expression (55), i.e. p_7 and m_9 has the common divisor k_2. It means that $p_7 = k_2 q_7, m_9 = k_2 q_9$ where q_7 and q_9 are polynomials of k_n too. The sixth component of (54) takes the form

$$-\frac{k_1^2 p_8 + k_2^2 q_9}{k_3}. \tag{56}$$

It is again the property (1) is satisfied for the expression (56), i.e. p_8 and q_9 has the common divisor k_3. Therefore, we have the representation $p_8 = k_3 r_8, q_9 = k_3 r_9$ where r_8, r_9 are polynomials. As a result, we obtain for the variant (1, 1, 1) the following form of **M**

$$\begin{aligned}\mathbf{M}^{(1,1,1)} = \big(0, 0, 0, -(k_2^2 q_7 + k_3^2 r_8), -(k_1^2 q_7 + k_3^2 r_9),\\ -(k_1^2 r_8 + k_2^2 r_9), k_1 k_2 q_7, k_1 k_3 r_8, k_2 k_3 r_9\big)^{\mathrm{T}}.\end{aligned} \tag{57}$$

Since q_7, r_8, r_9 are independent quantities then we can derive from (57) the following independent functions

$$\mathbf{M}_1 = \left(0, 0, 0, -k_2^2, -k_1^2, 0, k_1 k_2, 0, 0\right)^{\mathrm{T}}, \tag{58}$$

$$\mathbf{M}_2 = \left(0, 0, 0, -k_3^2, 0, -k_1^2, 0, k_1 k_3, 0\right)^{\mathrm{T}}, \tag{59}$$

$$\mathbf{M}_3 = \left(0, 0, 0, 0, -k_3^2, -k_2^2, 0, 0, k_2 k_3\right)^{\mathrm{T}}. \tag{60}$$

The variants (1, 0, 0), (0, 1, 0), (0, 0, 1) give the following solutions

$$\mathbf{M}_4 = \left(0, 0, 0, -2k_2 k_3, 0, 0, k_1 k_3, k_1 k_2, -k_1^2\right)^{\mathrm{T}}, \tag{61}$$

$$\mathbf{M}_5 = \left(0, 0, 0, 0, -2k_1 k_3, 0, k_2 k_3, -k_2^2, k_1 k_2\right)^{\mathrm{T}}, \tag{62}$$

$$\mathbf{M}_6 = \left(0, 0, 0, 0, 0, -2k_1 k_2, -k_3^2, k_2 k_3, k_1 k_3\right)^{\mathrm{T}}. \tag{63}$$

Consider now the variant (1, 1, 0). The first two steps for constructing the functions required completely repeat the previous procedure. The third step leads to the case (0), i.e. $k_1 m_8 + k_2 m_9 = k_1^2 p_8 + k_2^2 q_9 = 0$. Hence, we get that $p_8 = k_2^2 s$, $q_9 = -k_1^2 s$ where s is a polynomial. Therefore, we obtain for the variant (1, 1, 0) the following form of $\mathbf{M}$

$$\mathbf{M}^{(1,1,0)} = \left(0, 0, 0, -(k_2^2 q_7 + k_2^2 k_3 s), -(k_1^2 q_7 - k_1^2 k_3 s), 0, k_1 k_2 q_7, k_1 k_2^2 s, -k_1^2 k_2 s\right)^{\mathrm{T}}. \tag{64}$$

It enables us to derive more solutions. Specifically, we get

$$\mathbf{M}_7 = \left(0, 0, 0, -k_2^2 k_3, k_1^2 k_3, 0, 0, k_1 k_2^2, -k_1^2 k_2\right)^{\mathrm{T}}. \tag{65}$$

The variants (1, 0, 1) and (0, 1, 1) are investigated similarly and give in addition the new solutions:

$$\mathbf{M}_8 = \left(0, 0, 0, -k_2 k_3^2, 0, k_1^2 k_2, k_1 k_3^2, 0, -k_1^2 k_3\right)^{\mathrm{T}}, \tag{66}$$

$$\mathbf{M}_9 = \left(0, 0, 0, 0, -k_1 k_3^2, k_1 k_2^2, k_2 k_3^2, -k_2^2 k_3, 0\right)^{\mathrm{T}}. \tag{67}$$

Notice that not all solutions obtained are linearly independent. In order to show it, we consider the following matrixes

$$R_1 = (M_1, M_2, M_3), \quad R_2 = (M_4, M_5, M_6), \quad R_3 = (M_7, M_8, M_9). \tag{68}$$

By the direct verification, we can obtain that the following relationships hold

$$R_2 \begin{pmatrix} k_2 & k_3 & 0 \\ k_1 & 0 & k_3 \\ 0 & k_1 & k_2 \end{pmatrix} = 2R_1 \begin{pmatrix} k_3 & 0 & 0 \\ 0 & k_2 & 0 \\ 0 & 0 & k_1 \end{pmatrix}, \tag{69}$$

$$R_2 \begin{pmatrix} k_2 & k_3 & 0 \\ -k_1 & 0 & k_3 \\ 0 & -k_1 & -k_2 \end{pmatrix} = 2R_3. \tag{70}$$

The first equality means that R_1 and R_2 are linked by a differential constraint which contains the first-order derivatives. Therefore R_1 and R_2 are linear independent in the class of differential operators. It follows from the second equality that R_3 is a differential consequence of R_2. Thus, we proved that the equation $I\mathbf{M} = 0$ has 6 linearly independent solutions $M_1, \ldots, M_6$.

Linearly independent solutions of the equation in the space of differential operators read

$$
\begin{aligned}
\hat{M}_1 &= (0, 0, 0, \partial_{x_2}^2, \partial_{x_1}^2, 0, -\partial_{x_1}\partial_{x_2}, 0, 0)^{\mathrm{T}},\\
\hat{M}_2 &= (0, 0, 0, \partial_{x_3}^2, 0, \partial_{x_1}^2, 0, -\partial_{x_1}\partial_{x_3}, 0)^{\mathrm{T}},\\
\hat{M}_3 &= (0, 0, 0, 0, \partial_{x_3}^2, \partial_{x_2}^2, 0, 0, -\partial_{x_2}\partial_{x_3})^{\mathrm{T}},\\
\hat{M}_4 &= \left(0, 0, 0, 2\partial_{x_2}\partial_{x_3}, 0, 0, -\partial_{x_1}\partial_{x_3}, -\partial_{x_1}\partial_{x_2}, \partial_{x_1}^2\right)^{\mathrm{T}},\\
\hat{M}_5 &= \left(0, 0, 0, 0, 2\partial_{x_1}\partial_{x_3}, 0, -\partial_{x_2}\partial_{x_3}, \partial_{x_2}^2, -\partial_{x_1}\partial_{x_2}\right)^{\mathrm{T}},\\
\hat{M}_6 &= \left(0, 0, 0, 0, 0, 2\partial_{x_1}\partial_{x_2}, \partial_{x_3}^2, -\partial_{x_2}\partial_{x_3}, -\partial_{x_1}\partial_{x_3}\right)^{\mathrm{T}}.
\end{aligned}
\tag{71}
$$

In view of the symmetric form of $\hat{I}$, the conjugate operators $\hat{M}_i^* = \hat{M}_i^{\mathrm{T}}$ are solutions of the equation

$$
-\hat{M}_i^* \hat{I} = \hat{I}\hat{M}_i = 0. \tag{72}
$$

The left action of the conjugate operators $\hat{M}_i^{\mathrm{T}}$ on the system (32) gives the following identities

$$
\hat{M}_i^{\mathrm{T}} v_t = \hat{M}_i^{\mathrm{T}} \hat{I} \frac{\delta H[v]}{\delta v} \equiv 0, \quad i = 1, 2, 3. \tag{73}
$$

Then for an arbitrary Hamiltonian $H[v]$ the following relationships hold

$$
\begin{aligned}
\partial_t \psi_1 &\equiv 0, \quad \psi_1 = \partial_{x_2}^2 v_4 + \partial_{x_1}^2 v_5 - \partial_{x_1}\partial_{x_2} v_7,\\
\partial_t \psi_2 &\equiv 0, \quad \psi_2 = \partial_{x_3}^2 v_4 + \partial_{x_1}^2 v_6 - \partial_{x_1}\partial_{x_3} v_8,\\
\partial_t \psi_3 &\equiv 0, \quad \psi_3 = \partial_{x_3}^2 v_5 + \partial_{x_2}^2 v_6 - \partial_{x_2}\partial_{x_3} v_9,\\
\partial_t \psi_4 &\equiv 0, \quad \psi_4 = 2\partial_{x_2}\partial_{x_3} v_4 - \partial_{x_1}\partial_{x_3} v_7 - \partial_{x_1}\partial_{x_2} v_8 + \partial_{x_1}^2 v_9,\\
\partial_t \psi_5 &\equiv 0, \quad \psi_5 = 2\partial_{x_1}\partial_{x_3} v_5 - \partial_{x_2}\partial_{x_3} v_7 + \partial_{x_2}^2 v_8 - \partial_{x_1}\partial_{x_2} v_9,\\
\partial_t \psi_6 &\equiv 0, \quad \psi_6 = 2\partial_{x_1}\partial_{x_2} v_6 + \partial_{x_3}^2 v_7 - \partial_{x_2}\partial_{x_3} v_8 - \partial_{x_1}\partial_{x_3} v_9.
\end{aligned}
\tag{74}
$$

Thus for an arbitrary function $\Psi\,(x_1, x_2, x_3, \psi_1, \psi_2, \psi_3, \psi_4, \psi_5, \psi_6)$ and Hamiltonian $H[v]$, the system (32) conserves the following Casimir functional

$$
C_\Psi = \int \Psi\,(x_1, x_2, x_3, \psi_1, \psi_2, \psi_3, \psi_4, \psi_5, \psi_6)\, dx_1 dx_2 dx_3. \tag{75}
$$

The operator $\hat{I}$ can also be vanished yet on some function $N(x_1, x_2, x_3)$ which is defined by solving the system

$$\hat{I}N_i = \begin{pmatrix} 0 & 0 & 0 & \partial_{x_1} & 0 & 0 & \partial_{x_2} & \partial_{x_3} & 0 \\ 0 & 0 & 0 & 0 & \partial_{x_2} & 0 & \partial_{x_1} & 0 & \partial_{x_3} \\ 0 & 0 & 0 & 0 & 0 & \partial_{x_3} & 0 & \partial_{x_1} & \partial_{x_2} \\ \partial_{x_1} & 0 & 0 & 0 & 0 & 0 & 0 & 0 & 0 \\ 0 & \partial_{x_2} & 0 & 0 & 0 & 0 & 0 & 0 & 0 \\ 0 & 0 & \partial_{x_3} & 0 & 0 & 0 & 0 & 0 & 0 \\ \partial_{x_2} & \partial_{x_1} & 0 & 0 & 0 & 0 & 0 & 0 & 0 \\ \partial_{x_3} & 0 & \partial_{x_1} & 0 & 0 & 0 & 0 & 0 & 0 \\ 0 & \partial_{x_3} & \partial_{x_2} & 0 & 0 & 0 & 0 & 0 & 0 \end{pmatrix} \begin{pmatrix} n_1 \\ n_2 \\ n_3 \\ n_4 \\ n_5 \\ n_6 \\ n_7 \\ n_8 \\ n_9 \end{pmatrix} = 0. \tag{76}$$

To solve this system, we observe that (76) is decomposed into two subsystems with respect to the variables n_1–n_3 and n_4–n_9. Besides constant solutions, this system admits nontrivial solutions

$$\begin{aligned} \boldsymbol{N}_1 &= (0, x_3, -x_2, 0, 0, 0, 0, 0, 0)^{\mathrm{T}}, \\ \boldsymbol{N}_2 &= (-x_3, 0, x_1, 0, 0, 0, 0, 0, 0)^{\mathrm{T}}, \\ \boldsymbol{N}_3 &= (x_2, -x_1, 0, 0, 0, 0, 0, 0, 0)^{\mathrm{T}}, \end{aligned} \tag{77}$$

and

$$\begin{aligned} \boldsymbol{N}_1^f &= (0, 0, 0, f_{x_2x_2}, f_{x_1x_1}, 0, -f_{x_1x_2}, 0, 0)^{\mathrm{T}}, \\ \boldsymbol{N}_2^f &= (0, 0, 0, f_{x_3x_3}, 0, f_{x_1x_1}, 0, -f_{x_1x_3}, 0)^{\mathrm{T}}, \\ \boldsymbol{N}_3^f &= (0, 0, 0, 0, f_{x_3x_3}, f_{x_2x_2}, 0, 0, -f_{x_2x_3})^{\mathrm{T}} \end{aligned} \tag{78}$$

for an arbitrary function $f = f(x_1, x_2, x_3)$. Multiplying now (32) by the vector N_i^{T} and taking into account the form of variational derivative of the Hamiltonian (33), we get the equations

$$\begin{aligned} \partial_t(x_3v_2 - x_2v_3) = \; & \partial_{x_1}\left(x_3\frac{\delta H}{\delta v_7} - x_2\frac{\delta H}{\delta v_8}\right) + \partial_{x_2}\left(x_3\frac{\delta H}{\delta v_5} - x_2\frac{\delta H}{\delta v_9}\right) \\ & + \partial_{x_3}\left(x_3\frac{\delta H}{\delta v_9} - x_2\frac{\delta H}{\delta v_6}\right) \\ = \; & \partial_{x_2}\left(x_3\frac{\delta H}{\delta v_5}\right) - \partial_{x_3}\left(x_2\frac{\delta H}{\delta v_6}\right) + \partial_{x_1}\left(x_3\frac{\delta H}{\delta v_7}\right) \\ & - \partial_{x_1}\left(x_2\frac{\delta H}{\delta v_8}\right) + x_3\partial_{x_3}\frac{\delta H}{\delta v_9} - x_2\partial_{x_2}\frac{\delta H}{\delta v_9}, \end{aligned} \tag{79}$$

$$\begin{aligned} \partial_t(x_1v_3 - x_3v_1) = \; & -\partial_{x_1}\left(x_3\frac{\delta H}{\delta v_4}\right) + \partial_{x_3}\left(x_1\frac{\delta H}{\delta v_6}\right) - \partial_{x_2}\left(x_3\frac{\delta H}{\delta v_7}\right) \\ & + \partial_{x_2}\left(x_1\frac{\delta H}{\delta v_9}\right) - x_3\partial_{x_3}\frac{\delta H}{\delta v_8} + x_1\partial_{x_1}\frac{\delta H}{\delta v_8}, \end{aligned} \tag{80}$$

$$\partial_t(x_2v_1 - x_1v_2) = \partial_{x_1}\left(x_2\frac{\delta H}{\delta v_4}\right) - \partial_{x_2}\left(x_1\frac{\delta H}{\delta v_5}\right) + \partial_{x_3}\left(x_2\frac{\delta H}{\delta v_8}\right)$$

$$-\partial_{x_3}\left(x_1\frac{\delta H}{\delta v_9}\right)+x_2\partial_{x_2}\frac{\delta H}{\delta v_7}-x_1\partial_{x_1}\frac{\delta H}{\delta v_7}. \tag{81}$$

The following differential identities were applied for deriving (79)–(81):

$$x_i\partial_{x_i}F-x_j\partial_{x_j}F=\partial_{x_i}(x_iF)-\partial_{x_j}\left(x_jF\right). \tag{82}$$

If we integrate these equations over the spatial variables, we can see that the right-hand sides of the corresponding expressions obtained will be vanished under suitable boundary conditions which are independent of the form of Hamiltonian. As a result, we get the Casimir functionals

$$C_1=\int(x_3v_2-x_2v_3)dx_1dx_2dx_3, \tag{83}$$

$$C_2=\int(x_1v_3-x_3v_1)dx_1dx_2dx_3, \tag{84}$$

$$C_3=\int(x_2v_1-x_1v_2)dx_1dx_2dx_3. \tag{85}$$

Applying this procedure to N_i^f, we can find more Casimir functionals. Specifically, these are of the form

$$C_1^f=\int(f_{x_2x_2}v_4+f_{x_1x_1}v_5-f_{x_1x_2}v_7)dx_1dx_2dx_3, \tag{86}$$

$$C_2^f=\int(f_{x_3x_3}v_4+f_{x_1x_1}v_6-f_{x_1x_3}v_8)dx_1dx_2dx_3, \tag{87}$$

$$C_3^f=\int(f_{x_3x_3}v_5+f_{x_2x_2}v_6-f_{x_2x_3}v_9)dx_1dx_2dx_3 \tag{88}$$

for an arbitrary function f depending on the spatial variables. Integrating by parts in (86)–(88), we see that the Casimir functionals C_i^f can be derived from the functionals (75). Exemplarily, the Casimir functional (86) corresponds to the case which has been considered in (75) for

$$\Psi=f(x_1,x_2,x_3)\psi_1 \tag{89}$$

using that

$$\begin{aligned}&\int(f_{x_2x_2}v_4+f_{x_1x_1}v_5-f_{x_1x_2}v_7)dx_1dx_2dx_3\\&=\int f(x_1,x_2,x_3)\left(\partial_{x_2}^2v_4+\partial_{x_1}^2v_5-\partial_{x_1}\partial_{x_2}v_7\right)dx_1dx_2dx_3.\end{aligned} \tag{90}$$

Therefore, we found all Casimir functionals for the Poisson bracket defined by the operator $\hat{I}$. Moreover, the Casimir functionals C_1, C_2, C_3 coincide with the conservation laws of the momentum vector of the hydrodynamic equations for the constant density ρ_0, see, e.g. [10]. Notice that the conservation laws linked with the Casimir functionals obtained are not associated with the rotation symmetries in the frame of the Noether theorems [8, 10] since these functionals were resulted for arbitrary Hamiltonians.

6 Zero-Order Conservation Laws

The system (1) is written in the form of conservation laws. Furthermore, the system conserves the energy functional or the corresponding Hamiltonian (34). Besides we found for (1) the conservation laws presented by the Casimir functionals. Nevertheless, the question about the existence of additional conservation laws admitted is still open.

We are interested in finding zero-order conservation laws for the system (1), i.e. the conservation laws with the density which is independent of the derivatives of the dependent variables. We will seek conservation laws in the differential form

$$S \equiv \frac{\partial F}{\partial t} + \frac{\partial G_1}{\partial x_1} + \frac{\partial G_2}{\partial x_2} + \frac{\partial G_3}{\partial x_3} = 0, \tag{91}$$

where $F = F(t, x_1, x_2, x_3, \boldsymbol{u})$ is the density of the conservation law and $G_i = G_i(t, x_1, x_2, x_3, \boldsymbol{u})$ denotes the flow of one. Let us substitute these expressions into (91) and differentiate with respect to the variables t and x_i. Then changing the derivative with respect to t and taking into account the original system of equations, we get a system of 28 equations for the coefficients S_{ij} under the derivatives $\frac{\partial u_i}{\partial x_j}$. The coefficients under the derivative with respect to x_1 read

$$S_{11} = (\lambda + 2\mu)\frac{\partial F}{\partial u_4} + \lambda\left(\frac{\partial F}{\partial u_5} + \frac{\partial F}{\partial u_6}\right) + \frac{\partial G_1}{\partial u_1} = 0, \tag{92}$$

$$S_{21} = \mu\frac{\partial F}{\partial u_7} + \frac{\partial G_1}{\partial u_2} = 0, \qquad S_{31} = \mu\frac{\partial F}{\partial u_8} + \frac{\partial G_1}{\partial u_3} = 0, \tag{93}$$

$$S_{41} = \frac{1}{\rho_0}\frac{\partial F}{\partial u_1} + \frac{\partial G_1}{\partial u_4} = 0, \qquad S_{51} = \frac{\partial G_1}{\partial u_5} = 0, \tag{94}$$

$$S_{61} = \frac{\partial G_1}{\partial u_6} = 0, \qquad S_{71} = \frac{1}{\rho_0}\frac{\partial F}{\partial u_2} + \frac{\partial G_1}{\partial u_7} = 0, \tag{95}$$

$$S_{81} = \frac{1}{\rho_0}\frac{\partial F}{\partial u_3} + \frac{\partial G_1}{\partial u_8} = 0, \qquad S_{91} = \frac{\partial G_1}{\partial u_9} = 0. \tag{96}$$

Repeating the calculations, we find the following coefficients under the derivative with respect to x_2

$$S_{12} = \mu \frac{\partial F}{\partial u_7} + \frac{\partial G_2}{\partial u_1} = 0, \tag{97}$$

$$S_{22} = (\lambda + 2\mu) \frac{\partial F}{\partial u_5} + \lambda \Big(\frac{\partial F}{\partial u_4} + \frac{\partial F}{\partial u_6} \Big) + \frac{\partial G_2}{\partial u_2} = 0, \tag{98}$$

$$S_{32} = \mu \frac{\partial F}{\partial u_9} + \frac{\partial G_2}{\partial u_3} = 0, \qquad S_{42} = \frac{\partial G_2}{\partial u_4} = 0, \tag{99}$$

$$S_{52} = \frac{1}{\rho_0} \frac{\partial F}{\partial u_2} + \frac{\partial G_2}{\partial u_5} = 0, \qquad S_{62} = \frac{\partial G_2}{\partial u_6} = 0, \tag{100}$$

$$S_{72} = \frac{1}{\rho_0} \frac{\partial F}{\partial u_1} + \frac{\partial G_2}{\partial u_7} = 0, \qquad S_{82} = \frac{\partial G_2}{\partial u_8} = 0, \tag{101}$$

$$S_{92} = \frac{1}{\rho_0} \frac{\partial F}{\partial u_3} + \frac{\partial G_2}{\partial u_9} = 0. \tag{102}$$

The third group of equations is derived for the coefficients under the derivative with respect to x_3 correspondingly:

$$S_{13} = \mu \frac{\partial F}{\partial u_8} + \frac{\partial G_3}{\partial u_1} = 0, \qquad S_{23} = \mu \frac{\partial F}{\partial u_9} + \frac{\partial G_3}{\partial u_2} = 0, \tag{103}$$

$$S_{33} = (\lambda + 2\mu) \frac{\partial F}{\partial u_6} + \lambda \Big(\frac{\partial F}{\partial u_4} + \frac{\partial F}{\partial u_5} \Big) + \frac{\partial G_3}{\partial u_3} = 0, \tag{104}$$

$$S_{43} = \frac{\partial G_3}{\partial u_4} = 0, \qquad S_{53} = \frac{\partial G_3}{\partial u_5} = 0, \tag{105}$$

$$S_{63} = \frac{1}{\rho_0} \frac{\partial F}{\partial u_3} + \frac{\partial G_3}{\partial u_6} = 0, \qquad S_{73} = \frac{\partial G_3}{\partial u_7} = 0, \tag{106}$$

$$S_{83} = \frac{1}{\rho_0} \frac{\partial F}{\partial u_1} + \frac{\partial G_3}{\partial u_8} = 0, \qquad S_{93} = \frac{1}{\rho_0} \frac{\partial F}{\partial u_2} + \frac{\partial G_3}{\partial u_9} = 0. \tag{107}$$

The last equation establishes a connection between terms which are independent of the derivatives:

$$S_0 = \frac{\partial F}{\partial t} + \frac{\partial G_1}{\partial x_1} + \frac{\partial G_2}{\partial x_2} + \frac{\partial G_3}{\partial x_3} = 0. \tag{108}$$

Excepting the variables G_1, G_2, G_3 from the corresponding subsystems by the cross differentiation, we get 108 equations for the function F. Further, we use the notation $F_{ij} = \frac{\partial^2 F}{\partial u_i \partial u_j}$. The first group equations, which consists of 36 equations, is obtained by excepting G_1:

$$(\lambda + 2\mu)F_{24} + \lambda\,(F_{25} + F_{26}) - \mu F_{17} = 0, \tag{109}$$

$$(\lambda + 2\mu)F_{34} + \lambda\,(F_{35} + F_{36}) - \mu F_{18} = 0, \tag{110}$$

$$(\lambda + 2\mu)F_{44} + \lambda\,(F_{45} + F_{46}) - \frac{1}{\rho_0}F_{11} = 0, \tag{111}$$

$$(\lambda + 2\mu)F_{45} + \lambda\,(F_{55} + F_{56}) = 0, \qquad (\lambda + 2\mu)F_{46} + \lambda\,(F_{56} + F_{66}) = 0, \tag{112}$$

$$(\lambda + 2\mu)F_{47} + \lambda\,(F_{57} + F_{67}) - \frac{1}{\rho_0}F_{12} = 0, \tag{113}$$

$$(\lambda + 2\mu)F_{48} + \lambda\,(F_{58} + F_{68}) - \frac{1}{\rho_0}F_{13} = 0, \tag{114}$$

$$(\lambda + 2\mu)F_{49} + \lambda\,(F_{59} + F_{69}) = 0, \qquad \mu\,(F_{37} - F_{28}) = 0, \tag{115}$$

$$\mu F_{47} - \frac{1}{\rho_0}F_{12} = 0, \qquad \mu F_{57} = 0, \qquad \mu F_{67} = 0, \tag{116}$$

$$\mu F_{77} - \frac{1}{\rho_0}F_{22} = 0, \qquad \mu F_{78} - \frac{1}{\rho_0}F_{23} = 0, \qquad \mu F_{79} = 0, \tag{117}$$

$$\mu F_{48} - \frac{1}{\rho_0}F_{13} = 0, \qquad \mu F_{58} = 0, \qquad \mu F_{68} = 0, \tag{118}$$

$$\mu F_{78} - \frac{1}{\rho_0}F_{23} = 0, \qquad \mu F_{88} - \frac{1}{\rho_0}F_{33} = 0, \qquad \mu F_{89} = 0, \tag{119}$$

$$\frac{1}{\rho_0}F_{15} = 0, \qquad \frac{1}{\rho_0}F_{16} = 0, \qquad \frac{1}{\rho_0}\,(F_{17} - F_{24}) = 0, \tag{120}$$

$$\frac{1}{\rho_0}\,(F_{18} - F_{34}) = 0, \qquad \frac{1}{\rho_0}F_{19} = 0, \qquad 0 = 0, \tag{121}$$

$$-\frac{1}{\rho_0}F_{25} = 0, \qquad -\frac{1}{\rho_0}F_{35} = 0, \qquad 0 = 0, \qquad -\frac{1}{\rho_0}F_{26} = 0, \tag{122}$$

$$-\frac{1}{\rho_0}F_{36} = 0, \qquad 0 = 0, \qquad \frac{1}{\rho_0}\,(F_{28} - F_{37}) = 0, \tag{123}$$

$$\frac{1}{\rho_0}F_{29} = 0, \qquad \frac{1}{\rho_0}F_{39} = 0. \tag{124}$$

The exception of G_2 gives the following equations:

$$\mu F_{27} - (\lambda + 2\mu) F_{15} - \lambda (F_{14} + F_{16}) = 0, \tag{125}$$

$$\mu (F_{37} - F_{19}) = 0, \quad \mu F_{47} = 0, \quad \mu F_{57} - \frac{1}{\rho_0} F_{12} = 0, \tag{126}$$

$$\mu F_{67} = 0, \quad \mu F_{77} - \frac{1}{\rho_0} F_{11} = 0, \quad \mu F_{78} = 0, \quad \mu F_{79} - \frac{1}{\rho_0} F_{13} = 0, \tag{127}$$

$$(\lambda + 2\mu) F_{35} + \lambda (F_{34} + F_{36}) - \mu F_{29} = 0, \tag{128}$$

$$(\lambda + 2\mu) F_{45} + \lambda (F_{44} + F_{46}) = 0, \tag{129}$$

$$(\lambda + 2\mu) F_{55} + \lambda (F_{45} + F_{56}) - \frac{1}{\rho_0} F_{22} = 0, \tag{130}$$

$$(\lambda + 2\mu) F_{56} + \lambda (F_{46} + F_{66}) = 0, \tag{131}$$

$$(\lambda + 2\mu) F_{57} + \lambda (F_{47} + F_{67}) - \frac{1}{\rho_0} F_{12} = 0, \tag{132}$$

$$(\lambda + 2\mu) F_{58} + \lambda (F_{48} + F_{68}) = 0, \tag{133}$$

$$(\lambda + 2\mu) F_{59} + \lambda (F_{49} + F_{69}) - \frac{1}{\rho_0} F_{23} = 0, \tag{134}$$

$$\mu F_{49} = 0, \quad \mu F_{59} - \frac{1}{\rho_0} F_{23} = 0, \quad \mu F_{69} = 0, \tag{135}$$

$$\mu F_{79} - \frac{1}{\rho_0} F_{13} = 0, \quad \mu F_{89} = 0, \quad \mu F_{99} - \frac{1}{\rho_0} F_{33} = 0, \tag{136}$$

$$-\frac{1}{\rho_0} F_{24} = 0, \quad 0 = 0, \quad -\frac{1}{\rho_0} F_{14} = 0, \quad 0 = 0, \tag{137}$$

$$-\frac{1}{\rho_0} F_{34} = 0, \quad \frac{1}{\rho_0} F_{26} = 0, \quad \frac{1}{\rho_0} (F_{27} - F_{15}) = 0, \quad \frac{1}{\rho_0} F_{28} = 0, \tag{138}$$

$$\frac{1}{\rho_0} (F_{29} - F_{35}) = 0, \quad -\frac{1}{\rho_0} F_{16} = 0, \quad 0 = 0, \quad -\frac{1}{\rho_0} F_{36} = 0, \tag{139}$$

$$\frac{1}{\rho_0} F_{18} = 0, \quad \frac{1}{\rho_0} (F_{19} - F_{37}) = 0, \quad -\frac{1}{\rho_0} F_{38} = 0. \tag{140}$$

The exception of G_3 gives the third group of equations:

$$\mu\,(F_{28}-F_{19})=0, \tag{141}$$

$$\mu F_{38}-(\lambda+2\mu)F_{16}-\lambda\,(F_{14}+F_{15})=0, \tag{142}$$

$$\mu F_{48}=0,\qquad \mu F_{58}=0,\qquad \mu F_{68}-\frac{1}{\rho_0}F_{13}=0, \tag{143}$$

$$\mu F_{78}=0,\qquad \mu F_{88}-\frac{1}{\rho_0}F_{11}=0,\qquad \mu F_{89}-\frac{1}{\rho_0}F_{12}=0, \tag{144}$$

$$\mu F_{39}-(\lambda+2\mu)F_{26}-\lambda\,(F_{24}+F_{25})=0, \tag{145}$$

$$\mu F_{49}=0,\qquad \mu F_{59}=0,\qquad \mu F_{69}-\frac{1}{\rho_0}F_{23}=0, \tag{146}$$

$$\mu F_{79}=0,\qquad \mu F_{89}-\frac{1}{\rho_0}F_{12}=0,\qquad \mu F_{99}-\frac{1}{\rho_0}F_{22}=0, \tag{147}$$

$$(\lambda+2\mu)F_{46}+\lambda\,(F_{44}+F_{45})=0, \tag{148}$$

$$(\lambda+2\mu)F_{56}+\lambda\,(F_{45}+F_{55})=0, \tag{149}$$

$$(\lambda+2\mu)F_{66}+\lambda\,(F_{46}+F_{56})-\frac{1}{\rho_0}F_{33}=0, \tag{150}$$

$$(\lambda+2\mu)F_{67}+\lambda\,(F_{47}+F_{57})=0, \tag{151}$$

$$(\lambda+2\mu)F_{68}+\lambda\,(F_{48}+F_{58})-\frac{1}{\rho_0}F_{13}=0, \tag{152}$$

$$(\lambda+2\mu)F_{69}+\lambda\,(F_{49}+F_{59})-\frac{1}{\rho_0}F_{23}=0, \tag{153}$$

$$0=0,\qquad -\frac{1}{\rho_0}F_{34}=0,\qquad 0=0,\qquad -\frac{1}{\rho_0}F_{14}=0, \tag{154}$$

$$-\frac{1}{\rho_0}F_{24}=0,\qquad -\frac{1}{\rho_0}F_{35}=0,\qquad 0=0,\qquad -\frac{1}{\rho_0}F_{15}=0, \tag{155}$$

$$-\frac{1}{\rho_0}F_{25}=0,\qquad \frac{1}{\rho_0}F_{37}=0,\qquad \frac{1}{\rho_0}(F_{38}-F_{16})=0,\qquad \frac{1}{\rho_0}(F_{39}-F_{26})=0, \tag{156}$$

$$-\frac{1}{\rho_0}F_{17}=0,\qquad -\frac{1}{\rho_0}F_{27}=0,\qquad \frac{1}{\rho_0}(F_{19}-F_{28})=0. \tag{157}$$

Finally, excepting now G_1, G_2 and G_3 from S_0, we get 9 equations as yet:

$$F_{t1} = (\lambda + 2\mu)F_{x_1 4} + \lambda\left(F_{x_1 5} + F_{x_1 6}\right) + \mu\left(F_{x_2 7} + F_{x_3 8}\right), \tag{158}$$

$$F_{t2} = (\lambda + 2\mu)F_{x_2 5} + \lambda\left(F_{x_2 4} + F_{x_2 6}\right) + \mu\left(F_{x_1 7} + F_{x_3 9}\right), \tag{159}$$

$$F_{t3} = (\lambda + 2\mu)F_{x_3 6} + \lambda\left(F_{x_3 4} + F_{x_3 5}\right) + \mu\left(F_{x_1 8} + F_{x_2 9}\right), \tag{160}$$

$$F_{t4} = \frac{1}{\rho_0}F_{x_1 1}, \qquad F_{t5} = \frac{1}{\rho_0}F_{x_2 2}, \qquad F_{t6} = \frac{1}{\rho_0}F_{x_3 3}, \tag{161}$$

$$F_{t7} = \frac{1}{\rho_0}\left(F_{x_1 2} + F_{x_2 1}\right), F_{t8} = \frac{1}{\rho_0}\left(F_{x_1 3} + F_{x_3 1}\right), F_{t9} = \frac{1}{\rho_0}\left(F_{x_2 3} + F_{x_3 2}\right), \tag{162}$$

where $F_{yi} = \frac{\partial^2 F}{\partial y \partial u_i}$ for $y = t, x_1, x_2, x_3$.

From the system obtained, which consists of 117 equations, follows that

$$\begin{aligned} &F_{14} = F_{15} = F_{16} = F_{17} = F_{18} = F_{19} = F_{24} = F_{25} \\ &= F_{26} = F_{27} = F_{28} = F_{29} = F_{34} = F_{35} = F_{36} = F_{37} \\ &= F_{38} = F_{39} = F_{47} = F_{48} = F_{49} = F_{57} = F_{58} = F_{59} \\ &= F_{67} = F_{68} = F_{69} = F_{78} = F_{79} = F_{89} = 0. \end{aligned} \tag{163}$$

With the relationships obtained (163), we additionally obtain

$$F_{12} = F_{13} = F_{23} = 0 \tag{164}$$

and the following equations

$$\mu F_{77} - \frac{1}{\rho_0}F_{11} = 0, \quad \mu F_{77} - \frac{1}{\rho_0}F_{22} = 0, \quad \mu F_{88} - \frac{1}{\rho_0}F_{11} = 0, \tag{165}$$

$$\mu F_{88} - \frac{1}{\rho_0}F_{33} = 0, \quad \mu F_{99} - \frac{1}{\rho_0}F_{22} = 0, \quad \mu F_{99} - \frac{1}{\rho_0}F_{33} = 0, \tag{166}$$

$$(\lambda + 2\mu)F_{45} + \lambda\left(F_{55} + F_{56}\right) = 0, \quad (\lambda + 2\mu)F_{46} + \lambda\left(F_{56} + F_{66}\right) = 0, \tag{167}$$

$$(\lambda + 2\mu)F_{45} + \lambda\left(F_{44} + F_{46}\right) = 0, \quad (\lambda + 2\mu)F_{56} + \lambda\left(F_{46} + F_{66}\right) = 0, \tag{168}$$

$$(\lambda + 2\mu)F_{46} + \lambda\left(F_{44} + F_{45}\right) = 0, \quad (\lambda + 2\mu)F_{56} + \lambda\left(F_{45} + F_{55}\right) = 0, \tag{169}$$

$$(\lambda + 2\mu)F_{44} + \lambda\left(F_{45} + F_{46}\right) - \frac{1}{\rho_0}F_{11} = 0, \tag{170}$$

$$(\lambda + 2\mu)F_{55} + \lambda\left(F_{45} + F_{56}\right) - \frac{1}{\rho_0}F_{22} = 0, \tag{171}$$

$$(\lambda+2\mu)F_{66}+\lambda(F_{46}+F_{56})-\frac{1}{\rho_0}F_{33}=0. \tag{172}$$

We also derive from Eqs. (165)–(166) that

$$F_{11}=F_{22}=F_{33},\quad F_{77}=F_{88}=F_{99},\quad \frac{1}{\rho_0}F_{11}=\mu F_{77}. \tag{173}$$

The remaining Eqs. (167)–(169) give

$$F_{45}=F_{46}=F_{56},\quad F_{44}=F_{55}=F_{66}. \tag{174}$$

The last equations in (174) are derived for $\lambda\neq 0$. With the relationships obtained, we can conclude that Eqs. (167)–(169) and (170)–(172) consist of only two different equations:

$$\lambda F_{44}+2(\lambda+\mu)F_{45}=0,\quad \frac{1}{\rho_0}F_{11}=(\lambda+2\mu)F_{44}+2\lambda F_{45}. \tag{175}$$

Summarising, we get that, besides vanishing second-order partial derivatives (163)–(164), there exists nontrivial relationships (173)–(175) for differential quantities.

It follows from Eqs. (163) to (164) that

$$\begin{aligned} F=&f_1(t,x_1,x_2,x_3,u_1)+f_2(t,x_1,x_2,x_3,u_2)+f_3(t,x_1,x_2,x_3,u_3)\\ &+f_7(t,x_1,x_2,x_3,u_7)+f_8(t,x_1,x_2,x_3,u_8)+f_9(t,x_1,x_2,x_3,u_9)\\ &+f_{456}(t,x_1,x_2,x_3,u_4,u_5,u_6), \end{aligned} \tag{176}$$

where f_1, f_2, f_3, f_7, f_8 and f_{456} are arbitrary functions. To specify the above-mentioned functions, we consider at first the last equation in (173). Substituting (176) into this equation, we get the equation wherein the variables are separated since the functions f_7 and f_1 are independent of the variables u_1 and u_7 correspondingly:

$$\mu\frac{\partial^2 f_7}{\partial u_7^2}=\frac{1}{\rho_0}\frac{\partial^2 f_1}{\partial u_1^2}. \tag{177}$$

Integrating equation (177), we find that

$$f_1=\rho_0 C(t,x_1,x_2,x_3)u_1^2+B_1(t,x_1,x_2,x_3)u_1+C_{13}(t,x_1,x_2,x_3), \tag{178}$$

$$f_7=\frac{1}{\mu}C(t,x_1,x_2,x_3)u_7^2+B_7(t,x_1,x_2,x_3)u_7+C_{73}(t,x_1,x_2,x_3). \tag{179}$$

We put C_{13} and C_{73} equal zero since they are not included into the density of conservation laws. Further, using that $F_{11}=2\rho_0 C u_1^2$, we get the following expression for the function F:

$$
\begin{aligned}
F = \; & C(t, x_1, x_2, x_3)\Big\{\frac{\rho_0}{2}(u_1^2 + u_2^2 + u_3^2) + \frac{(3\tau)^2}{6(3\lambda + 2\mu)} \\
& + \frac{1}{4\mu}\left[(u_4 - \tau)^2 + (u_5 - \tau)^2 + (u_6 - \tau)^2 + 2(u_7^2 + u_8^2 + u_9^2)\right]\Big\} \\
& + B(t, x_1, x_2, x_3, \boldsymbol{u}),
\end{aligned}
$$

where $\tau = (1/3)(u_4 + u_5 + u_6)$ and B is of the form

$$B(t, x_1, x_2, x_3, \boldsymbol{u}) = \sum_{i=1}^{9} b_i(t, x_1, x_2, x_3)u_i = \boldsymbol{b} \cdot \boldsymbol{u}. \tag{180}$$

The linear form of B is a consequence of the form of the equations under consideration. Substituting now F into (162), we conclude that $C(t, x_1, x_2, x_3)$ is a constant function. Notice that the quadratic part of the function F in (180) coincides with $\boldsymbol{u}^{\mathrm{T}} A_{\boldsymbol{u}} \boldsymbol{u}$, where the operator $A_{\boldsymbol{u}}$ is defined in (24). Finally, the substitution of the linear function B into (158)–(162) gives the following system

$$\boldsymbol{b}_t = -\hat{L}^* \boldsymbol{b}, \quad \boldsymbol{b} = (b_1, \ldots, b_9)^{\mathrm{T}}, \tag{181}$$

where the symbol $*$ means the operation of conjugation. For any solution of (181) the following representation holds

$$\boldsymbol{b} = A_{\boldsymbol{u}} \boldsymbol{u}, \tag{182}$$

where $\boldsymbol{u}$ is a solution of the system (12).

Summarising the calculations performed, we put together the results obtained.

Theorem 1 *Zero-order conservation laws for the three-dimensional, linear-elastic model consist of only unique quadratic conservation law which coincides with the Hamiltonian admitted by the system (1) and linear conservation laws with the coefficients determined by solving Eq. (181).*

7 Concluding Remarks

The question about the existence of Hamiltonian structures (17) for the three-dimensional, linear-elastic model (2)–(11) was completely considered. We showed that this model admits the so-called non-canonical singular Poisson bracket and Casimir functionals. We presented two approaches to simplify the form of Eq. (17). The first one consists of performing the transformation of the Poisson bracket to the form (20). It follows from this that Casimir functionals are admitted by the system (2)–(11) and these functionals are independent of the form of Hamiltonian. Moreover, we found all Casimir functionals and showed that they are of the form (83)–

(88). In the three-dimensional case, the Casimir functional depends on 6 differential expressions, whereas in the two-dimensional case the Casimir functional depends on only one differential expression. The second approach was resulted in reducing the quadratic Hamiltonian obtained to the principal axes.

Notice that the existence of a Hamiltonian for the system (2)–(11) does not guarantee the stability of solutions to this system. In the norm given by the quadratic Hamiltonian obtained or the energy functional such property takes place provided that this Hamiltonian is a positive definite quadratic form. It was proven that the positive definiteness of the Hamiltonian occurs when the conditions (37) are fulfilled.

We found all zero-order conservation laws for the model wherein the derivatives are excluded from the list of variables. We used the direct method to derive these conservation laws and established that there is only unique quadratic conservation law with the density defined by (180) which actually coincides with the energy functional. Linear conservation laws were resulted with the density $B(t, x_1, x_2, x_3, \boldsymbol{u})$ defined by (180) where $b_i(t, x_1, x_2, x_3)$ for $i = 1, \dots, 9$ or the vector $\boldsymbol{b} = (b_1, \dots, b_9)^T$ satisfies the system (181) which can be presented in the form (182).

Appendix: The Energy Conservation Law

For the system (1) the energy conservation law reads [4]

$$\frac{\partial \Pi_0}{\partial t} + \frac{\partial \Pi_1}{\partial x_1} + \frac{\partial \Pi_2}{\partial x_2} + \frac{\partial \Pi_3}{\partial x_3} = 0, \tag{183}$$

where

$$\Pi_0 = \frac{\lambda}{2}(\varepsilon_{11} + \varepsilon_{22} + \varepsilon_{33})^2 + \mu\varepsilon_{ij}\varepsilon_{ji} + \rho_0\frac{u_1^2 + u_2^2 + u_3^2}{2} \tag{184}$$

$$= \rho_0 E + \rho_0\frac{u_1^2 + u_2^2 + u_3^2}{2} \tag{185}$$

is the density of the conservation law and

$$\Pi_i = -u_1\sigma_{1i} - u_2\sigma_{2i} - u_3\sigma_{3i} \tag{186}$$

denotes the components of the energy flow vector.

References

1. Landau LD, Lifshitz EM (2012) Theory of elasticity. Course of theoretical physics, vol 7. Butterworth-Heinemann

2. Hahn HG (1985) Elastizita Theorie Grundlagen der Linearen Theorie und Anwendungen auf Undimensionale. Ebene und Zaumliche Probleme, BG Teubner, Stuttgart
3. Slaughter WS (2001) The linearized theory of elasticity. Birkhausen, Boston
4. Godunov SK, Romenskii EI (2003) Elements of continuum mechanics and conservation laws. Springer, Berlin
5. Grebenev VN, Medvedev SB (2016) Hamiltonian structure and conservation laws of two-dimensional linear elasticity theory. ZAMM Z Angew Math Mech 96(10):1175–1183
6. Novikov SP (1982) The Hamiltonian formalism and a many-valued analogue of Morse theory. Russ Math Surv 37(5):1–56
7. Shepherd TG (1990) Symmetries, conservation laws, and Hamiltonian structure in geophysical fluid dynamics. In: Dmowska R (ed) Advances in geophysics. Elsevier, Amsterdam
8. Dubrovin BA, Fomenko TA, Novikov SP (1984) Modern geometry: methods and applications, part 1. Springer, Berlin
9. Sanz-Serna JM, Calvo MP (1994) Numerical Hamiltonian problems. Chapman and Hall, London
10. Ibragimov NH (1983) Transformation groups applied to mathematical physics. Reidel, Dordrech

Complex Methods for Lie Symmetry Analysis

Asghar Qadir and Fazal M. Mahomed

Abstract When Lie developed symmetry analysis, he took the equations to be defined in the complex domain but did not explicitly use the entailed complex analyticity. Making it explicit necessitates the incorporation of the Cauchy–Riemann equations into the original system of equations, which modifies the symmetries of the system. This point was followed up by us, and some of our students, in a series of papers (and theses). It was found that complex methods, when they are applicable, provide more powerful tools for obtaining solutions and integrals of differential equations, even enabling us to find solutions of systems of differential equations that possess no symmetries. In this chapter we review the methods developed and then pose the crucial question that was begged in saying "when they are applicable." When *would* they be applicable and why, or how, does the complex method work? We indicate some lines to pursue to try to find the answers, or at least partial answers, to these questions.

Dedicated to the memory of one of the most innovative workers in the field of Symmetry Analysis, after Sophus Lie, Nail Hairullovich Ibragimov, who initiated many new methods for using Lie Analysis to deal with differential equations.
FM is Visiting Professor at UNSW for 2020.

A. Qadir (✉)
Abdus Salam School for Mathematical Sciences, Government College University, Lahore, Pakistan

F. M. Mahomed
DSI-NRF Centre of Excellence in Mathematical and Statistical Sciences School of Computer Science and Applied Mathematics, University of the Witwatersrand, Johannesburg Wits 2050, South Africa
e-mail: fazal.mahomed@wits.ac.za

School of Mathematics and Statistics, The University of New South Wales, Sydney, NSW 2052, Australia

A. C. J. Luo and R. K. Gazizov (eds.), *Symmetries and Applications of Differential Equations*, Nonlinear Physical Science,
https://doi.org/10.1007/978-981-16-4683-6_4

1 Introduction

Before Lie, the usual method to solve differential equations (DEs) was by ad hoc approaches or by approximating it by a linear DE and solving the linear version. In general, the approximation will work well enough in some regime and become arbitrarily bad in other regimes. As such, it would be necessary to prove the existence of a solution and to determine the domain in which the approximation is good enough. Since these will be different for each DE, one is reduced to solving one DE at a time and not rely on any method for whole classes of DEs. Lie had wanted to do for DEs what Galois had done for polynomial equations, wherein he used the manifest symmetries of the roots of the equation to determine when the polynomial equation could be solved by means of radicals. Among the methods that had been used for solving some of the simpler DEs, there was the transformation of independent and dependent variables. Lie conceived of the idea of looking for invariance of the DE under such transformations [1–4] so that it could be determined when the DEs could be solved/reduced or transformed and then one can proceed to solve/reduce or transform the DEs. Lie used not only the groups of symmetries, but the algebra of the corresponding infinitesimal symmetry generators. The DEs are not necessarily single but could be systems (or we could say that they are *vector* DEs). Further, he did not restrict the domain of the DEs to be real, but allowed them to be complex.

Symmetry generators, inter alia, can be used to reduce the order of scalar ordinary differential equations (ODEs) or reduce the number of independent variables in partial differential equations (PDEs). Alternatively, the symmetries can be used to construct quantities that remain invariant under the transformation, thereby enabling a reduction of order or the number of independent variables [5, 6] by treating that combination as a new variable. If there are enough independent symmetries they can be used to fully solve the equation for scalar ODEs in the sense of providing an algebraic equation that constitutes an implicit solution, modulo quadratures or in the case of PDEs to construct invariant solutions. If the invariants contain derivatives in them, they are called *differential* invariants. If they depend only on dependent and/or independent variables, they are simply called "invariants." In effect, what Lie had done was to take the space of independent variables on which the dependent variables were defined and extended, or enlarged, it to include also the dependent variables. In this new, extended space, we could perform the equivalent of coordinate transformations called *point transformations*. In this way of looking at it, it is natural to require invariance in the higher dimensional manifold under point transformations. What Lie wanted was that the DEs remain invariant under these transformations, thus visualizing them as "living on the manifold."

To deal with DEs, we need to treat the derivatives *as if* they were independent variables and then constrain them in such a way that the DE is satisfied. The enlarged, or *prolonged* space of all the variables and their derivatives is also called the *jet space*. In this space, we restrict the transformations to be performed only in the original, non-prolonged, space. However, we could include any number of derivatives from the prolonged space that we choose. Thus, if we prolong to include the first derivative in our transformations but no more, we have "contact transformations." If we prolong further, we have "higher-order transformations," but there are no separate names for them. The name "contact" comes from the tangency requirement for the derivative to be met. Lie mainly restricted his analysis to point and contact transformations but subsequently others extended the Lie methods to higher order transformations. As such, the original transformations are called "Lie point transformations."

Among the various methods of using symmetries to solve DEs is the transformation of the DE to linear form, which is of special interest. The classic example for this is the Bernoulli equation, a first-order ODE in which the dependent variable appears to the nth power. However, there was no general procedure available for nonlinear DEs, especially higher order ones. If one can tell when a DE can, in principle, can be transformed to linear form, even without finding the required transformations and converting to linear form, we can say a lot about the solutions of the DE. For example, if it is an ODE, we know how many independent solutions there are, without having obtained the solutions. By requiring that the given DE transform to a chosen canonical form of a linear DE, we can arrive at conditions that the given DE must satisfy. Lie did this for second-order scalar ODEs [5] and demonstrated that the ODE must have eight infinitesimal symmetry generators that would constitute a Lie algebra as well as conditions on the DE. Then he looked at the maximal algebra admitted. He did not go further but others carried the work forward for higher order ODEs and for vector ODEs, using contact transformations and even Lie's original method. Equations that can be transformed to linear form are called *linearizable* and the process of transforming a DE to linear form by transformations of the dependent and independent variables is called *linearization* via point transformation. Note that this is *not* an approximation of the DE by a linear one, but a transformation that gives the exact solution of the DE. If it is a linearizable PDE, it has infinitely many linearly independent solutions. Consequently, there is no way that we can make the type of general statement that we could for ODEs. We then need boundary conditions to be able to arrive at a meaningful, unique, solution, or other invariant criteria. This entails that the conditions satisfy the symmetries of the PDE. Thus, for PDEs, invariants are especially useful although these needed further generalizations.

In this chapter we review work on a line that Lie did *not* take, namely making explicit use of complex analyticity. Recall that if a complex function of a complex variable is once differentiable in an open domain it is analytic in that domain, which entails infinite differentiability. This would not hold for real functions of real variables. While this fact simplifies statements of Lie's requirements for the DEs to be amenable to his symmetry methods, it is not obvious that it can make a fundamental difference to the procedures used to solve the equations. We demonstrate that it does so, provided certain additional conditions are met. Though there are explicit checks for when these complex methods *can be used*, there is no complete understanding of when they would be *useful*.

The plan of the chapter is as follows. In the next section, we give the preliminary background for Lie symmetry analysis and some basic geometry used in it, including contact and higher order symmetries. In section three, we review the fundamentals of the complex method. In the subsequent sections, we review its application for linearization and for Noether symmetries and their integrals. It is shown that the complex methods extend the applicability of symmetry analysis beyond the usual methods. In section seven, we present insights regarding the working of the complex methods obtained by iterative splitting of a scalar ODE. In the concluding section, we summarize the work reviewed and present the fundamental questions that need to be addressed so as to understand *why* complex methods work.

2 Preliminaries

For completeness, we give basic definitions despite the likelihood that the reader already knows them, in the hope that he/she will bear with us. At least they will be useful to establish notation. If there are l independent variables represented as a vector $\mathbf{x}$ and m dependent variables represented by $\mathbf{y}$, a *Lie point symmetry generator* is the operator

$$\mathbf{X} = \mathbf{A}(\mathbf{x}, \mathbf{y}).\nabla_{\mathbf{x}} + \mathbf{B}(\mathbf{x}, \mathbf{y}).\nabla_{\mathbf{y}} \,, \tag{1}$$

or using indices a for the independent variables and i for the dependent variables

$$\mathbf{X} = A^a(x^b, y^i)\frac{\partial}{\partial x^a} + B^i(x^a, y^j)\frac{\partial}{\partial y^i} \,, \tag{2}$$

where we have used the Einstein summation convention that repeated indices are summed over. Further, if the DE is of order n, we need to prolong the space and the generators to incorporate all the derivatives of the dependent variables with respect to the independent variables. For ODEs,

$$\mathbf{X}^{[n]} = A(x, y^i)\frac{\partial}{\partial x} + B^i(x, y^j)\frac{\partial}{\partial y^i} + B^{i\,[1]}(x, y^j, y^{j\,\prime})\frac{\partial}{\partial y^{i\,\prime}} + \cdots \,, \tag{3}$$

where

$$B^{i\,[p]} = D_x B^{i\,[p-1]} - y^{i\prime} D_x A \,, \tag{4}$$

$B^{i\,[0]}$ simply being B^i and D_x is the *total derivative in the prolonged space*,

$$D_x = \frac{\partial}{\partial x} + y^{i\,\prime}\frac{\partial}{\partial y^i} + \cdots + y^{i\,(p)}\frac{\partial}{\partial y^{i\,(p-1)}} \,. \tag{5}$$

For PDEs, the A would have to be replaced by $\mathbf{A}$ and the partial derivative with respect to x by $\nabla_{\mathbf{x}}$. While the former can be easily converted to index notation as A^i, the latter becomes somewhat involved in converting. The real problem in writing is for the $y^{i\,[p]}$, which would be a partial derivative with respect to x^a *to all orders* up to p. The set of all prolonged symmetry generators forms a Lie algebra and the symmetry group determines what reduction of the DE there can be. A system of

m ODEs of order n, $E^i(x, y^j; y^{j\,\prime}, \ldots, y^{j\,[n]}) = 0$, is said to be *symmetric* under the transformation generated by $\mathbf{X}$ if $\mathbf{X}^{[n]}E^i = 0$, when restricted to the solutions of $E^i = 0$. This is denoted by putting "$|_{\mathbf{E}=0}$" after the above equation. The generalization to PDEs is as before, with the corresponding complications.

A major activity arose of classifying the Lie point symmetry algebras of all second-order scalar ODEs. This was called the "classification problem for scalar second-order ODEs." The classification problem for higher order scalar ODEs rapidly becomes extremely difficult on account of the proliferation of possible cases and sub-cases of the allowed Lie algebras. Similarly, the classification problem for higher dimensional systems becomes even more complicated as the number of sub-cases proliferates even more. Increasing the order and dimension simultaneously makes the problem well high intractable. We do not go into this further here as the complex methods were not used for this purpose.

Since much of the complex work is motivated by considerations of linearization, it is necessary to very briefly review the key features of Lie's linearization procedure. By requiring that a scalar second-order semilinear ODE

$$y'' = f(x, y; y') \,, \tag{6}$$

be transformed under $p = p(x, y), q = q(x, y)$ to

$$q'' + A(p)q' + B(p)q + C(p) = 0 \,, \tag{7}$$

he showed that (6) would have to be of the form

$$y'' + a(x, y)y'^{\,3} + b(x, y)y'^{\,2} + c(x, y)y' + h(x, y) = 0 \,, \tag{8}$$

and would have to satisfy a system of four first-order conditions that the coefficients a, b, c, h and two auxiliary functions would have to satisfy. This is not as bad as it may sound, since one is not *solving* the coupled system of equations but merely verifying them. Nevertheless, the auxiliary functions complicate matters as they are arbitrary and would have to be guessed. Tressé [7] invariants eliminates the auxiliary functions via compatibility by taking derivatives and one obtains two second-order conditions to be satisfied by the coefficients, viz.

$$3(ac)_x + ha_y - 2bb_x - cb_y - 3a_{xx} - 2b_{xy} - c_{yy} = 0 \,, \tag{9}$$

$$3(hb)_y + ab_x - 2aa_y - ha_x - 3h_{yy} - 2c_{xy} - b_{xx} = 0 \,. \tag{10}$$

Tressé's formulation makes the application of the Lie conditions much easier.

Chern [8] did not use the Lie point transformations to linearize third-order scalar ODEs but incorporated the first derivative of the dependent variable in the coefficients of the operator, $A = A(x, y; y')$ and $B = B(x, y; y')$ in the scalar case for (2), to solve the problem. One must now ensure that under the transformation, the "derivative" used here corresponds to the derivative of the dependent variable by differentiation. Writing the transformation as $(x, y) \rightarrow (\overline{x}, \overline{y})$, the *contact* or *tangency* condition is

$$d\overline{y} - \overline{y}'d\overline{x} = \lambda(x, y; y')(dy - y'dx)\,, \tag{11}$$

where λ is an undetermined multiplier [5]. The contact transformations of Chern can be extended to systems of equations for several independent variables by re-inserting the indices, so that the condition becomes

$$d\overline{y}^i - \overline{y}^i_{,a}d\overline{x}^a = \lambda(x, y; y^i_{,a})(dy^i - y^i_{,a}dx^a)\,. \tag{12}$$

Lie had managed to prove that the second-order scalar ODE is linearizable if, and only if, it has eight Lie point symmetry generators. The Lie point symmetry algebra for order n scalar ODEs was obtained much later [9] and it was shown that Lie's theorem does not hold there, as there are three linearizable classes with $(n + 1)$, $(n + 2)$ or $(n + 4)$ generators. For the third-order case, the canonical forms associated with those symmetries were made explicit [10]. The classes of linearizable second-order systems was also achieved at around the same time, first for two-dimensional systems and then for arbitrary m [11]. The linear classes for the two-dimensional system have 5, 6, 7, 8, or 15 generators and for arbitrary m, $2m + 1, \ldots, (2m)^2 - 1$ symmetry generators [12, 13].

A question arises here, *why* is the $n = 2$ case special? The answer may lie in the geometric methods that had been developed and were used to linearize ODEs to which we now turn. To explain it, we need to establish the notation and concepts used there. For our purposes, we will be using a manifold with a metric tensor, g_{ab}, and inverse metric tensor, g^{ab}, defined on it, and assume that it is torsion-free so that the connection symbol is the Christoffel symbol in a coordinated basis (see Chaps. 2 and 3 of [14]),

$$\Gamma^a_{bc} = \frac{1}{2}g^{ad}(g_{bd,c} + g_{cd,b} - g_{bc,d})\,, \tag{13}$$

where "$_{,c}$" stands for the partial derivative relative to x^c. This object comes from the differentiation of the basis vectors relating the tensor quantity in the manifold to its components in the coordinate system chosen. As such this is *not* a tensor quantity or a fully coordinate quantity but hangs between the two. The *covariant derivative* of a contravariant vector, V^a, is $V^a_{;b} = V^a_{,b} + \Gamma^a_{bc}V^c$ and of a covariant vector W_a is $W_{a;b} = W_{a,b} - W_c\Gamma^c_{ab}$. The difference of the second derivative obtained by going first in one direction and then in another, or vice versa, gives a measure of the curvature of the space, measured by the *Riemann–Christoffel curvature tensor*

$$R^a{}_{bcd} = \Gamma^a_{bd,c} - \Gamma^a_{bc,d} + \Gamma^e_{bd}\Gamma^a_{ce} - \Gamma^e_{bc}\Gamma^a_{de}\,. \tag{14}$$

The trace of the curvature tensor is the Ricci tensor, $R_{bd} = R^a{}_{bad}$, whose trace $R = g^{bd}R_{bd}$ is the Ricci scalar. The Ricci tensor is symmetric and the curvature tensor is symmetric in the first and second pairs of indices and symmetric under the interchange of the two pairs of indices. Further, the skew part for any three of the indices is zero.

This reduces the number of linearly independent components. In m-dimensions, for $R^a{}_{bcd}$, there are $m^2(m^2-1)$, for R_{bd}, there are $m(m+1)$, and for R obviously just one.

As in flat spaces, so in curved spaces the "straightest" available path between two points is also the shortest. Such curves are called *geodesics* and satisfy the *geodesic equation*

$$y^{i\,\prime\prime} + \Gamma^i_{jk} y^{j\,\prime} y^{k\,\prime} = 0\,. \tag{15}$$

The above procedure relies on the differentiation of the basis vector, so how can it tell us about the curvature of the manifold? The point is that we define the vector on the manifold as a mapping of a point to a nearby point on the manifold, without reference to the coordinate system. Now we map the vector to the coordinate system and differentiate there using the covariant derivative as explained above and then map the quantity back to the manifold. This defines the derivative operator on the manifold without the coordinate system being involved, though an arbitrary coordinate system was used to be able to get the definition. The covariant and partial derivatives are identical for scalars and only differ for vectors or tensors. One is still left with the differentiation of the basis vector being "carried back" to the manifold. To eliminate this extra term, we can use one vector to move another on the manifold, which is called the *Lie derivative* of the vector moved, say $\mathbf{p}$ relative to the one along which it is moved, say $\mathbf{t}$. Going back to the covariant derivative, this is $[\mathscr{L}_{\mathbf{t}}\mathbf{p}]^a = t^b p^a_{;b} - p^b t^a_{;b}$. Since the Christoffel symbol is symmetric in the lower indices the two Christoffel symbols cancel and only the partial derivatives, $t^b p^a_{,b} - p^b t^a_{,b}$, are left in the expression. What has happened is that the derivatives of the basis vectors have cancelled out and so the worrying term is no longer present in the Lie derivative. We can transport one vector along the geodesic given by the other on the manifold by using the Taylor series, to obtain the other vector at the new point. Thus, if the geodesic lies between points P and Q on the manifold, and $\mathbf{t} : P \to Q$, then $\mathbf{p}|_Q$ is given in terms of $\mathbf{p}|_P$ by

$$\mathbf{p}|_Q = \exp[\mathscr{L}_{\mathbf{t}}]\mathbf{p}|_P\,. \tag{16}$$

Requiring that $\mathbf{p}$ be left invariant as it goes along $\mathbf{t}$, amounts to requiring that $\mathscr{L}_{\mathbf{t}}\mathbf{p}$ be zero. Consequently, the metric tensor, and hence the geometry, will be left invariant under Lie transport will if the Lie derivative of the metric tensor is zero. Such a direction is called an *isometry*, and is a generator of the symmetry implicit in the geometry.

Using the above definitions, in component form the equations for an isometric direction, k^a become,

$$g_{ab,c}k^c = g_{ac}k^c{}_{,b} + g_{bc}k^c{}_{,a}\,. \tag{17}$$

These are called *Killing's equations* and a vector satisfying this equation is called a *Killing vector*, or an isometry. Notice that (15) depends on the metric coefficients, which are functions of the dependent variable but do not depend explicitly on the independent variable. Thus the geodesic equations possess a translational symmetry along the independent variable.

It was noted by Aminova and Aminov [15], and independently, but later, by [16] that there is a direct connection between the symmetries of a system of geodesic equations and the isometries of the underlying manifold on which the solutions live. Aminova and Aminov further noted that projecting the geodesic equations down along x, one obtains a cubically semilinear system of second-order ODEs. We independently arrived at the same point [17]. We further noted that these are similar in the case of scalar ODE to the original Lie linearizable ODE. The projected equations then have the coefficients given as functions of the Christoffel symbols for the higher dimensional system. It turns out that the Lie conditions correspond exactly to the requirement that the curvature tensor constructed from those Christoffel symbols be zero, i.e., *the space is flat*. Further, there is a redundancy in the Christoffel symbols due to the freedom of choice of coordinates. When we project from two down to one-dimensional systems (i.e., the scalar equation) the redundancy is of two. These are the two arbitrary functions that Lie needed for his equations. As such, we have obtained the Lie linearization purely from Geometry. *This* is the sense in which Geometry explains what is special about order two. The requirement of flatness is natural as the shortest path between two points in a flat space is a straight line. We know the metric tensor in Cartesian coordinates and the equation of the straight line. It is possible to find the coordinate transformation that converts a flat metric locally to any given metric [18]. The coordinate transformation to get the metric tensor required to give the coefficients yields the linearizing transformation and one obtains the solution of the linearizable system in terms of the the original variables as a nonlinear superposition for the general solution. A more complete analysis of this linearization followed later [19]. A code was developed to determine if the system of second-order ODEs corresponds to a system of geodesics, and if it does to determine the metric tensor [20]. Put together, we could, in principle, feed in any system of the Lie type, check if it is a projected system of geodesics and then obtain the solution. It is this power of Geometry that we use at much of the base for the complex methods.

3 Complex Symmetry Analysis

The maximal Lie algebra for m-dimensional system of second-order ODEs is $sl(m+2)$, which is $sl(m+2, \mathbb{R})$ for real and $sl(m+2, \mathbb{C})$ for complex variables, with $[(m+2)^2-1]$ real or complex generators. Now, to obtain ODEs after splitting, the independent variable must be real and the dependent variables complex. In that case, to each generator containing the dependent variables in the complex ODE, there will be two after splitting. Thus, for $m = 1$ there should be 8 generators in the real case and 16 in the complex. However, splitting the scalar ODE into its real and imaginary parts yields a two-dimensional real system, which has an $sl(4, \mathbb{R})$ with 15 generators. This demonstrates that going from the real to the complex is non-trivial. The complication arises due to the fact that the complex ODE is not just the two-dimensional real system written explicitly but also the implicit Cauchy–Riemann conditions, which are two first-order equations. Thus the complex system regarded as a higher dimensional real system is constrained. This causes the reduction of symmetry of the complex system. Complex symmetry analysis follows up on the non-trivial consequences emerging from the constraints [21, 22].

To be more concrete, if the real independent variable is x and the complex dependent variable for a scalar ODE is $w = (u + \iota v)$, the complex translation operator, $\mathbf{W}_1$ splits as

$$\mathbf{W}_1 = \partial/\partial w = \frac{1}{2}(\partial/\partial u - \iota\partial/\partial v) = \mathbf{U}_1 + \iota\mathbf{V}_1 \, , \tag{18}$$

so that $\mathbf{W}_1 w = 1$ and $\mathbf{W}_1 \overline{w} = 0$. Hence the complex translation splits into two real orthogonal translations. Now there are no rotations for a single complex variable but there is a scaling symmetry $\mathbf{W}_2$,

$$\mathbf{W}_2 = w\partial/\partial w = \frac{1}{2}[(u\partial/\partial u + v\partial/\partial v) + \iota(v\partial/\partial u - u\partial/\partial v)] = \mathbf{U}_2 + \iota\mathbf{V}_2 \, . \tag{19}$$

Thus we get a real scaling, $\mathbf{U}_2$ *and* a rotation in two-dimensional, $\mathbf{V}_2$. In the context of our focus on "why complex methods are so effective" notice that, by definition, translations leave vector lengths invariant while scalings change lengths. The odd feature is that the complex scaling yields a rotation under the splitting, apart from the expected scaling. How did the complex scaling "know" that a real rotation was needed and had to be coded into the complex scaling?

The natural next step is to go to two complex dimensional systems, with the complex translation and rotation symmetry generators

$$\mathbf{W}_1 = \partial/\partial w_1 \, , \; \mathbf{W}_2 = \partial/\partial w_2 \, , \; \mathbf{R} = w_2\partial/\partial w_1 - w_1\partial/\partial w_2 \, , \tag{20}$$

which split into

$$\frac{1}{2}(\partial/\partial u_1 - \iota\partial/\partial v_1) = \mathbf{U}_1 - \iota\mathbf{V}_1 \, , \tag{21}$$

$$\frac{1}{2}(\partial/\partial u_2 - \iota\partial/\partial v_2) = \mathbf{U}_2 - \iota\mathbf{V}_2 \, , \tag{22}$$

$$\begin{aligned} &\frac{1}{2}[(u_2\partial/\partial u_1 - u_1\partial/\partial u_2) + (v_2\partial/\partial v_1 - v_1\partial/\partial v_2) \\ &+\iota\{(u_1\partial/\partial v_2 + v_1\partial/\partial u_2) - (u_2\partial/\partial v_1 + v_2\partial/\partial v_1)\}] \\ &= (\mathbf{R}_1 + \mathbf{R}_2) + \iota(\mathbf{L}_1 - \mathbf{L}_2) \, , \end{aligned} \tag{23}$$

where $\mathbf{R}_1$ and $\mathbf{R}_2$ are the expected rotations and $\mathbf{L}_1$ and $\mathbf{L}_2$ are two "Lorentz transformations," i.e., rotations through an imaginary angle. Notice that the rotations were arbitrarily identified. Instead of rotating between two u's and two v's, we could have "mixed" them to get rotations between the u's and v's, or broken the two Lorentz transformations differently to obtain two more cross rotations. Is *this* at the heart of the "unexpected effectiveness of the complex methods?"

What has happened is that we actually have the sixteen "quasi-scalings," which we can write, using the notation $k_1 = u_1, k_2 = u_2, k_3 = v_1, k_4 = v_2$, as $k_i\partial/\partial k_j$. To see the full significance of this point, let us proceed to the m complex dimensional system split into $2m$ real variables, k_i. Write the $2m$ translations as $\mathbf{X}_i$ and the $4m^2$ quasi-scalings as $\mathbf{Y}_{ij}$. Then the Lie algebra satisfied by these generators is:

$$[\mathbf{X}_i, \mathbf{X}_j] = 0\ , \ [\mathbf{X}_i, \mathbf{Y}_{jk}] = \delta_{ij}\mathbf{X}_k\ , \ [\mathbf{Y}_{ij}, \mathbf{Y}_{kl}] = \delta_{ik}\mathbf{Y}_{jl} - \delta_{jl}\mathbf{Y}_{ik}\ . \tag{24}$$

Thus we have $2m$ translations, $\mathbf{X}_i$, $m(2m-1)$ rotations $\mathbf{R}_{ij}$ given by the commutators of the quasi-scalings, $2m$ genuine scalings $\mathbf{S}_i = k_i\partial/\partial k_i$ and the remaining $m(2m-1)$ are proper "partial scaling." Somehow the Cauchy–Riemann (CR) conditions constrain the symmetries so that the quasi-scalings provide no new generators, and we are left with just the geometrically expected symmetries. The question remains, "How do the CR conditions get rid of the extra symmetries?"

For application to DEs, the prolongation of the generators proceeds in the usual way and the CR conditions do not need any further prolongation. Symmetry methods are used by enumerating all possible algebras of a given dimension. One-dimensional algebras are not in general sufficient for "group methods" to work. As such, we need at least a two-dimensional algebra. In the simpler cases of lower order and lower-dimensional systems, there are few higher dimensional algebras available and the classification problem is easy. For lower dimensions one gets whole classes of possible ODEs associated with each algebra of the given dimension. Thus, for the scalar ODE for two-dimensional algebras, there are four possible algebras, each with an ODE associated with it. On splitting the complex scalar ODE there is a much richer structure as one gets a two-dimensional, three three-dimensional and three four-dimensional algebras, each with its associated class of systems of two ODEs.

For every complex scalar ODE there is a system of two ODEs. However, the converse is obviously not true. Consequently there must be some compatibility conditions that the system satisfies for the correspondence to hold. The way they are obtained is to take the general relevant order complex scalar ODE and split it. The general form of the system corresponding to the general complex ODE is, thereby, obtained. What general form? This depends on the class of systems that is to be converted. The symmetries of the complex scalar ODE and the corresponding real system have been shown to be inequivalent [23, 24]. The procedure can, equally well, be applied to systems of complex ODEs being converted to systems of real ODEs of twice the dimension. A serious problem arises of being able to apply complex methods to odd-dimensional systems. The method used was to introduce an algebraic constraint, but that changes the system. Another method that could be explored would be to adjoin a real DE to a complex system, but it would entail additional complications, and may not be workable or worthwhile.

4 Complex Linearization

Every first-order complex scalar ODE

$$y'(x) = \omega(x, y)\,, \tag{25}$$

is linearizable [1, 4], where $y = f + \iota g$ and $\omega = \phi + \iota\gamma$. Thus the system

$$f'(x) = \phi(x, f(x), g(x))\,,\ g'(x) = \gamma(x, f(x), g(x))\,, \tag{26}$$

is linearizable, provided the CR equations

$$\frac{\partial\phi}{\partial f} = \frac{\partial\gamma}{\partial g}\,,\ \frac{\partial\phi}{\partial g} = -\frac{\partial\gamma}{\partial f}\,, \tag{27}$$

hold. We see that not every two-dimensional system of first-order ODEs is linearizable, but only those that satisfy the CR-equations as the linearization constraint equations. Notice that here the role of the CR-equations as a pair of first-order integrability conditions is obvious.

We now come to the general semilinear second-order complex scalar ODE,

$$y''(x) = \omega(x, y; y')\,. \tag{28}$$

Since it is not true that all second-order ODEs are linearizable, when we split the ODE, the CR-equations do not give linearization conditions, but only compatibility conditions,

$$f''(x) = \phi(x, f, g; f', g')\,,\ g''(x) = \gamma(x, f, g; f', g') \tag{29}$$

$$\frac{\partial\phi}{\partial f} = \frac{\partial\gamma}{\partial g}\,,\ \frac{\partial\phi}{\partial g} = -\frac{\partial\gamma}{\partial f}\,,\ \frac{\partial\phi}{\partial f'} = \frac{\partial\gamma}{\partial g'}\,,\ \frac{\partial\phi}{\partial g'} = -\frac{\partial\gamma}{\partial f'}\,. \tag{30}$$

For linearization there are further requirements that must be met. Lie's linearizable scalar second-order ODE, given by (8), can be split into its real and imaginary parts, bearing in mind that the four coefficients are also complex. The resulting two-dimensional system must be of the form

$$\begin{aligned} f'' + (a_1 f'^{\,3} - 3a_2 f'^{\,2}g' - 3a_1 f'g'^{\,2} + a_2 g'^{\,3})\\ +(b_1 f'^{\,2} - 2b_2 f'g' - b_1 g'^{\,2}) + (c_1 f' - c_2 g') + h_1 = 0,\\ g''(x) + (a_2 f'^{\,3} + 3a_1 f'^{\,2}g' - 3a_2 f'g'^{\,2} - a_1 g'^{\,3})\\ +(b_2 f'^{\,2} + 2b_1 f'g' - b_2 g'^{\,2}) + (c_2 f' + c_1 g') + h_2 = 0, \end{aligned} \tag{31}$$

subject not only to the CR-equations (30) written out explicitly for ϕ and γ but also to the CR-equations for each of the four complex coefficients, a, b, c, h. This is a system of two second-order real ODEs involving eight real functions to bring the system into the Lie form, which must satisfy a system of four first-order constraints that ensure integrability. We call such a system "complex linearizable."

It was proved by Goringe and Leach [12] that for a system of two second-order ODEs linear with constant coefficients there are 7, 8 or 15 symmetries. When extended to the fully general case 5 and 6 generators were added [13]. It may be recalled that the geometrically linearizable system has a Lie algebra of $sl(m+2, \mathbb{R})$, which yields 15 generators for $m = 2$. As such, only in the maximal symmetry case can we use the power of geometry at present to directly obtain the linearizing transformation and hence the solution. What is more disturbing is that the geometrical arguments for linearization are somehow bypassed in general – we are getting straight lines in a curved space. *How can that be?* Could it be that the space corresponding to the system of geodesics for 5, 6, 7 and 8 generators is like a higher dimensional cylinder, with some flat sections? It would be worth exploring this possibility.

The general second-order linear complex scalar equation

$$y''(z) + A(z)y'(z) + B(z)y(z) = 0\,, \tag{32}$$

where z is a complex variable, can be transformed to the form,

$$y''(w) + \alpha(w)y(w) = 0\,, \tag{33}$$

by re-scaling the dependent variable by a position dependent function or, equivalently, by transforming the independent variable, z, appropriately to an independent variable, w, to get rid of the first derivative term. This can then be split to

$$f''(x) + \alpha_1(x)f(x) - \alpha_2(x)g(x) = 0\,,\ g''(x) + \alpha_2(x)f(x) + \alpha_1(x)g(x) = 0\,, \tag{34}$$

where $\alpha = \alpha_1 + \iota\alpha_2$. When this was applied to the free particle equation (with $\alpha = 0$) [23, 25], the 15 generator Lie algebra case was recovered, which is amenable to geometric linearization. For the constant and the variable cases, the 7 and 6 generator algebras were also obtained. Though the system is not geometrically linearizable, the complex equation is and hence its power can be used to solve the scalar ODE and then convert to the system to get the solution of the system.

It is wonderful that that two more of the five classes of linearizable systems can be accessed by complex linearization, making them amenable to the geometric procedure that more-or-less writes down the solution for us, but now the question arises: "where did the other two go?" If the complex method works, "why does it work partially and not fully?" The answer may lie in a step that was glossed over. The scalar ODE was first transformed with the complex independent variable to obtain the simpler form (33) and then it was restricted to the real form. This procedure will not commute in general. If the independent variable is first restricted and then used, the reduction will not occur. There seems to be no good reason to take the reduced form (33) instead of the complete homogeneous liner form. Throwing away the first derivative term in the system may "throw the baby out with the bath-water." Can one not apply the Lie linearization procedure to the full (homogeneous) linear form to obtain the two-dimensional system? Perhaps that would provide the missing cases of 5 and 8 symmetry generators.

Notice that the second-order complex scalar ODE has eight complex symmetries to be linearizable but needs only two to be solvable by Lie's method. On the other hand the two-dimensional system needs at least five real symmetries to be linearizable and four to be solvable. Thus the minimum number of real symmetries required in both cases is four. We see that starting with linearizable second-order complex scalar ODEs, we can end up with two-dimensional systems with fewer symmetries. Is it possible to get a system with only four symmetries that is solvable? In that case, by the easy linearization of a complex ODE the more complicated process of solving the associated system can be bypassed. It was found that this could be done [23, 25–27]. In fact, not only could non-linearizable systems be solved by linearization (of a complex scalar ODE) but one could go further and find complex linearizable scalar ODEs corresponding to systems with *less* symmetries. Thus systems *not solvable* by symmetry methods in the usual way, could be solved by complex linearization. How much lower can one go? It turned out that there is an example with *no* symmetry. We cite the examples of four, one and zero here:

(a) Four symmetry case

$$f'' - f'^{\,3} + 3f'g'^{\,2} = 0\,,\ g'' - 3f'^{\,2}g' + g'^{\,3} = 0\,, \tag{35}$$

with the solution

$$\begin{aligned} f(x) &= c_1 \pm (\sqrt{(a-x)^2 + b^2} + a - x)^{1/2},\\ g(x) &= c_2 \pm (\sqrt{(a-x)^2 + b^2} - a + x)^{1/2}; \end{aligned} \tag{36}$$

(b) One symmetry case

$$\begin{aligned} f'' - xff'^{\,3} + 3xgf'^{\,2}g' + 3xff'g'^{\,2} - xgg'^{\,3} = 0\,,\\ g'' - xgf'^{\,3} - 3xff'^{\,2}g' + 3xgf'g'^{\,2} + xfg'^{\,3} = 0\,, \end{aligned} \tag{37}$$

with the implicit solution

$$\begin{aligned} \mathscr{R}[c_1 Ai(-f - \iota g) + c_2 Bi(-f - \iota g)] = x\,,\\ \mathscr{I}[c_1 Ai(-f - \iota g) + c_2 Bi(-f - \iota g)] = 0\,, \end{aligned} \tag{38}$$

where $\mathscr{R}$, $\mathscr{I}$ are the real and imaginary parts of the arguments and Ai, Bi are the two Airy functions;

(c) No symmetry case

$$\begin{aligned} f'' + (f^2 - g^2 - x^2)(f'^{\,3} - 3f'g'^{\,2}) - 2fg(3f'^{\,2}g' - g'^{\,3}) = 0\,,\\ g'' + (f^2 - g^2 - x^2)(3f'^{\,2}g' - g'^{\,3}) + 2fg(f'^{\,3} - 3f'g'^{\,2}) = 0\,, \end{aligned} \tag{39}$$

which corresponds to the complex scalar ODE

$$y'' - xy^2y'^3 = 0, \tag{40}$$

which is linearizable to $Y'' = 0$, yielding the solution directly.

5 Complex Noether Symmetries and Integrals

Noether's theorem [28] forms a basis of the use of symmetries in Mechanics and through it in all of Physics. It essentially generalizes Hamilton's principle of least action, which can be reformulated as saying that if there is time-translational invariance, energy will be conserved. The action, S, is a functional of a Lagrangian function, $\mathscr{L}[t, q^i(t), \dot{q}^i(t)]$, where q^i are the coordinates of a system of particles in the higher dimension and t is the time. If there is no explicit dependence on the time the action is minimized and a quantity associated with the Lagrangian, called the Hamiltonian, $\mathscr{H}$ is a conserved quantity. More generally, the theorem says that for every continuous symmetry, there is a conserved quantity. It was further generalized to extend to a continuum of "particles," i.e., a *field*, and thence to relativistic fields and further to quantum fields [29]. Hamilton's original method, used also by Noether, is to use the calculus of variations and require that the variation of the action be zero. This provides the necessary conditions for minimization. The sufficient condition, that the second variation be positive, is generally ignored or glossed over, but should be used to avoid getting spurious solutions.

Noether symmetries, as opposed to the usual symmetries, yield *double reduction* of the DEs for which they apply [30, 31], serving like two of the symmetries. Thus, if there is time translational invariance in an ODE (as for the time independent Schrödinger equation or the steady state heat equation) one can replace the derivative operator by a constant. Further, the energy conservation yields an invariant combination of the generalized coordinates and their derivatives, getting rid of another variable. Formally, $\mathbf{X}^{[1]}$, given by (3) for a single independent variable, is a *Noether symmetry* if there exists an appropriate (*gauge*) function, G, such that

$$\mathbf{X}^{[1]}\mathscr{L} + \mathscr{L}\frac{dA}{dx} = \frac{dG}{dx}, \tag{41}$$

where d/dx is the total derivative. This can be extended to PDEs by using several independent variables, x^a and the corresponding total derivatives with respect to each independent variable as well as introducing a vector gauge function, G^a.

An obvious problem of extending the variational principle to the complex domain arises: functionals map the space of functions into the *reals*, $\mathbb{R}$. Obviously the Lagrangian must be real for the action to be real, so that a minimum can be defined on it. This problem was "swept under the rug" at the time in [21, 32]. In defining distributions for complex arguments, the problem of defining functionals is addressed

[33–35] but not the problem of defining a minimum for a complex action. What is required is that the variations of the real and imaginary parts be separately zero and the minimum for both together requires that the magnitude of the action, $|S|$, be minimum. It is worth mentioning that on purely physical considerations Bender and Boettcher had also proposed complex Hamiltonians [36].

Let us now proceed with the complex Lagrangian [21, 32]. Let $\mathscr{L} = \mathscr{L}_1 + \iota\mathscr{L}_2$. Then the Euler–Lagrange equation splits into the pair of coupled equations:

$$\begin{aligned}
&\frac{\partial \mathscr{L}_1}{\partial f} + \frac{\partial \mathscr{L}_2}{\partial g} - \frac{d}{dx}\Big(\frac{\partial \mathscr{L}_1}{\partial f'} + \frac{\partial \mathscr{L}_2}{\partial g'}\Big) = 0 , \\
&\frac{\partial \mathscr{L}_2}{\partial f} - \frac{\partial \mathscr{L}_1}{\partial g} - \frac{d}{dx}\Big(\frac{\partial \mathscr{L}_2}{\partial f'} - \frac{\partial \mathscr{L}_1}{\partial g'}\Big) = 0 ,
\end{aligned} \tag{42}$$

which is not a pair of Euler–Lagrange equations. These were called "Euler–Lagrange-like" equations, but perhaps a better name would have been "*complex-EL*" equations.

The Noether operators, $\mathbf{X}_1^{[1]}, \mathbf{X}_2^{[1]}$ corresponding to the Lagrangians $\mathscr{L}_1, \mathscr{L}_2$

$$\begin{aligned}
\mathbf{X}_1^{[1]} &= \xi_1\partial_x + \frac{1}{2}(\eta_1\partial_f + \eta_2\partial_g + \eta_1'\partial_{f'} + \eta_2'\partial_{g'}) , \\
\mathbf{X}_2^{[1]} &= \xi_2\partial_x + \frac{1}{2}(\eta_2\partial_f - \eta_1\partial_g + \eta_2'\partial_{f'} - \eta_1'\partial_{f'}) ,
\end{aligned} \tag{43}$$

must satisfy the equation

$$\begin{aligned}
\mathbf{X_1}^{[1]}\mathscr{L}_1 - \mathbf{X_2}^{[1]}\mathscr{L}_2 + (D\xi_1)\mathscr{L}_1 - (D\xi_2)\mathscr{L}_2 = DG^1 , \\
\mathbf{X_1}^{[1]}\mathscr{L}_2 + \mathbf{X_2}^{[1]}\mathscr{L}_1 + (D\xi_1)\mathscr{L}_2 + (D\xi_2)\mathscr{L}_1 = DG^2 ,
\end{aligned} \tag{44}$$

for some gauge functions G^1, G^2, where $D = d/dx$. It might seem that the arbitrariness of the gauge functions allows infinitely many solutions and hence the "must satisfy" says nothing. This is not the case. In the scalar case one is requiring that the left side of the equation be an exact differential. For the coupled system, one is demanding that both left sides be total differentials, albeit of different "potentials." The resulting invariants are:

$$\begin{aligned}
I_1 &= \xi\mathscr{L}_1 - \xi_2\mathscr{L}_2 + \frac{1}{2}(\eta_1 - f'\xi_1 + g'\xi_2)\Big(\frac{\partial \mathscr{L}_1}{\partial f'} + \frac{\mathscr{L}_2}{\partial g'}\Big) \\
&\quad - \frac{1}{2}(\eta_2 - f'\xi_2 - g'\xi_1)\Big(\frac{\partial \mathscr{L}_2}{\partial f'} - \frac{\mathscr{L}_1}{\partial g'}\Big) - B_1 , \\
I_2 &= \xi\mathscr{L}_2 + \xi_2\mathscr{L}_1 + \frac{1}{2}(\eta_1 - f'\xi_1 + g'\xi_2)\Big(\frac{\partial \mathscr{L}_2}{\partial f'} - \frac{\mathscr{L}_1}{\partial g'}\Big) \\
&\quad + \frac{1}{2}(\eta_2 - f'\xi_2 - g'\xi_1)\Big(\frac{\partial \mathscr{L}_1}{\partial f'} + \frac{\mathscr{L}_2}{\partial g'}\Big) - B_2 .
\end{aligned} \tag{45}$$

The invariants of complex scalar second-order ODEs are often easier to obtain than those of a two-dimensional real system [37]. The question arises, as with complex linearization so with invariants, are they found for systems that could not be obtained for the two-dimensional system? With the question in mind of why the complex method is providing results that the real system did not, it is necessary to pursue the matter further.

In the simplest case, $y'' = xy'$, the complex method merely reproduces the results for the real system, albeit more simply. In the case of the complex simple harmonic oscillator it correctly gives a coupled system of harmonic oscillators [38] and provides the expression for the energy transferring back and forth between the two. As was put there, *one sees the energy in the field by putting on complex glasses*. It is found that new invariants arise for the complex Lagrangian in some cases. Unfortunately it was expressed in [37] in a way that misleadingly suggests that there are two Lagrangians for real two-dimensional systems arising from a variational principle. We present an example.

Example Consider the system of two second-order semi-linear ODEs

$$f'' + 3ff' - 3gg' + f^3 - 3fg^2 = 0\,,$$
$$g'' + 3fg' + 3gf' + 3f^2g - g^3 = 0\,. \tag{46}$$

It is not clear that it has any Lagrangian. If there is no ordinary conservation law arising from a variational principle, one can still get a conserved quantity (the generalization of the Hamiltonian) from what are called *partial Lagrangians* [39] or there may be nothing like a Lagrangian. It would be worth exploring which of the alternatives applies in this example and in general. This system corresponds to the complex ODE

$$y'' + 3yy' + y^3 = 0\,, \tag{47}$$

which has five infinitesimal Noether symmetry generators and the corresponding five invariants which split into ten real invariants for the system. It is noted in [37] that the two parts of the Lagrangian are equivalent Lagrangians for the system, which yields only one invariant. It would be interesting to explore if this spew of invariants is related to the spew of infinitesimal generators spawned by the split translation generator of the complex line.

6 Iterative Splitting of a Complex Scalar ODE

The idea of iterative splitting [40] is a strange one: (a) start with a complex (say) scalar ODE and obtain the split two-dimensional system of ODEs; (b) now get hit on the head and develop amnesia, so you forget where the split system came from and use the splitting procedure on it to get a four-dimensional system of ODEs; or (c) get a four-dimensional system of PDEs if you forgot that you intended to restrict yourself to ODEs. Why is the idea strange? To see this we have to get into what was not discussed before: the range of functions to which the procedures are applicable.

When we proceed for the splitting, we assume that the dependent variables are *complex analytic* functions and that the functions in the split system are *real analytic* functions. Now the ratio of the cardinality of the set of all complex analytic functions to the set of all complex functions is a second infinitesimal. Similarly, for the real analytic to all real functions. However, it does not appear that in going complex we have restricted our space "any more" than we have done for the real. In fact one feels that we have somehow made it "more general." This vague feeling lulls us into a false sense of security, as we see when we require the CR-equations. When we repeat the step of splitting we have required that the two dependent variables be *complex analytic functions themselves*. This obviously significantly restricts the space of permissible functions after the split. This will appear in the emergence of a second set of CR-equations. Obviously, there will be infinitely many functions that satisfy the requirement but the restriction on the space of permissible functions will make a big difference for what can be used in DEs. In our amnesia we have wandered into a cave with a narrow opening containing a magic lamp. To get out of the cave we may have to leave our magic lamp of splitting behind. The trick will be to bring a more constrained genie out without the lamp. Let us be more concrete. For complex symmetry analysis for ODEs, we need that f and g be n times differentiable functions of x and w for a complex analytic function $(f, g, f', g', \ldots, f^{(n)}, g^{(n)})$ and for PDEs that $y = f + \iota g$ be a complex analytic function of $z = u + \iota v$ and w be a complex analytic function of $(z, y, y', \ldots, y^{(n)})$.

One might have thought of generalizing the complexification of the DE to the quaternions, $q = 1 + ai + bj + ck$, subject to the requirements that $i^2 = j^2 = k^2 = -1$ and $ij = k = -ji$, $jk = i = -kj$, $ki = j = -ik$. It is easily verified that the requirements that $dq/dq = 1$ and $dq^2/dq = 2q$ are incompatible. Thus we cannot bound up the steps from one to four in a single leap and need to look elsewhere for a generalization. Why generalize? Apart from the search for simpler ways to get more powerful results, one wants to obtain insights into the working of the first step by going beyond. As mentioned earlier, the idea is to complexify twice over. In view of the important role of the CR-equations for double splitting, it is worthwhile to state them explicitly for the first splitting. For the initial complex scalar ODE (28), taking $z = x$ and proceeding with the split $y = f + \iota g$, we obtain a pair of ODEs as given before. Now we must also write

$$\omega(x; f, g; f', g') = w^r(x; f, g; f', g') + \iota w^i(x; f, g; f', g') . \tag{48}$$

Then the CR-equations are:

$$w^r_f = w^i_g \, , w^r_g = -w^i_f \, ; w^r_{f'} = f^i_{g'} \, ; w^r_{g'} = -w^i_{f'} \, . \tag{49}$$

It is easier to obtain four-dimensional systems of ODEs or PDEs by double splitting than a three-dimensional system because the number of equations would naturally be even. One can retain one of the functions of the split to be real and the other to be complex, so as to get the desired three-dimensional system, but the number of functions still remains even. To circumvent this problem, retain f as it is but split $g = h + \iota k$ in (28) and take all real terms that do not contain g or its derivative in one term and the rest in a second, complex, term

$$\omega(x, y, y') = w(x; f; f') + W(x; f, g; f', g') . \tag{50}$$

Now split W to write

$$W(x; f, g; f', g') = U(x; f, h, k; f', h', k') + \iota\, V(x; f, h, k; f', h', k') , \tag{51}$$

so that we obtain the three-dimensional second-order system:

$$\begin{aligned} f'' &= w(x; f; f') , \\ h'' &= U(x; f, h, k; f', h', k') , \\ k'' &= V(x; f, h, k; f', h', k') , \end{aligned} \tag{52}$$

subject to the CR-equations

$$U_h = V_k\ ,\ U_k = -V_h\ ;\ U_{h'} = V_{k'}\ ,\ U_{k'} = -V_{h'}\ . \tag{53}$$

Notice that the system of three coupled ODEs does not seem very general, as the first of (53) is independent of the other two dependent variables. However, it is not entirely clear how much of a restriction this is. We could try to take linear combinations of the three dependent variables so that in one equation we eliminate the other two. The problem is reminiscent of finding the Jordan canonical form and may need the symmetry structure of the system to be examined for the purpose. Incidentally, the second split given in [40] causes confusion by using ιW instead of W in (50) but is entirely equivalent to the one presented here.

We present an illustrative example here:

Example The system of generalized Emden–Fowler ODEs:

$$f'' = -2x^{-5}hk\ ,\ h'' = -2sx^{-5}fk\ ,\ k'' = 2x^{-5}fh\ , \tag{54}$$

corresponds to the completely integrable [41] scalar Emden–Fowler ODE

$$y'' = x^{-5}y^2\ , \tag{55}$$

subject to the algebraic constraint

$$f^2 + h^2 = k^2 , \tag{56}$$

which has the symmetry generators

$$\mathbf{X}_1 = x\frac{\partial}{\partial x} + 3y\frac{\partial}{\partial y} , \; \mathbf{X}_2 = x^2\frac{\partial}{\partial x} + xy\frac{\partial}{\partial y} \tag{57}$$

and is completely integrable. Double-splitting these symmetry generators yields eight Lie-like operators and *no* symmetries of the system. These are

$$Y_1 = x\partial_x + \frac{3}{2}f\partial_f + \frac{3}{4}h\partial_h + \frac{3}{4}k\partial_k , \; Y_2 = \frac{3}{4}k\partial_h - \frac{3}{4}h\partial_k ,$$

$$Y_3 = \frac{3}{2}k\partial_f + \frac{3}{4}f\partial_k , \; Y_4 = \frac{3}{2}h\partial_f - \frac{3}{4}f\partial_h,$$

$$Y_5 = x^2\partial_x + \frac{1}{2}xf\partial_f + \frac{1}{4}xh\partial_h + \frac{1}{4}xk\partial_k , \; Y_6 = \frac{1}{4}xk\partial_h - \frac{1}{4}xh\partial_k ,$$

$$Y_7 = \frac{1}{2}xk\partial_f + \frac{1}{4}xf\partial_k , \; Y_8 = \frac{1}{2}xh\partial_f - \frac{1}{4}xf\partial_h .$$

Our system (54) is completely integrable despite having too few symmetries. Cases with no symmetry were already seen above [26], but the first example of a two-dimensional integrable system with no symmetry was given in [42]. Here we have an integrable system of *three* ODEs with only two symmetries. The first integral of (55) is given by

$$I = \frac{1}{2}x^2y'^2 + \frac{1}{2}y^2 - \frac{1}{3}x^{-3}y^3 - xyy' ,$$

which has the symmetry X_2 [43]. The invariant obtained from it, $v = y/x$ reduces the equation to a simple quadrature as

$$x^2v' = \pm\sqrt{c + \frac{2}{3}v^3},$$

yielding the solution of (54). It is worth noting that the Lie-like operators have proliferated on double splitting and are likely to increase still more for further splits. The Lie symmetries seem lost in the abundance of Lie-like operators.

Put $f(x) = k(x) + \iota\, l(x)$ and $g(x) = m(x) + \iota\, n(x)$ in (48) to obtain the four-dimensional system of ODEs by double-splitting,

$$\begin{aligned} w^r(x; f, g; f', g') &= u^r(x; \mathbf{k}; \mathbf{k}') + \iota v^r(x; \mathbf{k}; \mathbf{k}') , \\ w^i(x; f, g; f', g') &= u^i(x; \mathbf{k}; \mathbf{k}') + \iota v^i(x; \mathbf{k}; \mathbf{k}') , \end{aligned} \tag{58}$$

where $\mathbf{k} := (k, l, m, n)$, yielding the system of four ODEs

$$k''(x) = u^r(x; \mathbf{k}; \mathbf{k}') \, , \ l''(x) = u^i(x; \mathbf{k}; \mathbf{k}') \, ,$$
$$m''(x) = v^r(x; \mathbf{k}; \mathbf{k}') \, , \ n''(x) = v^i(x; \mathbf{k}; \mathbf{k}') \, , \tag{59}$$

subject to the CR-conditions

$$u^r_k + v^r_l = u^i_m + v^i_n \, , \ u^r_l - v^r_k = u^i_n - v^i_m \, ,$$
$$u^r_m + v^r_n = -u^i_n - v^i_l \, , \ u^r_n - v^r_m = -u^i_l + v^i_k \, ,$$
$$u^r_{k'} + v^r_{l'} = u^i_{m'} + v^i_{n'} \, , \ u^r_{l'} - v^r_{k'} = u^i_{n'} - v^i_{m'} \, ,$$
$$u^r_{m'} + v^r_{n'} = -u^i_{k'} - v^i_{l'} \, , \ u^r_{n'} - v^r_{m'} = -u^i_{l'} + v^i_{n'} \, . \tag{60}$$

The prolonged symmetry generator can now be written as

$$\mathbf{X} = \xi(x, \mathbf{k})\frac{\partial}{\partial x} + \underline{\eta}(x, \mathbf{k}).\nabla_{\mathbf{k}} + \underline{\eta}^{[1]}(x; \mathbf{k}, \mathbf{k}').\nabla_{\mathbf{k}'} \, . \tag{61}$$

Writing this equation out in detail makes it too unwieldy to convey much wisdom.

Let us now come to the system of four PDEs. This is the most straightforward of the various possibilities considered. At the first step we regard both the independent and the dependent variables of (28) as complex, so that we take z instead of x there and do the usual split with $z = s + \iota t$, so that *both the independent and dependent variables are split*. This gives a system of two second-order PDEs for two functions of two variables. This is the standard complex symmetry analysis talked of earlier for PDEs. The double split repeats the process and yields a system of four PDEs of four variables. Due to the number of variables involved in the double split it becomes impossible to follow our notation above here and we copy the equations as given in [40], including the CR-conditions and the prolonged generator.

$$\begin{aligned}
& w_{ss} - w_{tt} + 2x_{st} - w_{uu} + w_{vv} - 2x_{uv} + 2y_{su} - 2y_{tv} \\
& +2z_{sv} + 2z_{tv} = 4g(\mathbf{s}; \mathbf{w}, \nabla_{\mathbf{s}}\mathbf{w}); \\
& x_{ss} - x_{tt} - 2w_{st} - x_{uu} + x_{vv} + 2w_{uv} + 2z_{su} - 2z_{tv} \\
& -2y_{sv} - 2y_{tv} = 4h(\mathbf{s}; \mathbf{w}, \nabla_{\mathbf{s}}\mathbf{w}); \\
& y_{ss} - y_{tt} + 2z_{st} - y_{uu} + y_{vv} - 2z_{uv} + 2w_{su} - 2w_{tv} \\
& +2x_{sv} + 2x_{tv} = 4k(\mathbf{s}; \mathbf{w}, \nabla_{\mathbf{s}}\mathbf{w}); \\
& z_{ss} - z_{tt} - 2y_{st} - z_{uu} + z_{vv} + 2y_{uv} + 2x_{su} - 2x_{tv} \\
& -2w_{sv} - 2w_{tv} = 4l(\mathbf{s}; \mathbf{w}, \nabla_{\mathbf{s}}\mathbf{w});
\end{aligned} \tag{62}$$

subject to the CR-conditions

$$
\begin{aligned}
w_s + x_t = y_u + z_v \, , \; w_t - x_s = y_v - z_u \, , \\
w_u + x_v = -y_s - z_t \, , \; w_v - x_u = -y_t + z_s \, ; \\
g_s + h_t = k_u + l_v \, , \; g_t - h_s = k_v - l_u \, , \\
g_u + h_v = -k_s - l_t \, , \; g_v - h_u = -k_t + l_s \, ; \\
g_w + h_x = k_y + l_z \, , \; g_x - h_w = k_z - l_y \, , \\
g_y + h_z = -k_w - l_x \, , \; g_z - h_y = -k_x + l_w \, .
\end{aligned}
\tag{63}
$$

The derivatives in the rest of the CR-conditions can be written in more familiar form using the variables

$$
\begin{aligned}
\alpha = w_s + x_t + y_u + z_v \, , \; \beta = w_t - x_s + y_v - z_u \, ; \\
\gamma = w_u + x_v - y_s - z_t \, , \; \delta = w_v - x_u - y_t + z_s \, ,
\end{aligned}
\tag{64}
$$

so that the rest of the CR-conditions are

$$
\begin{aligned}
g_\alpha - h_\beta = k_\gamma - l_\delta \, , \; g_\beta + h_\alpha = k_\delta + l_\gamma \, ; \\
g_\gamma - h_\delta = -k_\alpha + l_\beta \, , \; g_\delta + h_\gamma = -k_\beta - l_\alpha \, .
\end{aligned}
\tag{65}
$$

The prolonged symmetry generator for the system is

$$
\mathbf{X}^{[1]} = \underline{\xi}(\mathbf{s}, \mathbf{g}).\nabla_{\mathbf{s}} + \underline{\eta}(\mathbf{s}, \mathbf{g}).\nabla_{\mathbf{g}} + \underline{\eta}^{[1]}(\mathbf{s}, \mathbf{g}, \nabla_{\mathbf{s}}\mathbf{g}).\nabla_{\nabla_{\mathbf{s}}\mathbf{g}} \, . \tag{66}
$$

We again rely on an example to illustrate our system.

Example The free-particle system of equations is given by (62), with the right side set equal to zero. The CR-conditions are trivial. There are now 32 Lie-like operators, of which only 24 are symmetry generators. As before the local projective symmetries (eight in all) are lost, but the dilations are *not* lost. The generic problem of PDEs of an infinity of symmetry generators, persists but the 24 symmetries *do* form a Lie algebra, $\mathscr{A}_{24}$ say, which serves as a "core" for the system of PDEs, in that one could write the full Lie algebra $\mathscr{A} = \mathscr{A}_\infty \bigoplus \mathscr{A}_{24}$ and "throw away" the infinite dimensional algebra $\mathscr{A}_\infty$, to be left with a solution with 24 arbitrary constants.

Notice that for the three-dimensional system of ODEs, one could reverse the order of taking the split into two and one, to get a "dual" system. This would not be so simple for the split into a four-dimensional system, and there would be no obvious "dual." One could do the split first and then two "singles" after that; a "single," a split and a "single"; or two "singles" and a split. All would yield four ODEs for functions of one variable, but they would all be different. The first and third would, in some sense, be "duals." The same applies for the PDEs. In fact, the complex method for ODEs is not unique. Instead of first restricting the complex independent variable to be real, we could have first split and then restricted. The results would not be the same.

7 Discussion and Conclusion

In this chapter we reviewed the developments in Lie symmetry analysis that made explicit use of the complex analyticity of the solutions of complex differential equations. It might be recalled that Lie, himself, had assumed that the functions were complex analytic, but had not made the requirement explicit. So long as one remains entirely in the complex domain nothing new *can* emerge from the discussion. It is only when one splits the dependent and independent variables into their real and imaginary parts that the new features arise. In that case, a complex scalar ODE yields a pair of PDEs for two real functions of two real variables. As could be expected, the really new features arise when the independent variable is restricted to the real domain, which is needed to obtain ODEs. The areas where we particularly explored the consequences of the complex methods were linearization and Noether invariants. This is not to say that there are no consequences for more general situations, or that they would turn out to be less interesting, novel or useful. It is simply that these were the easiest to tackle, and hence provided a quick check on whether anything new would arise. In fact, there is reason to expect, as we shall discuss shortly, that the more general cases will lead to even more unexpected results. After all, if the complex linearizable ODE leads to the solution of ODEs not amenable to solution by symmetry methods, how many more may become solvable if the complex ODE is solvable, even if it is not linearizable?

In the applications to linearization, we discussed only the complex *scalar* ODE split to get a pair of real ODEs. By geometric methods one obtains the maximum symmetry case of linearizable systems. Using complex methods, two more of the five classes were accessed. Here, we have indicated that it should be possible to access the remaining two classes by *not* using the optimal canonical forms for the complex scalar ODE, as the restriction to the real independent variable (required for obtaining ODEs rather than PDEs by splitting) does not commute with the splitting procedure. This would be worth pursuing. However, the entire discussion is limited to a two-dimensional system. For higher (even) dimensional systems, we can split a higher, say m, dimensional complex system to a $2m$-dimensional system. This has been done for a two-dimensional system split to a four-dimensional system in [44]. It would be important to investigate if all the linearizable classes for the four-dimensional system are obtainable by complex methods, using the same point of avoiding the use of the optimal canonical form. More generally, if the m-d system split to the $2m$-d system covers all linearizable classes of the $2m$-d system. Further, one would need to see whether the double splitting of a scalar ODE yields the same results as the single split of the two-dimensional system. If not, is one of them more restrictive than the other, or is it that both methods give different extensions with some overlap? For odd dimensional systems, we have seen that one can appeal to iterative methods. However, it is worth exploring if direct algebraic constraints could also provide the desired linearization. Again, it would be fascinating to look for the connection between the two methods of enlarging the systems to which it is applied.

Complex methods have also been applied to the linearization of scalar third-order ODEs to deal with a two-dimensional system of third-order ODEs [45, 46] and a classification of two-dimensional linearizable systems of third-order ODEs has been obtained. This is a much harder problem as the ODEs and systems are not apparently connected to geodesic equations and hence to the geometrical methods. A method was developed to reduce the order of the ODE by defining a derivative of the dependent variable as a new dependent variable [47–49], thereby providing a possible connection with a system of geodesic equations, provided it satisfies the required criteria. Going one step further, one could ask if the second-order two-dimensional system could be obtained from a complex scalar ODE, so that the third-order ODE could be treated related to a second-order scalar ODE that could correspond to a geodesic equation? It is by no means clear that this could be done, but it seems very interesting to pursue this line of inquiry further. The procedure mentioned here was also used to reduce fourth-order ODEs to two-dimensional systems of second-order ODEs, albeit there is no classification for them. Of course, the above question would be as interesting for these equations as well. It was noted that the above procedure amounted to using contact transformations for third- or fourth-order ODEs and this provided the first classification for linearization by contact symmetries. It would be most interesting to see what would happen if one used complex methods for the contact transformations. The further ramifications involving iterative splitting may help provide insights into how the various methods, including contact and Lie–B*ä*cklund transformations are interconnected.

So far we have concentrated on reviewing the developments arising from complex methods that were useful but not really discussed the odd features, that turn up when we use the methods. This happens marginally in the first split, where the Lie operators are lost and what were called "Lie-like" which we called "complex-Lie," operators, replace them. Similarly for the Noether symmetries and integrals. This occurs more dramatically when iterative splitting is used. It is worthwhile to pursue this odd feature further. In this chapter we suggested that there may be a connection with the enormous proliferation of symmetry generators for the simple symmetries of the complex line: translation and scaling. It is worth pursuing precisely *how much* the proliferation is. Consider a complex scalar ODE, the translation splits into two and the rotation into four, giving a total of six. Now, at the second iteration, the two translations split into four and the four scaling-type operators split into sixteen, giving a total of twenty. In general, for n iterations we get the total number $N = 2^n + (2^n)^2 = 2^n(2^n + 1)$. Thus for $n = 3,\ N = 72$ and for $n = 4,\ N = 272$. Following the same logic, starting with an m-d complex system of ODEs, there will be initially m translations and m^2 scaling-type generators, which gives the general formula $N = 2^n m(2^n m + 1)$ after n iterations. Hence for a starting two-dimensional system, the effect is simply like increasing the number of iterations by one. Starting with $m = 3$ at two iterations, already $N = 156$. *This* is the extent to which the generators proliferate after n iterations. But the question then is "what difference does this proliferation make?" We will now consider the possibility that this may provide a clue to answer the big question that was mentioned at the start of this chapter: "*Why* does the complex method work?"

Take the case of linearization. Without the use of complex methods the geometric methods only give the *maximal* symmetry case. Complex methods have already provided two of the five linearizable classes for two-dimensional second-order systems and there is good reason to expect that the other two will also be found. To this extent, it is not all that strange that they work. After all, they only give what had been obtained by classical methods, albeit very much more explicitly. However, when complex linearization provides solutions for systems that are not linearizable, one really needs to explain how *that* could ever happen. Even stranger, where the number of symmetry generators of the system are inadequate to solve by symmetry methods, *how on Earth* can the complex methods work their magic there? The answer may lie in the much larger space of these "quasi-symmetry generators." One starts with the inadequately symmetric real system and "lifts" it as a lower-dimensional complex system, where the lifted equation is adequately symmetric and solves it there. However, the linearizing transformation that yields the solution changes the restricted real variable to a complex one, so that it cannot be used to linearize the system. So far it seems reasonable. The question now is, "why is it that there is at least one solution that is retained when one restricts the variable to be real? Why is it not that there is *no* solution for the real system?" An associated question is, "we have brought down one solution, but how do we know that we have not missed other solutions that could have been found?" For a $2m$-d real system of second-order ODEs, we need criteria that tell us precisely how many of the $2m$ complex solutions can be "brought down to the real world." No such criteria are available and they are crucial for using the complex methods to their full potential.

Now let us discuss the odd features of the complex variational principle. It has been noted that the real and imaginary parts of the complex Lagrangian are not the same but are equivalent, in that they satisfy the same Euler–Lagrange equations. Why should that be so? The final quantity that is minimized is the magnitude of the Lagrangian and not its real and imaginary parts. How is it that the two separately "know" that they must satisfy the same equations? Presumably it is because, apart from a constant value, the sum of the squares of the two parts has to become zero and that can only be if each is separately zero. However, the question then is "If they have to satisfy the same equations, why do they differ?" Again, the Noether invariants obtained from complex Lagrangians are many more than would have been expected at first sight. Essentially this must come from the tremendous enlargement of the quasi-symmetry operator space. The same point of obtaining many more invariants for the real system than should be possible, appears for the invariants. As such, the same questions as for linearization need to be answered. Again the explicit criteria are needed.

We have not discussed complex methods for PDEs. It is not that they cannot be used there. They *are* so used. They can be used to obtain systems of PDEs from ODEs. Nor is it that they are not useful. They pick out a "core" finite-dimensional symmetry algebra from an infinite dimensional Lie-algebra. The thing is that it is not so clear what relevance the PDEs would have for the big question asked here.

Acknowledgements We are grateful to Rimsha Khalil for pointing out some typographical errors in the chapter. FMM is grateful to the NRF of South Africa for support.

References

1. Lie S (1967) Differential equations. Chelsea, New York
2. Lie S (1883) Klassification und integration von gewönlichen differentialgleichungen zwischen x, y, die eine gruppe von transformationen gestaten. Arch Math
3. Lie S (1891) Lectures on differential equations with known infinitesimal transformations. Leipzig, Teubner (in German, Lie's lectures by Sheffers G)
4. Lie S (1893) Theorie der Transformationsgruppen. III. Teubner, Leipzig
5. Ibragimov NH (1999) Elementary Lie group analysis and ordinary differential equations. Wiley, Chichester
6. Stephani H (1996) Differential equations: their solutions using symmetries. Cambridge University Press, Cambridge
7. Tressé A (1894) Sur les invariants differentiels des groupes continus de transformations. Acta Math 18:1
8. Chern SS (1937) Sur la geometrie d'une equation differentielle du troiseme orde. CR Acad Sci Paris, 1227; (1940) The geometry of the differential equation $y''' = F(x, y, y', y'')$. Sci Rep Nat Tsing Hua Univ. 4, 97
9. Mahomed FM, Leach PGL (1990) Symmetry Lie algebra of nth. order ordinary differential equations. J Math Anal Appl 151:80
10. Grebot G (1997) The linearization of third order ODEs; the characterization of third order ordinary differential equations admitting a transitive fibre-preserving point symmetry group. J Math Anal Appl 206:364
11. Wafo Soh C, Mahomed FM (2001) Linearization criteria for a system of second-order ordinary differential equations. Int J Nonlinear Mech 36:671
12. Gorringe VM, Leach PGL (1988) Lie point symmetries for systems of 2nd order linear ordinary differential equations. Quaestiones Mathematicae 11:95
13. Wafo Soh C, Mahomed FM (2000) Symmetry Breaking for a System of Two Linear Second-Order Ordinary Differential Equation. Nonlinear Dyn 22:121
14. Qadir A (2020) Einstein's general theory of relativity. Cambridge Scholars Publishers, Cambridge
15. Aminova AV, Aminov NA-M (2000) Projective geometry of systems of differential equations: general conception. Tensor N S 62:65; (2006) Projective geometry of systems of second-order differential equations. Sbornik: Math 197:951
16. Feroze T, Mahomed FM, Qadir A (2006) The connection between isometries and symmetries of geodesic equations of the underlying spaces. Nonlinear Dyn 45:65
17. Mahomed FM, Qadir A (2007) Linearization criteria for a system of second order quadratically semi-linear ordinary differential equations. Nonlinear Dyn 48:417
18. Bokhari AH, Qadir A (1985) A prescription for n-dimensional vierbeins. ZAMP 36:184
19. Mahomed FM, Qadir A (2009) Invariant linearization criteria for systems of cubically semi-linear second order ordinary differential equations. J Nonlin Math Phys 16:1
20. Fredericks E, Mahomed FM, Momonia E, Qadir A (2008) Constructing a space from the system of geodesic equations. Comp Phys Commun 179:438
21. Ali S (2009) Complex symmetry analysis. PhD thesis, NUST CAMP
22. Ali S, Mahomed FM, Qadir A (2009) Complex Lie symmetries for scalar second-order ordinary differential equations. Nonlinear Anal: Real World App 10:3335
23. Safdar M (2003) Solvability of differential equations by complex symmetry analysis. PhD Thesis, NUST CAMP

24. Safdar M, Qadir A, Ali S (2011) Inequivalence of classes of linearizable systems of cubically semi linear ordinary differential equations obtained by real and complex symmetry analysis. Math Comp Appl 16:923
25. Safdar M, Qadir A, Ali S (2011) Linearizability of systems of ordinary differential equations obtained by complex symmetry analysis. Math Problems Eng 2011:171834
26. Ali S, Safdar M, Qadir A (2014) Linearization from complex Lie point transformations. J Appl Math 2014:793247
27. Dutt HM, Safdar M (2015) Linearization of two-dimensional complex linearizable system of second order ordinary differential equations. Appl Math Sci 9:2889
28. Noether E (1918) Invariante Variationsprobleme Nachrichten von der Gesellschaft der Wissenschaften zu Göttingen. Mathematisch-Physikalische Klasse. 1918:235
29. Kosmann-Schwarzbach Y (2010) The Noether theorems: invariance and conservation laws in the twentieth century. Sources and Studies in the History of Mathematics and Physical Sciences. Springer, Berlin
30. Kara AH, Vawda FE, Mahomed FM (1994) Symmetries of first integrals and solutions of differential equations. Lie Groups Appl 1:27
31. Bokhari AH, Al-Dweik AY, Zaman FD, Kara AH, Mahomed FM (2010) Generalization of the double reduction theory. Nonlinear Analysis B 11:3763
32. Ali S, Mahomed FM, Qadir A (2008) Complex Lie symmetries for variational problems. J Nonlin Math Phys 25:25
33. Zimmerman AH (1987) Generalized integral transformations. Dover Publications, Mineola
34. Tassaddiq A (2012) Some representations of the extended Fermi-Dirac and Bose-Einstein functions with application. NUST CAMP
35. Tassaddiq A, Qadir A (2011) Fourier transform and distributional representation of the generalized gamma function with some applications. Appl Math Comp 218:1084
36. Bender CM, Boetcher S (1998) Real spectra in non-Hermitian Hamiltonians having PT symmetry. Phys Rev Lett 80:5243
37. Safdar M, Qadir A, Farooq MU (2019) Comparison of Noether symmetries and first integrals of two-dimensional systems of second order ordinary differential equations by real and complex methods. Symmetry 11:1180
38. Farooq MU, Ali S, Qadir A (2010) Invariants of two-dimensional systems via complex Lagrangians with applications. Commun Nonlin Sci Num Sim 16:1804
39. Kara AH, Mahomed FM (2006) Noether-type symmetries and conservation laws via partial Lagrangians. Nonlinear Dyn 45:367
40. Qadir A, Mahomed FM (2015) Higher dimensional systems of differential equations obtainable by iterative use of complex methods. Int J Mod Phys (Conf Ser) 38:1560077. In: Ali S, Mahomed FM, Qadir A (eds)
41. Leach PGL, Maartens R, Maharaj SD (1992) Self-similar solutions of the generalized Emden-Fowler equation. Int J Non-Linear Mech 27:583
42. Gonzalez-Gascon F, Gonzalez-Lopez A (1988) Newtonian systems of differential equations, integrable via quadratures, with trivial group of point symmetry. Phys Lett A 129:153
43. Mahomed KS, Momoniat E (2014) Symmetry classification of first integrals for scalar dynamical equations. Int J Non-Linear Mech 59:52
44. Safdar M, Ali S, Mahomed FM (2011) Linearization of systems of four second order ordinary differential equations. Pramana - J Phys 77:171834
45. Dutt HM, Safdar M, Qadir A (2019) Linearization criteria for two dimensional systems of third order ordinary differential equations by complex approach. Arabian J Math 8:163
46. Dutt HM, Safdar M, Qadir A (2018) Classification of two-dimensional linearizable systems of third order ordinary differential equations by complex methods. In: Aslam MJ, Saifullah K (eds) Proceedings of the 14th regional conference in mathematical physics, pp 278–283, 2015. World Scientific
47. Dutt HM, Qadir A (2014) Reduction of fourth order ordinary differential equations to second and third Lie linearizable forms. Comm Nonlin Sci Numerical Sim 19:2653

48. Dutt HM, Qadir A (2017) Classification of scalar third order ordinary differential equations linearizable via generalized contact transformations. Quaestiones Mathematicae 2017:1
49. Dutt HM, Qadir A (2018) Classification of scalar fourth order ordinary differential equations linearizable via generalized Lie-Bäcklund transformations. In: Kac VG, Olver PJ, Winternitz P, Özer T (eds)Proceedings of the symmetries, differential equations and applications III, pp 67–74. Springer Proceedings in Mathematics & Statistics

Symmetry Analysis and Conservation Laws of a Family of Boussinesq Equations

M. S. Bruzón and M. L. Gandarias

Abstract This chapter presents a generalization depending on an arbitrary function $f(u)$ of a sixth-order Boussinesq equation which arises in shallow water waves theory from the point of view of the theory of symmetry reductions in partial differential equations. The reductions to ordinary differential equations are derived from the optimal system of subalgebras. In order to obtain exact solutions, we apply a direct method: a catalogue of exact solutions are given and a set of solitons, kinks, antikinks, and compactons are derived. Conservation laws for this equation are constructed. We have obtained a triple reduction to a fourth-order autonomous equation by combining first integrals, which are obtained from two of the conservation laws.

1 Introduction

In this chapter, we consider the sixth-order Boussinesq

$$u_{tt} - c_0^2 u_{xx} - (f'(u) + b_1 u_{tt} - b_2 u_{xx} + d u_{xxxx})_{xx} = 0 \tag{1}$$

with $f''(u) \neq 0$. This equation appeared in [16] where the authors agreed to call the Boussinesq paradigm a set of equations that simultaneously contains the following properties:

(i) bi-directionality of the wave solutions (propagation to the left and to the right; presence of a d'Alembertian operator);
(ii) nonlinearity of any order;
(iii) dispersion of any order, the latter resulting in the presence of combined space and time derivatives of the fourth order at least.

M. S. Bruzón · M. L. Gandarias (✉)
Department of Mathematics, Faculty of Sciences, University of Cádiz,
Puerto Real, 11510 Cádiz, Spain
e-mail: marialuz.gandarias@uca.es

M. S. Bruzón
e-mail: m.bruzon@uca.es

A. C. J. Luo and R. K. Gazizov (eds.), *Symmetries and Applications of Differential Equations*, Nonlinear Physical Science,
https://doi.org/10.1007/978-981-16-4683-6_5

Generally speaking, $f'(u)$ may be sought of as a polynomial in u, starting with second degree. The original equation had a positive sign in front of the dispersion connected with the fourth-order spatial derivative ($b_1 = 0$, $b_2 < 0$) and turned out to be mathematically improper, being incorrect in the sense of Hadamard (the initial value problem is ill-posed). The reason for this is that in the absence of nonlinearity, small perturbations would amplify as frequency may become imaginary. The equation, known nowadays as Boussinesq equation, is the incorrect equation (called sometimes bad Boussinesq equation), while if a sufficient strong surface tension is added in the model (making the coefficient b_2 positive), the equation is correct (called good Boussinesq equation).

Equation (1) with $c_0 = 1, b_1 = b_2 = 0, d = -1$, and $-f'(u)$ is $f(u)$ has the form

$$u_{tt} - u_{xx} + (f(u) + u_{xx})_{xx} = 0. \tag{2}$$

In [21] Gandarias and Bruzón derived classical and nonclassical symmetries of Eq. (2).

In [11, 12], Bruzón and Gandarias applied a new procedure for finding nonclassical symmetries of the following Boussinesq equation:

$$u_{tt} - uu_{xx} + u_x^2 + u_{xxxx} = 0. \tag{3}$$

The authors extended the algorithm described by Bîlă and Niesen to determine the nonclassical symmetries of a PDE. They observed that for any equation which can be expressed in the form

$$u_t = \mathscr{A}(x, t, u), \tag{4}$$

where $\mathscr{A}$ is an arbitrary function depending on x, t, and u, the nonclassical determining equation can be derived by substituting the corresponding functions $\mathscr{A}$ into the PDE. They applied the described algorithm to a Boussinesq equation and to a 2+1-dimensional shallow water wave equation. They proved that for the 2+1-dimensional shallow water wave equation the method yields a new symmetry reduction which is unobtainable by using Lie classical method.

In [13], Bruzón and Gandarias obtained a complete Lie group classification for the family of Boussinesq equations

$$u_{tt} = au_{xx} + \left(u^{m+1}\right)_{xx} + b\left[u\left(u^m\right)_{xx}\right]_{xx}, \tag{5}$$

where a and b are arbitrary constants. The authors made a full analysis of Eq. (5), by using classical symmetries, nonclassical symmetries and nonclassical potential symmetries, and they obtained new solutions. The authors also obtained some Type-II hidden symmetries [19] of Eq. (5) with $m = 1$ and $a = \lambda^2$.

Equation

$$u_{tt} = cu_{xx} + bu_{xxxx} + au_{xxxxxx} + (f(u))_{xx} \tag{6}$$

admits a Hamiltonian formulation when it is written as an equivalent system. For this equation, Recio, Gandarias, and Bruzón [37] established a point symmetry classification in terms of the function $f(u)$, determining the point symmetry group for all possible nonlinear differentiable functions $f(u)$. They also carried out an analogous classification of conservation laws by employing the multiplier method.

In [24], Gandarias and Bruzón derived all the low-order conservation laws of the Eq. (5) by using the multiplier method. Moreover, they considered potential and nonclassical potential symmetries for some of the associated systems. Taking into account the relationship between symmetries and conservation laws and applying the multiplier method to a reduced ordinary differential equation (ODE), they obtained a second-order ODE and two third-order ordinary differential equations.

Equation (1) with $c_0 = b_1 = b_2 = 0$, $d = -1$, where $-f'(u)$ becomes $f(u)$ and with the additional second term responsible for strong internal damping

$$u_{tt} - u_{txx} + u_{xxxx} - (f(u))_{xx} = g(x) \tag{7}$$

was studied by Gandarias and Rosa [23]. The authors gave the group classification as well as corresponding reduced ODEs.

In this chapter, we study Eq. (1) from the standpoint of the theory of symmetry reductions in PDEs. The fundamental basis of this method is that when a differential equation is invariant under a Lie group of transformations a reduction transformation does exist. This transformation reduces the number of independent variables of the partial differential equation, in particular, we might reduce the partial differential equation into ODEs. These ODEs may also have symmetries that allow us to reduce the order of the equation, and we can integrate to find exact solutions [9, 27, 28, 36].

In [3, 4], an algorithmic method was presented for finding the local conservation laws for partial differential equations with any number of independent and dependent variables. The method does not require the use or existence of a variational principle and reduces the calculation of conservation laws to solving a system of linear determining equations similar to that for finding symmetries, see [1–8]. Many papers using this method have been published (see [21–24]).

In [30], (see also [29]) a theorem on conservation laws for an arbitrary differential equation which does not require the existence of Lagrangian has been proved. This theorem is based on the concept of adjoint equations for nonlinear equations. The self-adjointness condition has been subsequently extended yielding to the nonlinear self-adjointnes condition in [18–20, 25, 26, 29]. After Ibragimov's results, several papers appeared concerned with the nonlinear self-adjointnes condition and its applications to PDEs [31–34].

In [14], Bruzón, Gandarias and Ibragimov by applying the algorithm of Ibragimov obtained conservation laws of a family of generalized thin film equations. In [35], Ibragimov et al. investigated the symmetries and conservation laws for the scalar nonlinear anisotropic wave equations with specific external sources, which involves two arbitrary functions, when the equations in question are nonlinearly self-adjoint. All conservation laws involving the first-order derivatives are constructed for the basic equation using the conservation laws theorem for nonlinearly self-adjoint differential equations.

Conservation laws that are symmetry invariant have some important applications. It is well known that when a differential equation admits a Noether symmetry, a conservation law is associated with this symmetry, and furthermore that a double reduction can be achieved as a result of this association. A more general double reduction method which applies to non-variational PDEs has been developed in [38, 39]. Examples about the method are in [10, 15, 17]. De la rosa, Gandarias, and Bruzón determined the subclasses of a generalized variable-coefficient Gardner equation which are nonlinearly self-adjoint, as well as the multipliers, of Anco and Bluman method. They derived conservation laws by using both methods. The authors showed that some of these conservation laws yields conserved integrals with physical meaning, such as mass and energy. As an example of another application of the conserved vectors, they applied the double reduction method to get exact solutions of the Gardner equation from solutions of a second-order reduced ODE. In [22], Gandarias and Rosa for Eq. (7) derived some non-trivial conservation laws by using the multipliers conservation laws method. Taking into account the relationship between symmetries and conservation laws and by applying the double reduction method, they derived a direct reduction of order of the ODEs and in particular they found a kink solution.

In [2, 5], the relationship between symmetries and conservation laws has been analyzed. By using the direct method of the multipliers, Anco developed symmetry properties of conservation laws of PDEs. In particular, the author provided a theoretical framework to comprehend and generalize the method proposed by Sjöberg. In this way, the author proved that conservation laws that are symmetry invariant or symmetry homogeneous have at least one important application: any symmetry-invariant conservation law will reduce to a first integral for the ODE obtained by symmetry reduction of the given PDE when symmetry-invariant solutions $u(x, t)$ are investigated. This provides a direct reduction of order of the ODE.

In [7], Anco and Gandarias provided an explicit algorithmic method to find all symmetry-invariant conservation laws that will reduce to first integrals for the ODE describing symmetry-invariant solutions of the PDE. This significantly generalizes the double reduction method known in the literature. They also proved that if the space of symmetry-invariant conservation laws has dimension $m \geq 1$, then the method yields m first integrals along with a check of which ones are non-trivial via their multipliers. In this work, for Eq. (1), we derive conservation laws by using the direct method of the multipliers. Moreover we will directly derive all the conservation laws which are invariant under translations, then a set of first integrals are obtained which allows for further reduction of the ODE. This method yields a direct triple reduction from a single symmetry.

The structure of the chapter is as follows. In Sect. 2, we have applied the Lie group method of infinitesimals transformations to the sixth-order Boussinesq (1) and we have reported its reductions obtained from the optimal system of subalgebras. In

Sect. 3 in order to obtain exact solutions, by applying a direct method we find the functions $f'(u)$ for which Eq. (1) admits traveling wave solutions. For these functions we obtain exact solutions of Eq. (1). Moreover, in Sect. 4 the double reduction method have been applied and, of course, their conservation laws have been obtained.

2 Lie Symmetries

To apply the classical method to (1) we consider the one-parameter Lie group of infinitesimal transformations in (x, t, u) given by

$$\begin{aligned} x^* &= x + \varepsilon\xi(x, t, u) + O(\varepsilon^2), \\ t^* &= t + \varepsilon\tau(x, t, u) + O(\varepsilon^2), \\ u^* &= u + \varepsilon\eta(x, t, u) + O(\varepsilon^2), \end{aligned} \tag{8}$$

where ε is the group parameter which is generated by the vector field

$$\mathbf{X} = \xi(x, t, u)\partial_x + \tau(x, t, u)\partial_t + \eta(x, t, u)\partial_u. \tag{9}$$

By Criterion of Invariance [36] we require to leave invariant the solution space of Eq. (1). This condition is given by

$$\mathrm{pr}^{(6)}\mathbf{X}\left(u_{tt} - c_0^2 u_{xx} - b_1 u_{ttxx} - b_2 u_{xxxx} - d u_{xxxxxx} - (f'(u))_{xx}\right) = 0 \tag{10}$$

when the equation is satisfied, where $\mathrm{pr}^{(6)}\mathbf{X}$ is the sixth-order prolongation of the vector field (9). This yields to an overdetermined, linear system of equations for the infinitesimals $\xi(x, t, u)$, $\tau(x, t, u)$ and $\eta(x, t, u)$, together with the function $f'(u)$ and the parameters b_1, b_2, d, c_0. Solving the system we obtain the point symmetries admitted by Eq. (1) with $d \neq 0$ and $f'(u), f''(u) \neq 0$. In Table 1, we list the cases for which (1) has symmetries.

In Table 1, $p \neq 0$, $k \neq 0$, a and b are arbitrary constants. On the other hand, symmetries $\mathbf{X}_1$ and $\mathbf{X}_2$ represent space and time-translations, which are admitted for any nonlinearity $f(u)$. The resulting form for invariant solutions is, in general, a traveling wave. In cases 2, 3 and 4 there is an additional admitted symmetry consisting of a scaling combined with a shift, where $f(u)$ is power nonlinearity in terms of the shift on u.

The corresponding symmetry transformation groups are given by

Table 1 Point symmetry classification of Eq. (1)

Case	$F(u) = f'(u)$	Conditions	Symmetry generators
1.	Arbitrary	–	$\mathbf{X}_1 = \partial_x, \mathbf{X}_2 = \partial_t$
2.	$k\,(a+u)^p - b - {c_0}^2 u$	$b_1 = b_2 = 0$	$\mathbf{X}_1 = \partial_x, \mathbf{X}_2 = \partial_t,$ $\mathbf{X}_3 = x\partial_x + 3t\partial_t$ $+ \frac{-4}{p-1}\,(a+u)\partial_u$
3.	$k\,\mathrm{e}^{pu} - {c_0}^2 u + b$	$b_1 = b_2 = 0$	$\mathbf{X}_1 = \partial_x, \mathbf{X}_2 = \partial_t,$ $\mathbf{X}_4 = x\partial_x + 3t\partial_t$ $- \frac{4}{p}\partial_u$
4.	$\ln\left({c_0}^2 + u\right) -$ ${c_0}^2\,(u+b)$	$b_1 = b_2 = 0$	$\mathbf{X}_1 = \partial_x, \mathbf{X}_2 = \partial_t,$ $\mathbf{X}_5 = x\partial_x + 3t\partial_t$ $+ 4(c_0^2 + u)\partial_u$

$$
\begin{aligned}
(t, x, u)_1 &\longrightarrow (t + \varepsilon, x, u),\\
(t, x, u)_2 &\longrightarrow (t, x + \varepsilon, u),\\
(t, x, u)_3 &\longrightarrow (t\mathrm{e}^{3\,\varepsilon}, x\mathrm{e}^{\varepsilon}, \mathrm{e}^{-4\,\frac{\varepsilon}{p-1}}u + \textstyle\int_0^{\varepsilon} -4\,\frac{a}{p-1}\mathrm{e}^{4\,\frac{z_1}{p-1}}\,\mathrm{d}z_1\mathrm{e}^{-4\,\frac{\varepsilon}{p-1}}),\\
(t, x, u)_4 &\longrightarrow (t\mathrm{e}^{3\,\varepsilon}, x\mathrm{e}^{\varepsilon}, -4\,\tfrac{\varepsilon}{p} + u),\\
(t, x, u)_5 &\longrightarrow (t\mathrm{e}^{3\,\varepsilon}, x\mathrm{e}^{\varepsilon}, \mathrm{e}^{4\,\varepsilon}u + \textstyle\int_0^{\varepsilon} 4\,{c_0}^2\mathrm{e}^{-4\,z_1}\,\mathrm{d}z_1\mathrm{e}^{4\,\varepsilon}),
\end{aligned}
$$

with ε the group parameter.

Each admitted point symmetry can be used to reduce Eq. (1) to an ordinary differential equation (ODE) whose solutions correspond to invariant solutions $u(x, t)$ of Eq. (1) under the point symmetry. These invariant solutions are naturally expressed in terms of similarity variables which are found by solving the invariance condition

$$\eta(t, x, u) - \tau(t, x, u)u_t - \xi(t, x, u)u_x = 0. \tag{11}$$

In Table 2, we present the similarity variables and form of the similarity solutions for an optimal set of point symmetry generators for each case in the symmetry classification, where $f'(u) = F(u)$.

In Table 3, we show the reduced equation.

3 Traveling Wave Solutions

In this section, we obtain traveling wave solutions of equation 1 of Table 3 by using a direct method. Integrating this equation twice with respect to z, we get

$$b_1\mu^2 h'''' + (\mu^2 b_1 - \lambda^2 b_2)h'' + (c_0^2\lambda^2 - \mu^2)h + \lambda^2 F(h) + Az + B = 0, \tag{12}$$

Table 2 Form of similarity solutions of Eq. (1)

Case	Optimal symmetry generators	Similarity variable z	Form of $u(t, x)$
1.	$\lambda \mathbf{X}_1 + \mu \mathbf{X}_2$	$\mu x - \lambda t$	$h(z)$
2.	$\lambda \mathbf{X}_1 + \mu \mathbf{X}_2$	$\mu x - \lambda t$	$h(z)$
	$\mathbf{X}_3$	$xt^{-1/3}$	$t^{-\frac{4}{3(p-1)}} h(z) - a$
3.	$\lambda \mathbf{X}_1 + \mu \mathbf{X}_2$	$x - \frac{\mu}{\lambda} t$	$h(z)$
	$\mathbf{X}_4$	$xt^{-1/3}$	$-\frac{4}{p} \ln x + h$
4.	$\lambda \mathbf{X}_1 + \mu \mathbf{X}_2$	$\mu x - \lambda t$	$h(z)$
	$\mathbf{X}_5$	$xt^{-1/3}$	$t^{4/3} h(z) - c_0^2$

Table 3 Reduced equations of Eq. (1)

1.	$-F''(h)\left(h'\right)^2 \lambda^2 - c_0{}^2 \left(h''\right) \lambda^2 - F'(h) h'' \lambda^2$ $- dh''''''\lambda^2 - b_1 h'''' \mu^2 + b_2 h'''' \lambda^2 + h'' \mu^2 = 0$
2.	$-dh''''''\lambda^2 - kp\lambda^2 \left(h'\right)^2 (p-1)(a+h)^{-2+p}$ $- h''\left((a+h)^{p-1} kp\lambda^2 - \mu^2\right) = 0$
	$h'\left(-4p^2 + 4\right) z^{\frac{7p-3}{p-1}}$ $+ \left(9d(p-1)^2 h'''''' + 9kp\left(h'\right)^2 (p-1)^3 (h(z))^{-2+p}\right.$ $\left.+9kph''(p-1)^2 h^{p-1} + (-12p-4)h\right) z^{\frac{6p-2}{p-1}}$ $-(p-1)^2 h'' z^{\frac{8p-4}{p-1}} = 0$
3.	$-d\lambda^2 (a+h)^2 h'''''' + h''\mu^2 (a+h)^2$ $-\lambda^2 p\left((a+h)h'' + \left(h'\right)^2 (p-1)\right) k (a+h)^p = 0$
	$h'' pz^8 + 4h' pz^7 - 9dh'''''' pz^6 - 9\left(h'\right)^2 kp^3 e^{hp} z^2$ $- 9h'' kp^2 e^{hp} z^2 d + 72h' kp^2 e^{hp} z$ $-180kpe^{hp} - 4320 = 0$
4.	$\left(-dh^2 - 2c_0{}^2 dh - c_0{}^4 d\right) h'''''' \mu^6$ $+ \left(\left(-h - c_0{}^2\right) h'' + \left(h'\right)^2\right) \mu^2$ $+ \left(\lambda^2 h^2 + 2\lambda^2 c_0{}^2 h + \lambda^2 c_0{}^4\right) h'' = 0$
	$h^2 h'' z^2 - 4h^2 h' z - 9h^2 h'''''' d + 4h^3 - 9hh''$ $+9\left(h'\right)^2 = 0$

where A and B are integrating constants.

We consider Eq. (12) with $A = B = 0$

$$h'''' + \frac{(\mu^2 b_1 - \lambda^2 b_2)}{b_1 \mu^2} h'' + \frac{(c_0^2 \lambda^2 - \mu^2)}{b_1 \mu^2} h + \frac{\lambda^2}{b_1 \mu^2} F(h) = 0. \tag{13}$$

Equation (13) can be written as

$$h'''' + bh'' + G(h) = 0, \tag{14}$$

where $b = \frac{\mu^2 b_1 - \lambda^2 b_2}{b_1 \mu^2}$ and

$$G(h) = \frac{(c_0^2\lambda^2 - \mu^2)}{b_1\mu^2} h + \frac{\lambda^2}{b_1\mu^2} F(h). \tag{15}$$

Equation (14) has solutions in the form

$$h = \alpha H^{\beta}(z), \tag{16}$$

where α and β are parameters and $H(z)$ is a solution of the Jacobi equation

$$(H')^2 = r + pH^2 + qH^4, \tag{17}$$

with r, p, and q constants. Substituting (16) into (14) and by using the identities which satisfy the standard Jacobi elliptic functions we can obtain an equation in h and $G(h)$. Since these equations are enormously long, we don't include them. From these equations, we can obtain the functions $G(h)$ for the which H is solution of Eq. (14):
Case (i): If $H(z) = \mathrm{sn}(z)$,

$$h(z) = \alpha\, \mathrm{sn}^{\beta}(z). \tag{18}$$

$$F(h) = \alpha_1 h^{\frac{4}{\beta}+1} + \alpha_2 h^{\frac{2}{\beta}+1} + \alpha_3 h^{-\frac{4}{\beta}+1} + \alpha_4 h^{-\frac{2}{\beta}+1} + \alpha_5 h, \tag{19}$$

where

$$\alpha_1 = -\alpha^{-\frac{4}{\beta}}\left[\beta^4 m^4 + \left(6\beta^3 + 8\beta^2 + 4\beta\right) m^3 + \left(3\beta^2 + 2\beta\right) m^2\right], \tag{20}$$

$$\begin{aligned}\alpha_2 = \alpha^{-\frac{2}{\beta}}\Big[&\left(2\beta^4 - 6\beta^3 + 8\beta^2 - 4\beta\right) m^4 + \left(12\beta^3 - 6\beta^2 + 8\beta\right) m^3 \\ &+ \left(2\beta^4 + (6-b)\beta^2\right) m^2 + \left(6\beta^3 + 8\beta^2 + (4-b)\beta\right) m\Big],\end{aligned} \tag{21}$$

$$\alpha_3 = -\alpha^{\frac{4}{\beta}}\beta^4 + 6\alpha^{\frac{4}{\beta}}\beta^3 - 11\alpha^{\frac{4}{\beta}}\beta^2 + 6\alpha^{\frac{4}{\beta}}\beta, \tag{22}$$

$$\alpha_4 = \left(2\alpha^{\frac{2}{\beta}}\beta^4 - 12\alpha^{\frac{2}{\beta}}\beta^3 + 22\alpha^{\frac{2}{\beta}}\beta^2 - 12\alpha^{\frac{2}{\beta}}\beta\right) m^2$$

$$+\left(\alpha^{\frac{2}{\beta}} b - 4\alpha^{\frac{2}{\beta}}\right)\beta + 2\alpha^{\frac{2}{\beta}}\beta^4 - 6\alpha^{\frac{2}{\beta}}\beta^3 + \left(8\alpha^{\frac{2}{\beta}} - \alpha^{\frac{2}{\beta}} b\right)\beta^2 \tag{23}$$

$$+\left(6\alpha^{\frac{2}{\beta}}\beta^3 - 14\alpha^{\frac{2}{\beta}}\beta^2 + 8\alpha^{\frac{2}{\beta}}\beta\right) m, \tag{24}$$

$$\alpha_5 = \left(-\beta^4 + 6\beta^3 - 11\beta^2 + 6\beta\right) m^4 + \left(-6\beta^3 + 14\beta^2 - 8\beta\right) m^3$$

$$+\left(-4\beta^4 + 12\beta^3 + (b-19)\beta^2 + (10-b)\beta\right) m^2 \tag{25}$$

$$+\left(-12\beta^3 + 6\beta^2 + (b-8)\beta\right) m - \beta^4 + b\beta^2. \tag{26}$$

Case (ii): If $H(z) = \mathrm{cn}(z)$,

$$h(z) = \alpha\, \mathrm{cn}^{\beta}(z). \tag{27}$$

$$F(h) = \beta_1 h^{\frac{4}{\beta}+1} + \beta_2 h^{\frac{2}{\beta}+1} + \beta_3 h^{-\frac{4}{\beta}+1} + \beta_4 h^{-\frac{2}{\beta}+1} + \beta_5 h, \tag{28}$$

where

$$\beta_1 = \alpha^{-\frac{4}{\beta}} \beta m^2 \left(\beta^3 m^2 + 6\beta^2 m + 8\beta m + 4m + 3\beta + 2\right), \tag{29}$$

$$\beta_2 = -\alpha^{-\frac{2}{\beta}} \left[\beta m \left(4\beta^3 m^3 - 6\beta^2 m^3 + 8\beta m^3 - 4m^3 + 18\beta^2 m^2 + b \right.\right.$$
$$\left.\left. +2\beta m^2 + 12m^2 - 2\beta^3 m + b\beta m + 6\beta m - 6\beta^2 - 8\beta - 4\right)\right], \tag{30}$$

$$\beta_3 = \alpha^{\frac{4}{\beta}} (\beta - 3)(\beta - 2)(\beta - 1)\beta (m - 1)^2 (m + 1)^2, \tag{31}$$

$$\beta_4 = -\alpha^{\frac{2}{\beta}} (\beta - 1)\beta (m - 1)(m + 1)\left(4\beta^2 m^2 - 14\beta m^2 + 16m^2 \right.$$
$$\left. +6\beta m - 8m - 2\beta^2 + 4\beta + b - 4\right), \tag{32}$$

$$\beta_5 = \beta \left(6\beta^3 m^4 - 18\beta^2 m^4 + 27\beta m^4 - 14m^4 + 18\beta^2 m^3 - 20\beta m^3 \right.$$
$$+16m^3 - 6\beta^3 m^2 + 12\beta^2 m^2 + 2b\beta m^2 - 13\beta m^2 - bm^2 + 6m^2$$
$$\left. -12\beta^2 m + 6\beta m + bm - 8m + \beta^3 - b\beta\right). \tag{33}$$

Case (iii): If $H(z) = \mathrm{dn}(z)$,

$$h(z) = \alpha\, \mathrm{dn}^{\beta}(z). \tag{34}$$

$$F(h) = \gamma_1 h^{-\frac{2}{q}+1} + \gamma_2 h + \gamma_3 h^{\frac{2}{q}+1} + \gamma_4 h^{-\frac{4}{q}+1} + \gamma_5 h^{\frac{4}{q}+1}, \tag{35}$$

where

$$\gamma_1 = -\alpha^{-\frac{4}{\beta}} m^{-4} \beta \left(8m^3 + 28\beta m^2 - 12m^2 + 12\beta^2 m - 28\beta m \right.$$
$$\left. +16m + \beta^3 - 6\beta^2 + 11\beta - 6\right), \tag{36}$$

$$\gamma_2 = -\alpha^{-\frac{2}{\beta}} m^{-4} \beta \left(4m^5 + 28\beta m^4 - 12m^4 + 18\beta^2 m^3 - 42\beta m^3 \right.$$
$$-2bm^3 + 16m^3 + 2\beta^3 m^2 - 12\beta^2 m^2 - b\beta m^2 - 34\beta m^2 + bm^2 \tag{37}$$
$$\left. +12m^2 - 36\beta^2 m + 84\beta m - 48m - 4\beta^3 + 24\beta^2 - 44\beta + 24\right)$$

$$\gamma_3 = -\alpha^{\frac{4}{\beta}} m^{-4} (\beta - 3)(\beta - 2)(\beta - 1)\beta (m - 1)^2 (m + 1)^2, \tag{38}$$

$$\gamma_4 = \alpha^{\frac{2}{\beta}} m^{-4} (\beta - 1)\beta (m - 1)(m + 1)\left(6\beta m^3 - 8m^3 + 2\beta^2 m^2 \right. \tag{39}$$
$$\left. -10\beta m^2 - bm^2 + 12m^2 - 12\beta m + 16m - 4\beta^2 + 20\beta - 24\right)$$

$$\gamma_5 = -\beta \left(3\beta m^6 - 2m^6 + 6\beta^2 m^5 - 14\beta m^5 - bm^5 + 8m^5 + \beta^3 m^4 \right.$$
$$-6\beta^2 m^4 - b\beta m^4 - 17\beta m^4 + bm^4 + 6m^4 - 36\beta^2 m^3 + 84\beta m^3$$
$$+2bm^3 - 48m^3 - 6\beta^3 m^2 + 36\beta^2 m^2 + 2b\beta m^2 - 38\beta m^2 \tag{40}$$
$$-2bm^2 + 24m^2 + 36\beta^2 m - 84\beta m + 48m + 6\beta^3 - 36\beta^2$$
$$\left. +66\beta - 36\right) m^{-4}. \tag{41}$$

Substituting (19), (28) and (35) into (15), we obtain the functions $f(h)$ for which h is solution of Eq. (13):
Case (i): Substituting (19) into (15), we obtain that

$$F(h) = \alpha_1' h^{\frac{4}{\beta}+1} + \alpha_2' h^{\frac{2}{\beta}+1} + \alpha_3' h^{-\frac{4}{\beta}+1} + \alpha_4' h^{-\frac{2}{\beta}+1} + \alpha_5' h, \tag{42}$$

where $\alpha_i' = -\mu^2\lambda^2\alpha_i, i = 1, \ldots, 4, \alpha_5' = -\mu^2\lambda^2\alpha_5 + \mu^2(\lambda^2 - \mu^2), \alpha_i, i = 1, \ldots, 5,$ are given in (20)–(26) and a solution of Eq. (13) is

$$h(z) = \alpha \,\mathrm{sn}^{\beta}(z|m).$$

Consequently, an exact solution of Eq. (1), where $F(u)$ is obtained substituting h by u in (42), is

$$u(x,t) = \alpha \,\mathrm{sn}^{\beta}(\mu x - \lambda t|m). \tag{43}$$

In the following, we give some solutions with physical interest:

- For $m = 0$, $\mu = \lambda = \frac{1}{2}\sqrt{\frac{5}{12}}$, $\alpha = 1$ and $\beta = 2$, substituting in (42) we obtain

$$F(h) = -\tfrac{5}{288}\left(12\,c^2 - 17\right)(2\,h - 1)\,. \tag{44}$$

and, as $\mathrm{sn}(z, 0) = \sin(z)$, we can obtain the particular solution (Fig. 1)

$$h(z) = \sin^2(z).$$

Consequently, an exact solution of Eq. (1), where $F(u)$ is obtained substituting h by u in (56), is

$$u(x,t) = \begin{cases} \sin^2\left[\frac{1}{2}\sqrt{\frac{5}{12}}(x-t)\right] & |x-t| \le \frac{2\pi}{k}, \\ 0 & |x-t| > \frac{2\pi}{k}, \end{cases} \tag{45}$$

with $k = \sqrt{\frac{5}{12}}$.

- For $m = \mu = \beta = 1$, $\lambda = \frac{1}{2}$ and $\alpha = \frac{1}{4}$, substituting into (42), we obtain

$$F(h) = 1536h^5 + 32c\,h^3 - 168h^3 + \left(\frac{15}{4} - 2c\right)h, \tag{46}$$

and, as $\mathrm{sn}(z|1) = \tanh(z)$, a solution of Eq. (12) (Fig. 2) is

$$h(z) = \frac{1}{4}\tanh(z).$$

Consequently, an exact solution of Eq. (1), where $F(u)$ is obtained substituting h by u in (46), is

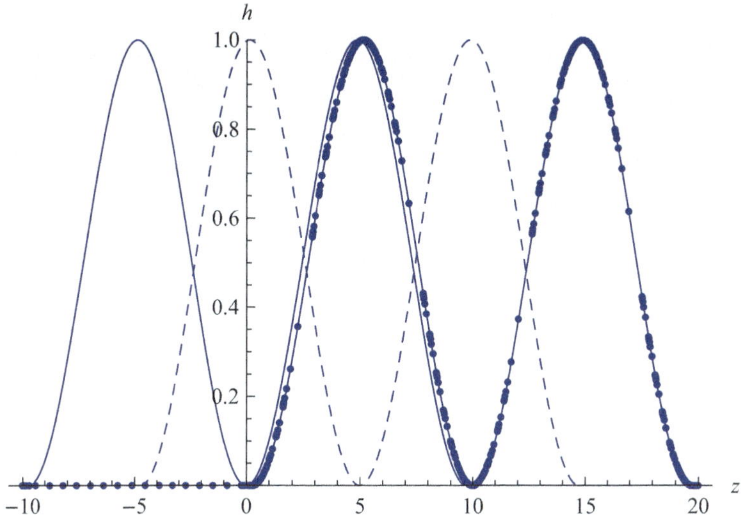

Fig. 1 Solution $h(z) = \sin^2(z)$ of Eq. (13)

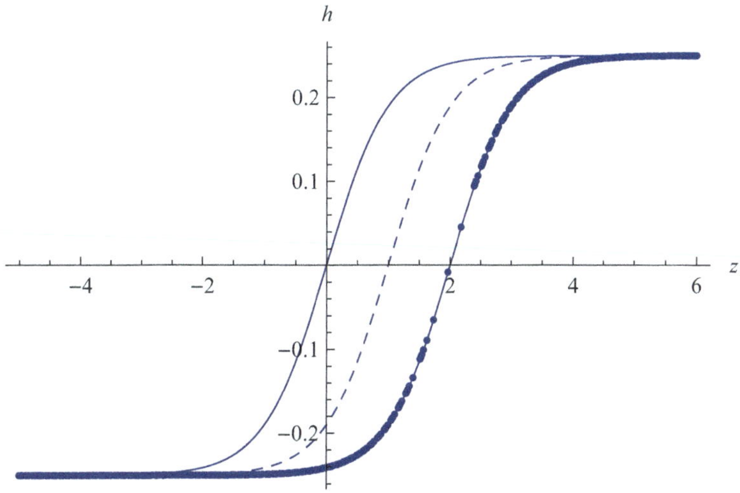

Fig. 2 Solution $h(z) = \frac{1}{4}\tanh(z)$ of Eq. (13)

$$u(x,t) = \frac{1}{4}\tanh\left(x - \frac{t}{2}\right). \tag{47}$$

- For $m = \mu = \alpha = 1$, $\lambda = \frac{1}{2}$, and $\beta = 3$, substituting into (42), we obtain

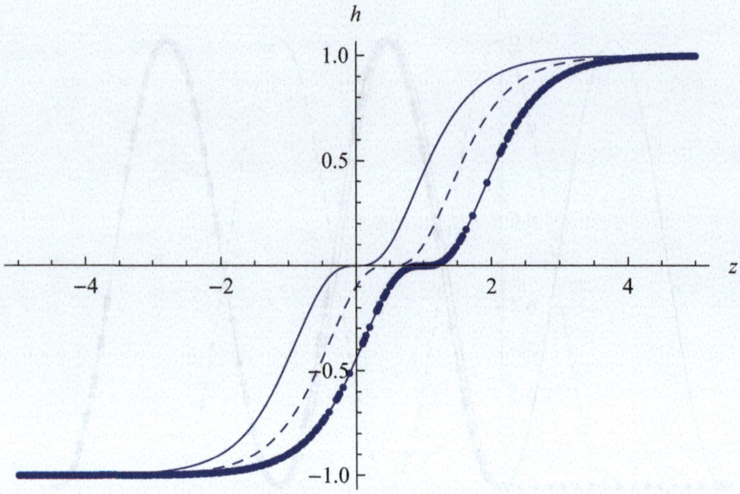

Fig. 3 Solution $h(z) = \tanh^3(z)$ of Eq. (13)

$$F(h) = 90h^{\frac{7}{3}} + 3\left(4c^2 - 69\right)h^{\frac{5}{3}} + \frac{3}{2}\left(4c^2 - 21\right)\sqrt[3]{h} - \frac{3}{4}\left(24c^2 - 197\right)h^{-\frac{1}{3}}, \tag{48}$$

and a solution of Eq. (12) is

$$h(z) = \tanh^3 z.$$

Consequently, an exact solution of Eq. (1), where $f(u)$ is obtained substituting h by u in (48), is (Fig. 3)

$$u(x,t) = \tanh^3\left(x - \frac{t}{2}\right). \tag{49}$$

Case (ii): Substituting (28) into (15) we obtain that

$$F(h) = \beta_1' h^{\frac{4}{\beta}+1} + \beta_2' h^{\frac{2}{\beta}+1} + \beta_3' h^{-\frac{4}{\beta}+1} + \beta_4' h^{-\frac{2}{\beta}+1} + \beta_5' h, \tag{50}$$

where $\beta_i' = -\mu^2\lambda^2\beta_i$, $i = 1, \ldots, 4$, $\beta_5' = -\mu^2\lambda^2\beta_5 + \mu^2(\lambda^2 - \mu^2)$, β_i, $i = 1, \ldots,$ 5, are given in (29)–(33) and a solution of Eq. (13) is

$$h(z) = \alpha\, \mathrm{cn}^\beta(z|m).$$

Consequently, an exact solution of Eq. (1), where $F(u)$ is obtained substituting h by u in (50), is

$$u(x,t) = \alpha\, \mathrm{cn}^\beta(\mu x - \lambda t|m). \tag{51}$$

- For $m = 0$, $\mu = \lambda = \frac{1}{2}\sqrt{\frac{5}{7}}$, $\alpha = 1$ and $\beta = 2$, substituting in (50) we obtain

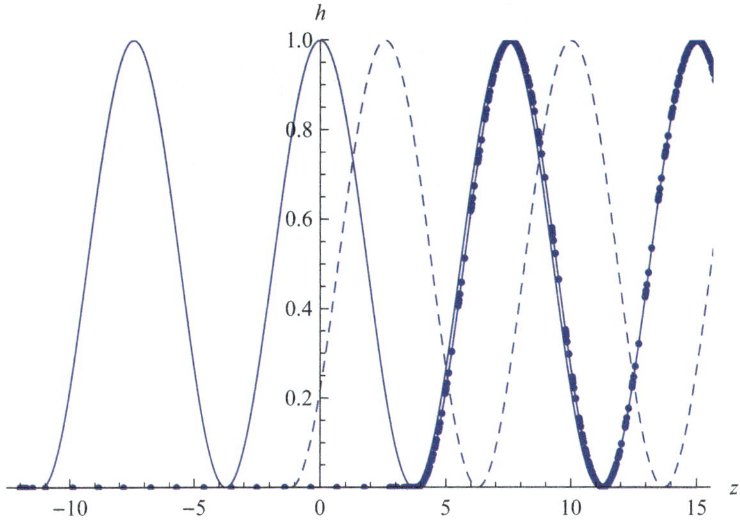

Fig. 4 Solution $h(z) = \cos^2(z)$ of Eq. (13)

$$F(h) = \frac{5}{288}\left(12\,c^2 - 17\right)(2\,h - 1)\,. \tag{52}$$

Due that $\mathrm{cn}(z|0) = \cos(z)$, we can obtain the particular solution Fig. 4

$$h(z) = \cos^2(z)$$

Consequently, an exact solution of Eq. (1), where $F(u)$ is obtained substituting h by u in (56), is

$$u(x,t) = \begin{cases} \cos^2(\mu x - \lambda t) & |x - t| \le \frac{3\pi}{k}, \\ 0 & |x - t| > \frac{3\pi}{k}, \end{cases} \tag{53}$$

with $k = \sqrt{\frac{5}{7}}$.

Case (iii): Substituting (35) in (15), we obtain that

$$F(h) = \gamma_1' h^{\frac{4}{\beta}+1} + \gamma_2' h^{\frac{2}{\beta}+1} + \gamma_3' h^{-\frac{4}{\beta}+1} + \gamma_4' h^{-\frac{2}{\beta}+1} + \gamma_5' h, \tag{54}$$

where $\gamma_i' = -\mu^2\lambda^2\gamma_i, i = 1, \ldots, 4, \gamma_5' = -\mu^2\lambda^2\gamma_5 + \mu^2(\lambda^2 - \mu^2), \gamma_i, i = 1, \ldots, 5,$ are given in (36)–(41) and a solution of Eq. (13) is

$$h(z) = \alpha\, \mathrm{dn}^{\beta}(z|m).$$

Consequently, an exact solution of Eq. (1), where $F(u)$ is obtained substituting h by u in (54), is

$$u(x,t) = \alpha\, \mathrm{dn}^{\beta}(\mu x - \lambda t|m). \tag{55}$$

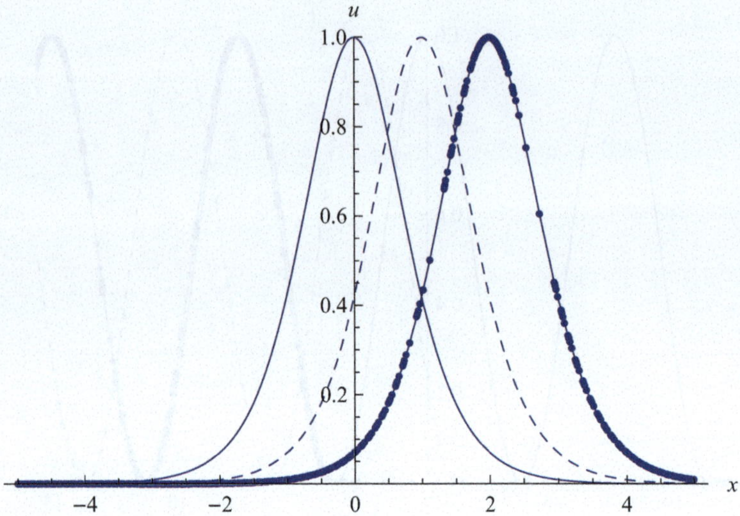

Fig. 5 Solution $h(z) = \text{sech}^2(z)$ of Eq. (13)

- For $m = \lambda = \mu = \alpha = 1$ and $\beta = 2$, substituting into (54) we obtain

$$F(h) = 120\,h^3 - 6\,\left(c^2 + 19\right)\,h^2 + 4\,\left(c^2 + 3\right)\,h. \tag{56}$$

Due that $\text{dn}(z, 1) = \text{sech}(z)$, we can obtain the particular solution (Fig. 5)

$$h(z) = \text{sech}^2(z).$$

Consequently, an exact solution of Eq. (1), where $F(u)$ is obtained substituting h by u in (56), is

$$u(x, t) = \text{sech}^2(x - t). \tag{57}$$

4 Conservation Laws

One interesting applications of symmetry reduction is the reduction to a traveling wave ODE. It happens that in most of the applications of double reduction to PDEs the considered conservation laws are invariant under translations. In a recent paper, a new method has been proposed and one of advantages is that we start from a symmetry to be used for reduction of a PDE, and then find all conservation laws that are invariant under the symmetry. Each one will be inherited by the reduced differential equation. This extension is more interesting when a PDE in two independent variables, such as the Boussinesq equation, is being reduced to an ODE, as then a set of first integrals can be obtained which allows for further reduction of the ODE.

A local conservation law of a scalar PDE $G(t, x, u, u_t, u_x, \ldots) = 0$ for $u(t, x)$ is a continuity equation $D_t T + D_x \Phi^x = 0$ holding on the space ε of solutions of the PDE, where T is the conserved density and $\Phi = (\Phi^x)$ is the spatial flux, which are functions of t, x, u, and derivatives of u. The conserved current is (T, Φ).

Every non-trivial conservation law of the PDE $G = 0$ arises from a multiplier, and there is a one-to-one correspondence between non-trivial conserved currents $(T, \Phi)|_\varepsilon$ modulo trivial ones and non-zero multipliers $Q|_\varepsilon$, with $QG = D_t T + D_x \Phi^x$ holding as an identity. Here Q is a function of t, x, u, and derivatives of u, such that $Q|_\varepsilon$ is non-singular. All multipliers are given by the solutions of the determining equation. For each solution Q, a conserved current $(T, \Phi)|_\varepsilon$ can be obtained by several explicit methods.

A traveling wave has the form

$$u(t, x) = U(x - \nu t) \tag{58}$$

where ν =const.

Invariance of a PDE $G(t, x, u, u_t, u_x, \ldots) = 0$ under the translation symmetry

$$X = \partial_t + c\partial_x, \tag{59}$$

gives rise to traveling wave solutions, with $z = x - ct$ and $u = U$ being the invariants.

In this chapter, we will focus on the conservation laws of Eq. (1) which are invariant under the translation symmetry (59).

We will consider the following low-order multipliers:

$$Q(t, x, u, u_x, u_t). \tag{60}$$

The determining equations split into an over-determined linear system which is straightforward to solve for Q $f'' \neq 0$, with $c_0 \neq 0, d <> 0$ $b_i <> 0$ $i = 1, \ldots, 2$.

Proposition 1 *The low-order multipliers (60) admitted by the generalized sixth-order Boussinesq equation (1) with* $c_0 \neq 0$, $d <> 0$, $b_i <> 0$ $i = 1 \ldots 2$. *invariant under the translations group with* $c_0 \neq 0$, $d <> 0$, $b_i <> 0$, $i = 1, \ldots, 2$ *are given by*

(i) $c_0 \neq 0$, $d <> 0$ $b_i <> 0$ $i = 1, \ldots, 2$, $f'' <> 0$

$$Q_1 = 1,$$
$$Q_2 = x - ct;$$

(ii) $f'''(u) = 0,$

$$Q_1 = 1,$$
$$Q_2 = x - ct$$
$$Q_3 = u_t,$$
$$Q_4 = u_x$$

These multipliers yield non-trivial conservation laws of low order, summarized as follows.

Theorem 1 *(i) The low-order conservation laws for the sixth-order Boussinesq equation (1) with* $c_0 \neq 0$, $d <> 0$, $b_i <> 0$, $i = 1 \ldots 2$ *for* $f'' \neq 0$ $f''' = 0$ *are given by (up to equivalence):*

case 1:

$$\begin{aligned} T_1 &= u_t \\ X_1 &= -du_{x,x,x,x,x} + b_2 u_{xxx} - b_1 u_{ttx} + (-c_0^2 - f'')u_x \end{aligned} \quad (61)$$

$$\begin{aligned} T_2 &= (-ct + x)u_t + cu + u_{txx} b_1 (ct - x) \\ X_2 &= -b_2(ct - x)u_{xxx} - b_1 cutx - b_2 u_{xx} + (ct - x)(c_0^2 + f'')f'' u_x + c_0^2 u \\ &\quad + d(ctu_{xxxxx} - xu_{xxxxx} + u_{xxxx}) + f' \end{aligned} \quad (62)$$

case2: $c_0 \neq 0$, $d <> 0$, $b_i <> 0$ $i = 1 \ldots 2$ *for* $f(u) = (1/2)c_1 u^2 + c_2 u + c_3$

$$\begin{aligned} T_3 &= (1/2)du_{xxx}^2 + (1/2)b_2 u_{xx}^2 + (1/2)b_1 u_{tx}^2 + (1/2)u_t^2 + (1/2)c_0^2 u_x^2 + (1/2)c_1 u_x^2 \\ X_3 &= (b_2 u_t - du_{txx})u_{xxx} + (-b_2 u_{xx} + du_{xxxx})u_{tx} \\ &\quad + (-du_{xxxxx} + (-c_0^2 - c_1)u_x - b_1 u_{ttx})u_t \end{aligned} \quad (63)$$

$$\begin{aligned} T_4 &= -b_1 u_x u_{txx} + u_t u_x \\ X_4 &= -(1/2)du_{xxx}^2 + b_2 u_x u_{xxx} + (1/2)b_1 u_{tx}^2 - (1/2)b_2 u_{xx}^2 \\ &\quad + du_{xx}u_{xxxx} - (1/2)u_t^2 + (-(1/2)c_0^2 - (1/2)c_1)u_x^2 - du_x u_{xxxxx} \end{aligned} \quad (64)$$

A traveling wave has the form

$$u(t, x) = U(x - ct) \quad (65)$$

where c =const.

Invariance of a PDE $G(t, x, u, u_t, u_x, \ldots) = 0$ under the translation symmetry

$$X = \partial_t + c\partial_x, \quad (66)$$

gives rise to traveling wave solutions, with $z = x - ct$ and $u = U$ being the invariants.

Substitution of the traveling wave expression

$$u(t,x) = U(x-ct) \tag{67}$$

into Eq. (1) yields a nonlinear sixth-order ODE

$$-dU'''''' + (-b_1c^2 + b_2)U'''' - \left(U'^2 c_1(p-1)(p-2)U(z)^{-3+p}(-c_1(p-1)U(z)^{p-2} \right.$$
$$\left. +c^2 - c0^2\right) U'' = 0 \tag{68}$$

By using the two conservation laws (61), (62), we get the following two first integrals

$$\Psi_1 = (-(c-c_0)(c+c_0) + c_1(p-1)U(z)^{p-2}U') + U''')b_1c^2 - U'''b_2 + U'''''d$$
$$= C_1 \tag{69}$$
$$\Psi_2 = (-z(c-c0)(c+c0) + zc1(p-1)U(z)^{p-2}U') + (c-c_0)(c+c_0)U$$
$$+U'''b_1c^2z - b_1c^2U''$$
$$-U'''b_2z + U'''''dz - U''''d + b_2U'' - c_2 - U^{p-1}c_1 = C_2 \tag{70}$$

By combining these first integrals, we have obtained a triple reduction to a fourth-order autonomous equation

$$U'''' = (1/d)(-U^{p-1}c_1 + ((-b_1c^2 + b_2)U'' + (c^2 - c_0^2)U^{-1} + C_1zU^{-2} - (C_2 - c_2)U^{-2} \tag{71}$$

The remaining conservation laws are also invariant under the translation group yielding first integrals, setting $C_1 = C_2 = 0$.

5 Conclusion

We have considered a nonlinearly generalized sixth-order Boussinesq equation (1), depending on an arbitrary nonlinear differential function $f(u)$ and the parameters b_1, b_2, c_0. By using a direct method we have derived traveling wave solutions for Eq. (1). Among them, we found solitons, kinks, antikinks, and compactons.

We have established a point symmetry classification of Eq. (1)in terms of the function $f(u)$ and the parameters b_1, b_2, c_0, determining the point symmetry group for all possible nonlinear differential functions $f(u)$ and the parameters b_1, b_2, c_0. We have also carried out an analogous classification of conservation laws of Eq. (1) by employing the multiplier method. By using the two conservation laws, we construct two first integrals. By combining these first integrals, we have obtained a triple reduction to a fourth-order autonomous equation.

Acknowledgements The support of Junta de Andalucía FQM-201 group and University of Cádiz is gratefully acknowledged.

References

1. Anco SC (2017) Generalization of Noether's theorem in modern form to non-variational partial differential equations. In: Recent progress and modern challenges in applied mathematics. Modeling and computational science. Fields Inst Commun 79:119–182
2. Anco SC (2016) Symmetry properties of conservation laws. Int J Modern Phys B 30:1640003–1640014
3. Anco SC, Bluman G (2002) Direct construction method for conservation laws of partial differential equations. Part I: examples of conservation law classifications. Euro J Appl Math 13:545–566
4. Anco SC, Bluman G (2002) Direct construction method for conservation laws of partial differential equations. Part II: General treatment Examples of conservation law classifications. Euro J Appl Math 13: 567–585
5. Anco SC, Kara AH (2018) Symmetry invariance of conservation laws. Euro J Appl Math 29(1):78–117
6. Anco SC, Rosa M, Gandarias ML (2017) On conservation laws for a generalized Boussinesq equation. AIP Conf Proc 1863:280003–28007
7. Anco SC, Gandarias ML, Multi-reduction theory for PDEs with conservation laws. Comun Nonlinear Sci Numer Sim (Accepted)
8. Bluman G, Temuerchaolu, Anco SC (2006) New conservation laws obtained directly from symmetry action on known conservation laws. J Math Anal Appl 322:233–250
9. Bluman G, Kumei S (1989) Symmetries and differential equations. Springer, Berlin
10. Bokhari AH, Dweik AY, Zaman FD, Kara AH, Mahomed FM (2010) Generalization of the double reduction theory. Nonlinear Anal: Real World Appl 11(5):3763–3769
11. Bruzón MS, Gandarias ML (2006) Applying a new algorithm method to derive nonclassical symmetries. Nonlinear Sci Compl 7–12
12. Bruzón MS, Gandarias ML (2008) Applying a new algorithm to derive nonclassical symmetries. Nonl Sci Num Sim 13:517–523
13. Bruzón MS, Gandarias ML (2009) Symmetries for a family of Boussinesq equations with nonlinear dispersion. Commun Nonlinear Sci Numer Simulat 14:3250–3257
14. Bruzón MS, Gandarias ML, Ibragimov NH (2009) Self-adjoint sub-classes of generalized thin film equations. J Math Anal Appl 357:307–313
15. Caraffini GL, Galvani M (2012) Symmetries and exact solutions via conservation laws for some partial differential equations of Mathematical Physics. Appl Math Comput 219(4):1474–1484
16. Christov CI, Maugin GA, Porubov AV (2007) On Boussinesq's paradigm in nonlinear wave propagation. C R Mecanique 335:521–535
17. de la Rosa R, Gandarias ML, Bruzón MS (2016) On symmetries and conservation laws of a Gardner equation involving arbitrary functions. Appl Math Comput 290:125–134
18. Freire IL (2013) New classes of nonlinearly self-adjoint evolution equations of third- and fifth-order. Commun Nonlin Sci Numer Sim 18:493–499
19. Gandarias ML (2011) Weak self-adjoint differential equations. J Phys A Math Theor 44:262001–262006
20. Gandarias ML (2014) Nonlinear self-adjointness through differential substitutions. Commun Nonlinear Sci Numer Sim 19:3523–3528
21. Gandarias ML, Bruzón MS (1998) Classical and Nonclassical Symmetries of a Generalized Boussinesq Equation. J Nonlinear Math Phys 5(1):8–12
22. Gandarias ML, Rosa M (2016) On double reductions from symmetries and conservation laws for a damped Boussinesq equation. Chaos Solitons Fractals 89:560–565
23. Gandarias ML, Rosa M (2016) Symmetries and conservation laws of a damped Boussinesq equation. Int J Modern Phys B 30:1640012–1640023
24. Gandarias ML, Bruzón MS (2017) Conservation laws for a Boussinesq equation. Appl Math Nonlin Sci 2(2):465–472
25. Ibragimov NH (2010) Nonlinear self-adjointness in constructing conservation laws. Arch ALGA 7(8):1–86

26. Ibragimov NH (2011) Nonlinear self-adjointness and conservation laws. J Phys A Math Theor 44:432002–432009
27. Ibragimov NH (1994–1996) CRC handbook of lie group analysis of differential equations, vol I, II, III. CRC Press
28. Ibragimov, N.H.: Elementary Lie Group Analysis and Ordinary Differential Equations. John Wiley & Sons (1998)
29. Ibragimov NH (2007) Quasi self-adjoint differential equations. Arch ALGA 4:55–60
30. Ibragimov NH (2007) A new conservation theorem. J Math Anal Appl 333:311–328
31. Ibragimov NH, Torrisi M, Tracinà R (2010) Quasi self-adjoint nonlinear wave equations. J Phys A: Math Theor 43:442001–442009
32. Ibragimov NH (2006) Integrating factors, adjoint equations and Lagrangians. J Math Anal Appl 318:742–757
33. Ibragimov NH (2006) The answer to the question put to me by LV Ovsiannikov 33 years ago. Arch ALGA 3:53–80
34. Ibragimov NH (2011) Nonlinear self-adjointness in constructing conservation laws. Arch ALGA 8:59–63
35. Ibragimov NH, Gandarias ML, Galiakberova LR, Bruzón MS, Avdonina ED (2016) Group classification and conservation laws of anisotropic wave equations with a source. J Math Phys 57:083504–083513
36. Olver PJ (1993) Applications of Lie groups to differential equations. Springer, Berlin
37. Recio E, Gandarias ML, Bruzón MS (2016) Symmetries and conservation laws for a sixth-order Boussinesq equation. Chaos Solitons Fractals 89:572–577
38. Sjöberg A (2007) Double reduction of PDEs from the association of symmetries with conservation laws with applications. Appl Math Comput 184:608–616
39. Sjöberg A (2009) On double reduction from symmetries and conservation laws. Nonlinear Anal: Real World Appl 10:3472–3477

Group Analysis of the Guéant and Pu Model of Option Pricing and Hedging

Khristofor V. Yadrikhinskiy, Vladimir E. Fedorov, and Mikhail M. Dyshaev

Abstract The Guéant and Pu model of option pricing and hedging with the execution costs and the market impact is analyzed with different stationary execution costs functions. The group classification of the model with a nonlinear execution costs functions is obtained. Symmetries of concrete models are used for invariant solutions and invariant submodels search. The model with the linear execution costs function is reduced to the heat equation. Known results on this equation are applied to the study of the linear function case.

1 Introduction

Classical option pricing models are based on the perfect market hypothesis. Under this hypothesis, there are no execution costs and market participants use only the prevailing market prices and cannot influence the prices by their operations, either temporarily, or permanently. These assumptions, despite the obvious contradiction with the market practice, are quite widely used and the resulting models give useful results when the underlying asset is liquid and the transaction amount is not too large for the market.

However, in the case of options for an illiquid asset or large denominations relative to the normally traded volume in the market, the market impact and execution costs can no longer be excluded from the consideration.

K. V. Yadrikhinskiy
M.K. Ammosov North-Eastern Federal University, Belinsky St. 58, Yakutsk, Russia
e-mail: ghdsfdf@yandex.ru

V. E. Fedorov (✉)
Chelyabinsk State University, Kashirin Brothers St. 129, Chelyabinsk, Russia
e-mail: kar@csu.ru

South Ural State University (National Research University), Lenin Av. 76, Chelyabinsk, Russia

M. M. Dyshaev
Chelyabinsk State University, Kashirin Brothers St. 129, Chelyabinsk, Russia

A. C. J. Luo and R. K. Gazizov (eds.), *Symmetries and Applications of Differential Equations*, Nonlinear Physical Science,
https://doi.org/10.1007/978-981-16-4683-6_6

Perhaps one of the first works on the pricing of options was the thesis of Bachelier [1] in 1900. L. Bachelier calculated the prices of stock options, assuming a change in the price of the underlying asset (stock) according to the laws of the Brownian motion, and compared them with the current prices.

In 1965, Samuelson [2] proposed to use the so-called geometric (economic) Brownian motion to describe the dynamics of the stock price. The geometric Brownian motion served as the basis for the Black–Scholes–Merton model (1973) [3–5] and the well-known Black–Scholes formula.

Due to the fact that the Black–Scholes model does not take into account execution costs and the impact of market participants' transactions on current prices, researchers are actively studying changes in the model, which can be taken into account. There are two approaches to account for the impact of transactions on prices.

Kyle [6] introduced the concept that trades made by market participants can influence the market price. The market impact is a direct consequence of the order size effect. There is plenty of empirical evidence for this. As a rule, the main task when executing orders is to have the least possible impact on the current prices.

The first approach is usually called the ≪supply curve≫ approach. It takes into account the impact on the price of a traded asset of operations of high volume or insufficient liquidity. This approach was created and further developed in the works of Bank and Baum [7], Çetin, Jarrow and Protter [8, Sect. 4], Çetin and Rogers [9, Sect. 6].

The second approach examines the situations observed in practice related to the influence of delta hedging (dynamic hedging) on the dynamics of the underlying asset and the resulting feedback effect on the option price. Grossman [10] wrote one of the first works in this direction. There are also works, devoted to the study of this approach, of Platen and Schweizer [11], Sircar and Papanicolaou [12], Schönbucher and Wilmott [13].

The work of Magill and Constantinides [14] was one of the earliest studies to investigate the effect of transaction costs on portfolio pricing. This article has shown a number of fundamental qualitative changes that occur in the behavior of an investor's portfolio when trading opportunities must be paid in one form or another.

Leland [15] proposed one of the first models which took transactions into account when determining the price of options. The model of Barles and Soner [16] takes into account transaction costs and the risk aversion factor of hedgers, it was obtained using the asymptotic analysis methods. In the work of Cvitanić and Karatzas [17] by means of the martingale approach, a formula is obtained for calculating the minimum of the initial capital required to hedge an arbitrary conditional claim in a continuous-time model, taking into account proportional transaction costs.

In addition, an approach which takes into account transaction costs and is based on "optimal execution theory" has been introduced recently. In this approach, Rogers and Singh [18], Li and Almgren [19] considered execution costs that are not linear relative to the executed volume, but convex to account for the impact of liquidity.

These models have been studied by many authors both numerically and analytically. Analytical study of the Black–Scholes equation by the group analysis methods was carried out in the work of Ibragimov and Gazizov [20]. In the works of Bordag with co-authors [21–25], of Dyshaev and Fedorov [26–32] and other authors group properties of various nonlinear Black–Scholes type models were studied, their invariant solutions and submodels were calculated.

1.1 A Brief Description of the Guéant and Pu Model

In the work of Guéant and Pu [33] (see [34] also), devoted to the analysis of options pricing taking into account transaction costs and the impact of operations on the market, the problem of call option selling by a bank or trader in the market to a client with a maturity of T is solved under the next assumptions:

(1) the constant risk-free rate r, the absolute risk aversion parameter γ and the volatility σ are considered;
(2) the process of market trading volume V_t is considered deterministic, non-negative, and bounded;
(3) trading is limited to the maximum degree of participation ρ_m and, therefore, processes v are considered from the set of valid strategies A, which has a restriction $|v_t| \leq \rho_m V_t$ almost everywhere on $(0, T) \times \Omega$;
(4) the number of shares in the hedged portfolio is modeled as $q_t = q_0 + \int_0^t v_s ds$;
(5) the price process is modeled as $dS_t = \mu dt + \sigma dW_t + kv_t dt$, where μ is the trend forecast, expected return of the underlying asset, and k linearly models the permanent market impact; the authors considered the dynamics of the price of the underlying asset, as in the Bachelier model instead of the classic Black–Scholes model;
(6) to model execution costs a continuous non-negative function $L : \mathbb{R} \to \mathbb{R}_+$ is used, which is even, increasing on $\mathbb{R}_+$, $L(0) = 0$, strictly convex and coercive, i.e., $\lim_{\rho\to+\infty} \frac{L(\rho)}{\rho} = +\infty$;
(7) for any $v \in A$ the account X is changed as $dX_t = rX_t dt - v_t S_t dt - V_t L(v_t/V_t)dt$;
(8) the penalty function $\mathcal{L}(q, q')$ models the liquidity price when moving from a portfolio with q shares to a portfolio with q' shares. Its form is specified as $\mathcal{L}(q, q') = l(|q - q'|) + \frac{1}{2}k(q - q')^2$, where l is a convex and increasing function (possible variants of its form are suggested in [33], note 4).

Under these conditions, the problem of optimal stochastic control is set

$$\sup_{v\in A} \mathbb{E}\left[-\exp(-\gamma(X_T + q_T S_T - \Pi(q_T, S_T)))\right],$$

where $\mathbb{E}$ is the mathematical expectation. The value function and the associated Hamilton–Jacobi–Bellman equation are defined for it.

When $k = 0$ and, therefore, in the absence of a permanent market impact, the function $\theta(t, S, q)$ is obtained, which models the price of indifference of a call option, and by introducing the function $H(p) = \sup_{|\rho|\le\rho_m} [p\rho - L(\rho)]$ the differential equation associated with θ is derived

$$-\theta_t + r\theta + (\mu - rS)q - \mu\theta_S - \frac{1}{2}\sigma^2\theta_{SS} - \frac{1}{2}\gamma\sigma^2 e^{r(T-t)}(\theta_S - q)^2 + V_t H(\theta_q) = 0.$$

In this chapter, the group classification of this equation with a constant V_t is obtained, and for different specifications of the free element H from the classification invariant solutions and submodels are found.

2 The Group Classification of the Model

In this section, we will obtain the group classification for the Guéant–Pu equation with stationary nonlinear free element.

2.1 Continuous Groups of the Equivalence Transforms

Consider the Guéant–Pu equation with a constant market trading volume

$$\theta_t = r\theta + (\mu - rS)q - \mu\theta_S - \frac{1}{2}\sigma^2\theta_{SS} - \frac{1}{2}\gamma\sigma^2 e^{r(T-t)}(\theta_S - q)^2 + F, \quad (1)$$

where $\theta = \theta(t, S, q)$ is unknown function, $F = F(\theta_q)$ is so-called free element, i. e., it is a given function of an arbitrary form. Hereafter, as before,

$$\theta_t = \frac{\partial\theta}{\partial t}, \quad \theta_q = \frac{\partial\theta}{\partial q}, \quad \theta_S = \frac{\partial\theta}{\partial S}, \quad \theta_{SS} = \frac{\partial^2\theta}{\partial S^2}.$$

For the search of equivalence transformations groups, we consider the free element F and its derivatives as independent variables [35]. Generators of such groups will be searched in the form $Y = \tau\partial_t + \xi\partial_S + \alpha\partial_q + \eta\partial_\theta + \zeta\partial_F$, where τ, ξ, α, η depend on t, S, q, θ, and ζ depends on $t, S, q, \theta, F, \theta_t, \theta_S, \theta_q$. Additional equations

$$F_t = 0, \quad F_q = 0, \quad F_S = 0, \quad F_\theta = 0, \quad F_{\theta_S} = 0, \quad F_{\theta_t} = 0, \quad (2)$$

meaning the dependence of F on θ_q only, will be considered together with (1) as a manifold M in the extended space of the corresponding variables. Let us act on Eq. (1) by the prolonged operator

$$\underset{2}{Y} = Y + \eta^t \partial_{\theta_t} + \eta^S \partial_{\theta_S} + \eta^q \partial_{\theta_q} + \eta^{SS} \partial_{\theta_{SS}} + \zeta^t \partial_{F_t} + \zeta^S \partial_{F_S} + \zeta^q \partial_{F_q} + \zeta^\theta \partial_{F_\theta} + \zeta^{\theta_t} \partial_{F_{\theta_t}} + \zeta^{\theta_S} \partial_{F_{\theta_S}} + \zeta^{\theta_q} \partial_{F_{\theta_q}}$$

and after the restriction of the result on the manifold M obtain the equation by virtue of the invariance criterion

$$- \eta^t + r\eta - rS\alpha - rq\xi + \left(\mu + \gamma\sigma^2 e^{r(T-t)}(\theta_S - q)\right)(-\eta^S + \alpha) + \frac{r}{2}\gamma\sigma^2 e^{r(T-t)}(\theta_S - q)^2\tau + \zeta - \frac{1}{2}\sigma^2\eta^{SS}|_M = 0. \tag{3}$$

Analogously from (2), we obtain

$$\zeta^t|_M = 0, \;\; \zeta^q|_M = 0, \;\; \zeta^S|_M = 0, \;\; \zeta^\theta|_M = 0, \;\; \zeta^{\theta_S}|_M = 0, \;\; \zeta^{\theta_t}|_M = 0. \tag{4}$$

The coefficients of the operator $\underset{2}{Y}$ are calculated through the full derivative operators, for example

$$D_S = \frac{\partial}{\partial S} + \theta_S \frac{\partial}{\partial \theta} + \theta_{SS} \frac{\partial}{\partial \theta_S} + \cdots, \quad \tilde{D}_t = \frac{\partial}{\partial t} + F_t \frac{\partial}{\partial F} + \cdots,$$
$$\tilde{D}_\theta = \frac{\partial}{\partial \theta} + F_\theta \frac{\partial}{\partial F} + \cdots, \quad \tilde{D}_{\theta_t} = \frac{\partial}{\partial \theta_t} + F_{\theta_t} \frac{\partial}{\partial F} + \cdots,$$

and the prolongation formulas

$$\begin{aligned}
\eta^t &= D_t\eta - \theta_t D_t\tau - \theta_S D_t\xi - \theta_q D_t\alpha, \\
\eta^S &= D_S\eta - \theta_t D_S\tau - \theta_S D_S\xi - \theta_q D_S\alpha, \\
\eta^{SS} &= D_S\eta^S - \theta_{tS} D_S\tau - \theta_{SS} D_S\xi - \theta_{Sq} D_S\alpha, \quad \ldots, \\
\zeta^q &= \tilde{D}_q\zeta - F_t\tilde{D}_q\tau - F_S\tilde{D}_q\xi - F_q\tilde{D}_q\alpha - F_\theta\tilde{D}_q\eta - F_{\theta_t}\tilde{D}_q\eta^t \\
&\quad - F_{\theta_S}\tilde{D}_q\eta^S - F_{\theta_q}\tilde{D}_q\eta^q, \\
\zeta^\theta &= \tilde{D}_\theta\zeta - F_t\tilde{D}_\theta\tau - F_S\tilde{D}_\theta\xi - F_q\tilde{D}_\theta\alpha - F_\theta\tilde{D}_\theta\eta - F_{\theta_t}\tilde{D}_\theta\eta^t \\
&\quad - F_{\theta_S}\tilde{D}_\theta\eta^S - F_{\theta_q}\tilde{D}_\theta\eta^q, \\
\zeta^{\theta_t} &= \tilde{D}_{\theta_t}\zeta_{\theta_t} - F_t\tilde{D}_{\theta_t}\tau - F_S\tilde{D}_{\theta_t}\xi - F_q\tilde{D}_{\theta_t}\alpha - F_\theta\tilde{D}_{\theta_t}\eta - F_{\theta_t}\tilde{D}_{\theta_t}\eta^t \\
&\quad - F_{\theta_S}\tilde{D}_{\theta_t}\eta^S - F_{\theta_q}\tilde{D}_{\theta_t}\eta^q, \quad \ldots
\end{aligned}$$

We calculate them

$$
\begin{aligned}
\eta^t &= \eta_t + \theta_t \eta_\theta - \theta_t(\tau_t + \theta_t \tau_\theta) - \theta_S(\xi_t + \theta_t \xi_\theta) - \theta_q(\alpha_t + \theta_t \alpha_\theta),\\
\eta^S &= \eta_S + \theta_S \eta_\theta - \theta_t(\tau_S + \theta_S \tau_\theta) - \theta_S(\xi_S + \theta_S \xi_\theta) - \theta_q(\alpha_S + \theta_S \alpha_\theta),\\
\eta^q &= \eta_q + \theta_q \eta_\theta - \theta_t(\tau_q + \theta_q \tau_\theta) - \theta_S(\xi_q + \theta_q \xi_\theta) - \theta_q(\alpha_q + \theta_q \alpha_\theta),\\
\eta^{SS} &= \eta_{SS} + 2\theta_S \eta_{S\theta} + \theta_S^2 \eta_{\theta\theta} - 2\theta_{tS}(\tau_S + \theta_S \tau_\theta)\\
&\quad + \theta_{SS}(\eta_\theta - \theta_t \tau_\theta - \theta_q \alpha_\theta - 2\xi_S - 3\theta_S \xi_\theta) - 2\theta_{Sq}(\alpha_S + \theta_S \alpha_\theta)\\
&\quad - \theta_t(\tau_{SS} + 2\theta_S \tau_{S\theta} + \theta_S^2 \tau_{\theta\theta}) - \theta_S(\xi_{SS} + 2\theta_S \xi_{S\theta} + \theta_S^2 \xi_{\theta\theta})\\
&\quad - \theta_q(\alpha_{SS} + 2\theta_S \alpha_{S\theta} + \theta_S^2 \alpha_{\theta\theta}),
\end{aligned}
$$

substitute the coefficients into Eqs. (3), (4) and get the system

$$
\begin{aligned}
\zeta^t|_M &= \zeta_t - F_{\theta_q}(\eta_{tq} + \theta_q \eta_{t\theta} - \theta_S(\xi_{tq} + \theta_q \xi_{t\theta}) - \theta_q(\alpha_{tq} + \theta_q \alpha_{t\theta})\\
&\quad - \left(r\theta + (\mu - rS)q - \mu\theta_S - \frac{1}{2}\sigma^2\theta_{SS} - \frac{1}{2}\gamma\sigma^2 e^{r(T-t)}(\theta_S - q)^2 + F\right)\\
&\quad \times (\tau_{tq} + \theta_q \tau_{t\theta})) = 0, \qquad (5)\\
\zeta^S|_M &= \zeta_S - F_{\theta_q}(\eta_{Sq} + \theta_q \eta_{S\theta} - \theta_S(\xi_{Sq} + \theta_q \xi_{S\theta}) - \theta_q(\alpha_{Sq} + \theta_q \alpha_{S\theta})\\
&\quad - (r\theta + (\mu - rS)q - \mu\theta_S - \frac{1}{2}\sigma^2\theta_{SS} - \frac{1}{2}\gamma\sigma^2 e^{r(T-t)}(\theta_S - q)^2 + F)\\
&\quad \times (\tau_{Sq} + \theta_q \tau_{S\theta})) = 0, \qquad (6)\\
\zeta^q|_M &= \zeta_q - F_{\theta_q}(\eta_{qq} + \theta_q \eta_{q\theta} - \theta_S(\xi_{qq} + \theta_q \xi_{q\theta}) - \theta_q(\alpha_{qq} + \theta_q \alpha_{q\theta})\\
&\quad - \left(r\theta + (\mu - rS)q - \mu\theta_S - \frac{1}{2}\sigma^2\theta_{SS} - \frac{1}{2}\gamma\sigma^2 e^{r(T-t)}(\theta_S - q)^2 + F\right)\\
&\quad \times (\tau_{qq} + \theta_q \tau_{q\theta})) = 0, \qquad (7)\\
\zeta^\theta|_M &= \zeta_\theta - F_{\theta_q}(\eta_{q\theta} + \theta_q \eta_{\theta\theta} - \theta_S(\xi_{q\theta} + \theta_q \xi_{\theta\theta}) - \theta_q(\alpha_{q\theta} + \theta_q \alpha_{\theta\theta})\\
&\quad - (r\theta + (\mu - rS)q - \mu\theta_S - \frac{1}{2}\sigma^2\theta_{SS} - \frac{1}{2}\gamma\sigma^2 e^{r(T-t)}(\theta_S - q)^2 + F)\\
&\quad \times (\tau_{q\theta} + \theta_q \tau_{\theta\theta})) = 0, \qquad (8)\\
\zeta^{\theta_t}|_M &= \zeta_{\theta_t} + F_{\theta_q}(\tau_q + \theta_q \tau_\theta) = 0, \qquad (9)\\
\zeta^{\theta_S}|_M &= \zeta_{\theta_S} + F_{\theta_q}(\xi_q + \theta_q \xi_\theta) = 0, \qquad (10)
\end{aligned}
$$

$$
\begin{aligned}
& r\eta - rS\alpha - rq\xi + \frac{r}{2}\gamma\sigma^2 e^{r(T-t)}(\theta_S - q)^2\tau + (\mu + \gamma\sigma^2 e^{r(T-t)}(\theta_S - q))\\
& \times (-\eta_S - \theta_S\eta_\theta + \theta_S(\xi_S + \theta_S\xi_\theta) + \theta_q(\alpha_S + \theta_S\alpha_\theta) + \alpha) - \eta_t + \theta_S\xi_t\\
& + \theta_q\alpha_t + \zeta - \frac{1}{2}\sigma^2(\eta_{SS} + 2\theta_S\eta_{S\theta} + \theta_S^2\eta_{\theta\theta} - 2\theta_{tS}(\tau_S + \theta_S\tau_\theta)\\
& + \theta_{SS}(\eta_\theta - \theta_q\alpha_\theta - 2\xi_S - 3\theta_S\xi_\theta) - 2\theta_{Sq}(\alpha_S + \theta_S\alpha_\theta)\\
& - \theta_S(\xi_{SS} + 2\theta_S\xi_{S\theta} + \theta_S^2\xi_{\theta\theta}) - \theta_q(\alpha_{SS} + 2\theta_S\alpha_{S\theta} + \theta_S^2\alpha_{\theta\theta}))\\
& + (\tau_t - \eta_\theta + \theta_S\xi_\theta + \theta_q\alpha_\theta + \frac{1}{2}\sigma^2(\tau_{SS} + 2\theta_S\tau_{S\theta} + \theta_S^2\tau_{\theta\theta} + \theta_{SS}\tau_\theta)\\
& + (\mu + \gamma\sigma^2 e^{r(T-t)}(\theta_S - q))(\tau_S + \theta_S\tau_\theta))
\end{aligned}
$$

$$\times\left(r\theta+(\mu-rS)q-\mu\theta_S-\frac{1}{2}\sigma^2\theta_{SS}-\frac{1}{2}\gamma\sigma^2e^{r(T-t)}(\theta_S-q)^2+F\right)$$
$$+\tau_\theta\left(r\theta+(\mu-rS)q-\mu\theta_S-\frac{1}{2}\sigma^2\theta_{SS}-\frac{1}{2}\gamma\sigma^2e^{r(T-t)}(\theta_S-q)^2+F\right)^2=0. \quad (11)$$

Splitting the last equation with respect to the variables θ_{tS}, θ_{Sq} we obtain that $\tau_S=0$, $\tau_\theta=0$, $\alpha_S=0$, $\alpha_\theta=0$. By splitting with respect F_{θ_q} Eqs. (9) and (10) we get the equalities $\tau_q=0$, $\xi_q=0$, $\xi_\theta=0$. So, (5)–(10) will have the form

$$\zeta_t-F_{\theta_q}(\eta_{tq}+\theta_q\eta_{t\theta}-\theta_q\alpha_{tq})=0, \quad (12)$$
$$\zeta_S-F_{\theta_q}(\eta_{Sq}+\theta_q\eta_{S\theta})=0, \quad (13)$$
$$\zeta_q-F_{\theta_q}(\eta_{qq}+\theta_q\eta_{q\theta}-\theta_q\alpha_{qq})=0, \quad (14)$$
$$\zeta_\theta-F_{\theta_q}(\eta_{q\theta}+\theta_q\eta_{\theta\theta})=0, \quad (15)$$
$$\zeta_{\theta_t}=0, \quad \zeta_{\theta_S}=0. \quad (16)$$

Split with respect to F_{θ_q} in (15) and obtain $\eta_{q\theta}=0$, then split with $F_{\theta_q}\theta_q$ in (14) and get $\alpha_{qq}=0$. Thus, due to (12)–(16), we have

$$\eta_{t\theta}=\alpha_{tq}, \quad \eta_{tq}=\eta_{Sq}=\eta_{S\theta}=\eta_{qq}=\eta_{q\theta}=\eta_{\theta\theta}=\alpha_{qq}=0, \quad (17)$$
$$\zeta_t=\zeta_S=\zeta_q=\zeta_\theta=\zeta_{\theta_t}=\zeta_{\theta_S}=0. \quad (18)$$

Hence, $\zeta=\zeta(\theta_q,F)$.

Substitute the obtained functions into (11)

$$r\eta-rS\alpha-rq\xi+\frac{r}{2}\gamma\sigma^2e^{r(T-t)}(\theta_S-q)^2\tau$$
$$+(\mu+\gamma\sigma^2e^{r(T-t)}(\theta_S-q))(-\eta_S-\theta_S\eta_\theta+\theta_S\xi_S+\alpha)-\eta_t+\theta_S\xi_t+\theta_q\alpha_t$$
$$+\zeta-\frac{1}{2}\sigma^2(\eta_{SS}+\theta_{SS}(\eta_\theta-2\xi_S)-\theta_S\xi_{SS})$$
$$+(\tau_t-\eta_\theta)\left(r\theta+(\mu-rS)q-\mu\theta_S-\frac{1}{2}\sigma^2\theta_{SS}-\frac{1}{2}\gamma\sigma^2e^{r(T-t)}(\theta_S-q)^2+F\right)=0. \quad (19)$$

Split with respect to θ_{SS}

$$\tau_t=2\xi_S. \quad (20)$$

Therefore, $\xi_{SS}=0$.

Now split (19) with respect to θ_S and obtain due to (20)

$$r\tau-\eta_\theta=0, \quad (21)$$

$$\xi_t + (\mu - \gamma\sigma^2 e^{r(T-t)} q)(\xi_S - \tau_t) + \gamma\sigma^2 e^{r(T-t)}(-qr\tau - \eta_S + \alpha) = 0, \tag{22}$$

$$\begin{aligned}\zeta &= \gamma\sigma^2 e^{r(T-t)} q\big(-\frac{r}{2}q\tau + \alpha - \eta_S\big) + rq\xi - (\mu - rS)\alpha - \theta_q\alpha_t - r\eta + \eta_t \\ &\quad + \mu\eta_S + \frac{1}{2}\sigma^2\eta_{SS} - \big(r\theta + (\mu - rS)q - \frac{1}{2}\gamma\sigma^2 e^{r(T-t)}q^2 + F\big)(\tau_t - \eta_\theta).\end{aligned} \tag{23}$$

Differentiate (22) with respect to q, taking into account (17), then

$$\alpha_q = r\tau + \xi_S - \tau_t = r\tau - \tau_t/2. \tag{24}$$

Differentiate the last equation by t and use (17) and (21), then $r\tau_t = r\tau_t - \tau_{tt}/2$, $\tau_{tt} = 0$. Now differentiate (23) with respect to θ_q and q and obtain by virtue of (24) $\alpha_{tq} = 0 = r\tau_t - \tau_{tt}/2$. Hence, $\tau_t = 0$, $\alpha_{tq} = 0$. Therefore, due to (20) $\xi_S = 0$. Differentiate (22) by S and get

$$\eta_{SS} = 0. \tag{25}$$

Denote $\tau = A$, $\xi = B(t)$. Since $\alpha_q = r\tau = rA$, we have $\alpha = rAq + C(t)$. From (17), (21), and (25), it follows that $\eta = rA\theta + Dq + E(t)S + G(t)$.

Substitute them in (23) and obtain

$$\begin{aligned}\zeta &= \gamma\sigma^2 e^{r(T-t)} q(C(t) - E(t)) + rqB(t) - (\mu - rS)C(t) - \theta_q C'(t) - rDq \\ &\quad - rE(t)S - rG(t) + E'(t)S + G'(t) + \mu E(t) + rAF = 0.\end{aligned}$$

Differentiating the both sides of this equality by θ_q and t, we get that $C''(t) = 0$ due to (18), $C = C_0 + C_1 t$. After differentiating with respect to S, we get $E(t) = C_0 + C_1 t + \frac{C_1}{r} + C_2 e^{rt}$. The differentiation by q implies the equality

$$B(t) = D + \frac{\gamma}{r}\sigma^2 e^{r(T-t)}\big(\frac{C_1}{r} + C_2 e^{rt}\big)$$

and after differentiation by t we have $G''(t) - rG'(t) = -\frac{C_1}{r} - C_2 e^{rt}$, therefore,

$$G(t) = \big(C_3 - \frac{\mu}{r}C_2 t\big)e^{rt} + \frac{\mu}{r^2}C_1 t + C_4.$$

After substitution of the obtained functions into (22) we get $C_1 = C_2 = 0$, therefore

$$\begin{gathered}\tau = A, \quad \xi = D, \quad \alpha = rAq + C_0, \\ \eta = rA\theta + Dq + C_0 S + C_3 e^{rt} + C_4, \quad \zeta = rAF - rC_4.\end{gathered}$$

So, we have obtained the next result

Theorem 1 *The Lie algebra for the continuous group of equivalence transforms to Eq. (1) is generated by the operators*

$$Y_1 = e^{rt}\partial_\theta, \quad Y_2 = \partial_S + q\partial_\theta, \quad Y_3 = \partial_q + S\partial_\theta,$$
$$Y_4 = \partial_\theta - r\partial_F, \quad Y_5 = \partial_t + rq\partial_q + r\theta\partial_\theta + rF\partial_F.$$

Consequently, all the equivalence transforms for the free element F and its argument θ_q have the form

$$\overline{F} = kF + l, \quad \overline{\theta}_{\overline{q}} = k\theta_q + m, \quad k, l, m \in \mathbb{R}.$$

2.2 The Defining System of Equations

Act on the equation

$$\theta_t = r\theta + (\mu - rS)q - \mu\theta_S - \frac{1}{2}\sigma^2\theta_{SS} - \frac{1}{2}\gamma\sigma^2 e^{r(T-t)}(\theta_S - q)^2 + F(\theta_q) \quad (26)$$

by the second prolongation of the operator $X = \tau\partial_t + \xi\partial_S + \alpha\partial_q + \eta\partial_\theta$

$$\underset{2}{X} = X + \eta^t\partial_{\theta_t} + \eta^S\partial_{\theta_S} + \eta^q\partial_{\theta_q} + \eta^{SS}\partial_{\theta_{SS}}.$$

Here the functions τ, ξ, α, η depend on t, S, q, θ. Then we have

$$\begin{aligned}
&-\eta^t + r\eta + (\mu - rS)\alpha - rq\xi - \mu\eta^S - \frac{1}{2}\sigma^2\eta^{SS} - \gamma\sigma^2 e^{r(T-t)}(\theta_S - q)\eta^S \\
&+ \gamma\sigma^2 e^{r(T-t)}(\theta_S - q)\alpha + \frac{r}{2}\gamma\sigma^2 e^{r(T-t)}(\theta_S - q)^2\tau + F'\eta^q|_M \\
&= -\eta^t + r\eta - rS\alpha - rq\xi + (\mu + \gamma\sigma^2 e^{r(T-t)}(\theta_S - q))(-\eta^S + \alpha) \\
&+ \frac{r}{2}\gamma\sigma^2 e^{r(T-t)}(\theta_S - q)^2\tau + F'\eta^q - \frac{1}{2}\sigma^2\eta^{SS}|_M = 0.
\end{aligned}$$

The substitution of the prolongation formulas (5) implies the equality

$$
\begin{aligned}
&-(\eta_t+\theta_t\eta_\theta-\theta_t(\tau_t+\theta_t\tau_\theta)-\theta_S(\xi_t+\theta_t\xi_\theta)-\theta_q(\alpha_t+\theta_t\alpha_\theta))\\
&+r\eta-rS\alpha-rq\xi+(\mu+\gamma\sigma^2e^{r(T-t)}(\theta_S-q))\\
&\times(-(\eta_S+\theta_S\eta_\theta-\theta_t(\tau_S+\theta_S\tau_\theta)-\theta_S(\xi_S+\theta_S\xi_\theta)-\theta_q(\alpha_S+\theta_S\alpha_\theta))+\alpha)\\
&+\frac{r}{2}\gamma\sigma^2e^{r(T-t)}(\theta_S-q)^2\tau\\
&+F'(\eta_q+\theta_q\eta_\theta-\theta_t(\tau_q+\theta_q\tau_\theta)-\theta_S(\xi_q+\theta_q\xi_\theta)-\theta_q(\alpha_q+\theta_q\alpha_\theta))\\
&-\frac{1}{2}\sigma^2(\eta_{SS}+2\theta_S\eta_{S\theta}+\theta_S^2\eta_{\theta\theta}-2\theta_{St}(\tau_S+\theta_S\tau_\theta)\\
&+\theta_{SS}(\eta_\theta-\theta_t\tau_\theta-\theta_q\alpha_\theta-2\xi_S-3\theta_S\xi_\theta)-2\theta_{Sq}(\alpha_S+\theta_S\alpha_\theta)\\
&-\theta_t(\tau_{SS}+2\theta_S\tau_{S\theta}+\theta_S^2\tau_{\theta\theta})-\theta_S(\xi_{SS}+2\xi_{S\theta}\theta_S+\theta_S^2\xi_{\theta\theta})\\
&-\theta_q(\alpha_{SS}+2\theta_S\alpha_{S\theta}+\theta_S^2\alpha_{\theta\theta}))|_M=0.
\end{aligned}
\tag{27}
$$

Express θ_t from Eq. (26), substitute it in (27) and obtain

$$
\begin{aligned}
&\Big(r\theta+(\mu-rS)q-\mu\theta_S-\frac{1}{2}\sigma^2\theta_{SS}-\frac{1}{2}\gamma\sigma^2e^{r(T-t)}(\theta_S-q)^2+F\Big)^2\tau_\theta\\
&+\Big(r\theta+(\mu-rS)q-\mu\theta_S-\frac{1}{2}\sigma^2\theta_{SS}-\frac{1}{2}\gamma\sigma^2e^{r(T-t)}(\theta_S-q)^2+F\Big)\\
&\times\Big((\mu+\gamma\sigma^2e^{r(T-t)}(\theta_S-q))(\tau_S+\theta_S\tau_\theta)-\eta_\theta+\tau_t+\theta_S\xi_\theta+\theta_q\alpha_\theta\\
&+\frac{1}{2}\sigma^2(\tau_{SS}+2\theta_S\tau_{S\theta}+\theta_S^2\tau_{\theta\theta}+\theta_{SS}\tau_\theta)-F'(\tau_q+\theta_q\tau_\theta)\Big)\\
&+r\eta-rS\alpha-rq\xi+(\mu+\gamma\sigma^2e^{r(T-t)}(\theta_S-q))\\
&\times(-(\eta_S+\theta_S\eta_\theta-\theta_S(\xi_S+\theta_S\xi_\theta)-\theta_q(\alpha_S+\theta_S\alpha_\theta))+\alpha)\\
&-(\eta_t-\theta_S\xi_t-\theta_q\alpha_t)+\frac{r}{2}\gamma\sigma^2e^{r(T-t)}(\theta_S-q)^2\tau\\
&+F'(\eta_q+\theta_q\eta_\theta-\theta_S(\xi_q+\theta_q\xi_\theta)-\theta_q(\alpha_q+\theta_q\alpha_\theta))\\
&-\frac{1}{2}\sigma^2(\eta_{SS}+2\theta_S\eta_{S\theta}+\theta_S^2\eta_{\theta\theta}-2\theta_{St}(\tau_S+\theta_S\tau_\theta)\\
&+\theta_{SS}(\eta_\theta-\theta_q\alpha_\theta-2\xi_S-3\theta_S\xi_\theta)-2\theta_{Sq}(\alpha_S+\theta_S\alpha_\theta)\\
&-\theta_S(\xi_{SS}+2\xi_{S\theta}\theta_S+\theta_S^2\xi_{\theta\theta})-\theta_q(\alpha_{SS}+2\theta_S\alpha_{S\theta}+\theta_S^2\alpha_{\theta\theta}))=0.
\end{aligned}
\tag{28}
$$

By splitting with respect to θ_{Sq} and θ_{St} of Eq. (28), we obtain

$$
\tau_S=0,\quad \tau_\theta=0,\quad \alpha_S=0,\quad \alpha_\theta=0. \tag{29}
$$

Consequently, (28) has the form

$$
\begin{aligned}
&\big(r\theta + (\mu - rS)q - \mu\theta_S - \frac{1}{2}\sigma^2\theta_{SS} - \frac{1}{2}\gamma\sigma^2 e^{r(T-t)}(\theta_S - q)^2 + F\big)\\
&\times(-\eta_\theta + \tau_t + \theta_S\xi_\theta - F'\tau_q) + r\eta - rS\alpha - rq\xi\\
&+ (\mu + \gamma\sigma^2 e^{r(T-t)}(\theta_S - q))(-(\eta_S + \theta_S\eta_\theta - \theta_S(\xi_S + \theta_S\xi_\theta)) + \alpha)\\
&- (\eta_t - \theta_S\xi_t - \theta_q\alpha_t) + \frac{r}{2}\gamma\sigma^2 e^{r(T-t)}(\theta_S - q)^2\tau\\
&+ F'(\eta_q + \theta_q\eta_\theta - \theta_S(\xi_q + \theta_q\xi_\theta) - \theta_q\alpha_q)\\
&- \frac{1}{2}\sigma^2(\eta_{SS} + 2\theta_S\eta_{S\theta} + \theta_S^2\eta_{\theta\theta} + \theta_{SS}(\eta_\theta - 2\xi_S - 3\theta_S\xi_\theta)\\
&- \theta_S(\xi_{SS} + 2\xi_{S\theta}\theta_S + \theta_S^2\xi_{\theta\theta})) = 0.
\end{aligned}
\tag{30}
$$

Using the splitting of (30) by θ_{SS}, we obtain

$$
\tau_t - F'\tau_q - 2\xi_S - 2\theta_S\xi_\theta = 0. \tag{31}
$$

Therefore, $\xi_\theta = 0$. Differentiate (31) with respect to S and get $\xi_{SS} = 0$ due to (29), hence

$$
\tau_t - F'\tau_q - 2\xi_S = 0, \quad \xi_{SS} = 0, \quad \xi_\theta = 0. \tag{32}
$$

Taking into account these equalities in (30), we have

$$
\begin{aligned}
&\big(r\theta + (\mu - rS)q - \mu\theta_S + \frac{1}{2}\gamma\sigma^2 e^{r(T-t)}(2q\theta_S - \theta_S^2 - q^2) + F\big)\\
&\times(-\eta_\theta + \tau_t - F'\tau_q) + r\eta - rS\alpha - rq\xi + \mu(-\eta_S - \theta_S\eta_\theta + \theta_S\xi_S + \alpha)\\
&+ \gamma\sigma^2 e^{r(T-t)}(\theta_S^2(\xi_S - \eta_\theta) + \theta_S(q\eta_\theta - q\xi_S - \eta_S + \alpha) + q(\eta_S - \alpha))\\
&- \eta_t + \theta_S\xi_t + \theta_q\alpha_t + \frac{r}{2}\gamma\sigma^2 e^{r(T-t)}(-2q\theta_S + \theta_S^2 + q^2)\tau\\
&+ F'(\eta_q + \theta_q\eta_\theta - \theta_S\xi_q - \theta_q\alpha_q) - \frac{1}{2}\sigma^2(\eta_{SS} + 2\theta_S\eta_{S\theta} + \theta_S^2\eta_{\theta\theta}) = 0.
\end{aligned}
$$

Split it with respect to θ_S and get due to (32)

$$
\gamma e^{r(T-t)}(-\eta_\theta + r\tau) - \eta_{\theta\theta} = 0, \tag{33}
$$

$$
\xi_t - \mu\xi_S + \gamma\sigma^2 e^{r(T-t)}(q\xi_S - rq\tau - \eta_S + \alpha) - F'\xi_q - \sigma^2\eta_{S\theta} = 0, \tag{34}
$$

$$
\begin{aligned}
&\big(r\theta + (\mu - rS)q - \frac{1}{2}\gamma\sigma^2 e^{r(T-t)}q^2 + F\big)(-\eta_\theta + 2\xi_S) + r\eta - rS\alpha\\
&- rq\xi + (\mu - \gamma\sigma^2 e^{r(T-t)}q)(-\eta_S + \alpha) - \eta_t + \theta_q\alpha_t + \frac{r}{2}\gamma\sigma^2 e^{r(T-t)}q^2\tau\\
&+ F'(\eta_q + \theta_q\eta_\theta - \theta_q\alpha_q) - \frac{1}{2}\sigma^2\eta_{SS} = 0.
\end{aligned}
\tag{35}
$$

From (32) and (33), it follows that

$$\begin{aligned} \xi &= A(t,q)S + B(t,q), \\ \eta &= r\tau\theta + C(t,S,q)e^{-\gamma e^{r(T-t)}\theta} + D(t,S,q). \end{aligned} \tag{36}$$

Substitute it in (34) and obtain

$$A_t S + B_t - \mu A + \gamma\sigma^2 e^{r(T-t)}(Aq - rq\tau - D_S + \alpha) - F'(A_q S + B_q) = 0,$$

hence, $D_{SSS} = 0$,

$$D = D_2(t,q)S^2 + D_1(t,q)S + D_0(t,q), \tag{37}$$

$$A_t S + B_t - \mu A + \gamma\sigma^2 e^{r(T-t)}(Aq - rq\tau - 2D_2 S - D_1 + \alpha) - F'(A_q S + B_q) = 0.$$

After splitting by S of this equality, we have

$$B_t - \mu A + \gamma\sigma^2 e^{r(T-t)}(Aq - rq\tau - D_1 + \alpha) - F' B_q = 0, \tag{38}$$

$$A_t - 2\gamma\sigma^2 e^{r(T-t)} D_2 - F' A_q = 0. \tag{39}$$

Now (35) has the form, taking into account (32),

$$\begin{aligned} &\left((\mu - rS)q - \frac{1}{2}\gamma\sigma^2 e^{r(T-t)}q^2 + F\right)(-r\tau + \gamma C e^{r(T-t)} e^{-\gamma e^{r(T-t)}\theta} + 2A) \\ &+ r(Ce^{-\gamma e^{r(T-t)}\theta} + D_2 S^2 + D_1 S + D_0) - rS\alpha \\ &- rq(AS + B) + (\mu - \gamma\sigma^2 e^{r(T-t)}q)(-C_S e^{-\gamma e^{r(T-t)}\theta} - 2D_2 S - D_1 + \alpha) \\ &- C_t e^{-\gamma e^{r(T-t)}\theta} - D_{2t} S^2 - D_{1t} S \\ &- D_{0t} + \theta_q \alpha_t + \frac{r}{2}\gamma\sigma^2 e^{r(T-t)} q^2 \tau - \frac{1}{2}\sigma^2(C_{SS} e^{-\gamma e^{r(T-t)}\theta} + 2D_2) \\ &+ F'(C_q e^{-\gamma e^{r(T-t)}\theta} + D_{2q} S^2 + D_{1q} S + D_{0q} \\ &+ \theta_q(r\tau - \gamma C e^{r(T-t)} e^{-\gamma e^{r(T-t)}\theta}) - \theta_q \alpha_q) = 0. \end{aligned} \tag{40}$$

The differentiation by θ implies

$$\begin{aligned} &\left((\mu - rS)q - \frac{1}{2}\gamma\sigma^2 e^{r(T-t)}q^2 + F\right)\gamma e^{r(T-t)} C + rC \\ &- (\mu - \gamma\sigma^2 e^{r(T-t)}q)C_S - C_t + F'(C_q - \gamma\theta_q C e^{r(T-t)}) - \frac{1}{2}\sigma^2 C_{SS} = 0, \end{aligned} \tag{41}$$

so, (40) has the form

$$((\mu - rS)q + F)(-r\tau + 2A) + r(D_2S^2 + D_1S + D_0)$$
$$- rS\alpha - rq(AS + B) + (\mu - \gamma\sigma^2 e^{r(T-t)}q)(-2D_2S - D_1 + \alpha)$$
$$+ \gamma\sigma^2 e^{r(T-t)}q^2(r\tau - A) - D_{2t}S^2 - D_{1t}S - D_{0t} + \theta_q\alpha_t - \sigma^2 D_2$$
$$+ F'(D_{2q}S^2 + D_{1q}S + D_{0q} + r\theta_q\tau - \theta_q\alpha_q) = 0.$$

Split with respect to S and obtain the equations

$$rD_2 - D_{2t} + F'D_{2q} = 0, \tag{42}$$

$$- rq(-r\tau + 2A) + rD_1 - r\alpha - rqA \tag{43}$$
$$- 2D_2(\mu - \gamma\sigma^2 e^{r(T-t)}q) - D_{1t} + F'D_{1q} = 0,$$
$$(\mu q + F)(-r\tau + 2A) + rD_0 - rqB + (\mu - \gamma\sigma^2 e^{r(T-t)}q)(-D_1 + \alpha) - D_{0t}$$
$$+ \theta_q\alpha_t + \gamma\sigma^2 e^{r(T-t)}q^2(r\tau - A) + F'(D_{0q} + r\theta_q\tau - \theta_q\alpha_q) - \sigma^2 D_2 = 0. \tag{44}$$

Thus, we have the system of Eqs. (29), (32), (36), (37), (38), (39), (41), (42), (43), (44).

2.3 *The Case of a Nonlinear Function F*

We will assume in further arguments that $F''(\theta_q) \neq 0$.

Differentiate with respect to θ_q Eq. (41) and obtain

$$F''(C_q - \gamma\theta_q C e^{r(T-t)}) = 0,$$

hence, $C = 0$. After the differentiation by θ_q of Eqs. (32), (38), (39), (42), (43), we get that due to (32)

$$\tau_q = A_q = B_q = D_{1q} = D_{2q} = 0, \quad \tau_t = 2A.$$

Differentiate (38) by q and obtain

$$\alpha_q = r\tau - A, \quad \alpha = (r\tau - A)q + G(t).$$

Equality (42) implies that $D_2 = Ee^{rt}$, therefore, due to (39), we have that A_t is constant, $A = A_0 + A_1 t$,

$$D_2 = \frac{e^{r(t-T)}A_1}{2\gamma\sigma^2}.$$

The derivative of (43) with respect to q gives the equality $-2r(A_1t + A_0) + A_1 = 0$. Consequently, $A_1 = A_0 = 0$, $A \equiv D_2 \equiv 0$, τ is constant.

Now differentiate (44) by θ_q and obtain $G' - r\tau F' + F'' D_{0q} = 0$. Therefore, $D_{0qq} = 0$, $D_0 = H(t)q + J(t)$. Thus, $\xi = B(t)$, $\alpha = r\tau q + G(t)$, $\eta = r\tau\theta + D(t)S + H(t)q + J(t)$ after reassignment $D_1 := D$. Now Eqs. (38), (43), (44) have the form

$$B'(t) + \gamma\sigma^2 e^{r(T-t)}(G(t) - D(t)) = 0, \tag{45}$$

$$rD(t) - D'(t) - rG(t) = 0, \tag{46}$$

$$\begin{aligned} &- r\tau F + rqH + rJ - rqB + (\mu - \gamma\sigma^2 e^{r(T-t)} q)(G - D) \\ &- H'q - J' + \theta_q G' + F'H = 0. \end{aligned}$$

Split the last equation with respect to q and get

$$rH - rB - \gamma\sigma^2 e^{r(T-t)}(G - D) - H' = 0, \tag{47}$$

$$-r\tau F + rJ + \mu(G - D) - J' + \theta_q G' + F'H = 0. \tag{48}$$

Equations (45) and (47) imply that

$$\begin{gathered} rH - rB - H' + B' = r(H - B) - (H - B)' = 0, \\ \xi(t) = B(t) = H(t) + Ke^{rt}. \end{gathered} \tag{49}$$

From Eq. (47), it follows that

$$G(t) = -\frac{(H'(t) + rKe^{rt})e^{r(t-T)}}{\gamma\sigma^2} + D(t),$$

$$G'(t) = -r\frac{(H'(t) + rKe^{rt})e^{r(t-T)}}{\gamma\sigma^2} - \frac{(H''(t) + r^2Ke^{rt})e^{r(t-T)}}{\gamma\sigma^2} + D'(t). \tag{50}$$

Due to (46) and (50), we have

$$D'(t) = r\frac{(H'(t) + rKe^{rt})e^{r(t-T)}}{\gamma\sigma^2}, \quad G'(t) = -\frac{(H''(t) + r^2Ke^{rt})e^{r(t-T)}}{\gamma\sigma^2}. \tag{51}$$

So, we have (48), (49), (50), (51). Equation (48) defines the function F, it has the form

$$\beta F' + \lambda F + \delta\theta_q + \varepsilon = 0. \tag{52}$$

1. Let $\beta = \lambda = 0$ in (52), then $\delta = \varepsilon = 0$ and F may be arbitrary, hence, $\tau = 0$, $H = G' = 0$, hence, due to (51) $K = 0$, $G(t) \equiv D(t) \equiv D_0$, $J(t) = J_0 e^{rt}$. Consequently, $\xi = 0$, $\alpha \equiv D_0$, $\eta = D_0 S + J_0 e^{rt}$. So, we have the symmetries $X_1 = e^{rt}\partial_\theta$ and $X_2 = \partial_q + S\partial_\theta$ for arbitrary F.

2. If $\beta = 0$, $\lambda \neq 0$, then F does not satisfy the condition $F'' \neq 0$. Suppose $\beta \neq 0$, $\lambda = 0$. Then due to (52)

$$F(\theta_q) = -\frac{\delta}{2\beta}\theta_q^2 - \frac{\varepsilon}{\beta}\theta_q + F_0.$$

At $\delta = 0$ we have the contradiction again, therefore, we allow that $\delta \neq 0$. If $\varepsilon \neq 0$, after the equivalence transform $\overline{\theta}_{\overline{q}} = \theta_q + c$, we obtain $F = \delta_1\theta_q^2 + F_0$. By means of another equivalence transform $\overline{F} = aF + b$, we get $F = \theta_q^2$. Substitute this function into (48) and after splitting with respect to θ_q obtain $\tau = 0$

$$H'' - 2\gamma\sigma^2 H(t)e^{r(T-t)} = -r^2 K e^{rt}, \tag{53}$$

$$rJ(t) - J'(t) - \frac{\mu(H'(t) + rKe^{rt})e^{r(t-T)}}{\gamma\sigma^2} = 0. \tag{54}$$

3. Let $\beta \neq 0$, $\lambda \neq 0$, $\delta = 0$. Using the equivalence transforms, we get $F = e^{r\nu\theta_q}$, where $\nu \neq 0$ can not be changed by such transforms. By splitting (48) with respect to θ_q, we obtain that $\tau = \nu H(t)$, hence $H' = 0$, $G' = 0$, and $K = 0$, $G(t) \equiv D(t) \equiv D_0$, $J = J_0 e^{rt}$, $\xi = \text{const}$, $\tau = \nu\xi$, $\alpha = r\nu\xi q + D_0$, $\eta = r\nu\xi\theta + D_0 S + \xi q + J_0 e^{rt}$. Thus, we obtain the symmetries $X_1 = e^{rt}\partial_\theta$, $X_2 = \partial_q + S\partial_\theta$, and the third symmetry $X_3 = \nu\partial_t + \partial_S + r\nu q\partial_q + (r\nu\theta + q)\partial_\theta$. The parameter ν is the ratio of τ and ξ and may be arbitrary. For $F = e^{\nu\theta_q}$ it will be $X_3 = \nu\partial_t + r\partial_S + r\nu q\partial_q + (r\nu\theta + rq)\partial_\theta$.

4. At $\beta \neq 0$, $\lambda \neq 0$, $\delta \neq 0$ by means of the equivalence transforms, we obtain the equation $F' - r\nu F + \theta_q = 0$, $\nu \neq 0$, which has a unique solution

$$F(\theta_q) = F_1 e^{r\nu\theta_q} + \frac{\theta_q}{r\nu} + \frac{1}{r^2\nu^2}.$$

It can be transformed to the equivalent form $F(\theta_q) = e^{r\nu\theta_q} + F_0\theta_q$, $\nu \neq 0$, $F_0 \neq 0$. Substitute it into (48) and obtain $\tau = \nu H(t)$, $H' = 0$, $G' = r\tau F_0$, $K = 0$, hence $G' = 0$ and F_0. We get the contradiction.

2.4 The Case of a Quadratic Function F

So, for the case $F = \theta_q^2$, we have the system of equations

$$\tau = 0, \quad \xi(t) = H(t) + Ke^{rt}, \tag{55}$$

$$\alpha = G(t) = -\frac{(H'(t) + rKe^{rt})e^{r(t-T)}}{\gamma\sigma^2} + D(t),$$

$$\eta = D(t)S + H(t)q + J(t), \tag{56}$$

$$D'(t) = r\frac{(H'(t) + rKe^{rt})e^{r(t-T)}}{\gamma\sigma^2}, \tag{57}$$

$$H''(t) - 2\gamma\sigma^2 e^{r(T-t)}H(t) = -r^2Ke^{rt}, \tag{58}$$

$$rJ(t) - J'(t) - \mu\frac{(H'(t) + rKe^{rt})e^{r(t-T)}}{\gamma\sigma^2} = 0. \tag{59}$$

By integrating Eq. (57), we get

$$\begin{aligned} D(t) &= \int r\frac{(H'(t) + rKe^{rt})e^{r(t-T)}}{\gamma\sigma^2}dt \\ &= \frac{(H'(t) + rKe^{rt})e^{r(t-T)}}{\gamma\sigma^2} - \int \frac{(H''(t) + r^2Ke^{rt})e^{r(t-T)}}{\gamma\sigma^2}dt \\ &= \frac{(H'(t) + rKe^{rt})e^{r(t-T)}}{\gamma\sigma^2} - \int 2H(t)dt, \end{aligned}$$

therefore, due to (56)

$$\alpha = G(t) = -\int 2H(t)ds. \tag{60}$$

Multiply by e^{-rt} the both sides of Eq. (59) and obtain

$$(e^{-rt}J(t))' = -\mu\frac{H'(t) + rKe^{rt}}{\gamma\sigma^2}e^{-rT},$$

hence

$$J(t) = J_0e^{rt} - \frac{\mu}{\gamma\sigma^2}(H(t) + Ke^{rt})e^{r(t-T)}. \tag{61}$$

Make the change of the unknown function $H(t) = L(t) - Ke^{rt}$. Then Eq. (53) can be rewritten as

$$L''(t) - 2\gamma\sigma^2 e^{r(T-t)}L(t) = -2K\gamma\sigma^2 e^{rT}. \tag{62}$$

After the replacement $x = \frac{2}{r}\sqrt{2\gamma\sigma^2 e^{r(T-t)}}$, $e^{-rt} = \frac{r^2x^2e^{-rT}}{8\gamma\sigma^2}$, $k := -8r^{-2}K\gamma\sigma^2 e^{rT}$, (62) has the form

$$x^2 L_{xx} + xL_x - x^2 L = k, \ x > 0. \tag{63}$$

It is the inhomogeneous modified Bessel equation of the index $\nu = 0$. The fundamental system of solutions for such homogeneous equation consists of the modified Bessel functions of the first kind $I_0(x)$ and of the second kind $K_0(x)$.

A partial solution of Eq. (63) has the form

$$L_{\rm p}(x) = kI_0(x)\int \frac{K_0(x)}{x}dx - kK_0(x)\int \frac{I_0(x)}{x}dx.$$

Note that, for example,

$$\int \frac{K_0(x)}{x}dx\Big|_{x=\frac{2}{r}\sqrt{2\gamma\sigma^2 e^{r(T-t)}}} = -\frac{r}{2}\int K_0\big(\frac{2}{r}\sqrt{2\gamma\sigma^2 e^{rT}}e^{-rt/2}\big)dt,$$

$$\int \frac{I_0(x)}{x}dx\Big|_{x=\frac{2}{r}\sqrt{2\gamma\sigma^2 e^{r(T-t)}}} = -\frac{r}{2}\int I_0\big(\frac{2}{r}\sqrt{2\gamma\sigma^2 e^{rT}}e^{-rt/2}\big)dt,$$

hence the general form of a solution of (58) is

$$H(t) = \alpha_1\varphi_1(t) + \alpha_2\varphi_2(t) + K\psi(t) - Ke^{rt}, \tag{64}$$

where α_1 and α_2 are arbitrary constants and

$$\begin{aligned}\varphi_1(t) =& I_0\big(\frac{2}{r}\sqrt{2\gamma\sigma^2 e^{rT}}e^{-rt/2}\big), \quad \varphi_2(t) = K_0\big(\frac{2}{r}\sqrt{2\gamma\sigma^2 e^{rT}}e^{-rt/2}\big),\\ \psi(t) =& \frac{4\gamma\sigma^2 e^{rT}}{r}\big(\varphi_1(t)\int\varphi_2(t)dt - \varphi_2(t)\int\varphi_1(t)dt\big).\end{aligned} \tag{65}$$

Remark 1 Using the Meijer G-function [36], it can be shown that

$$\begin{aligned}\psi(t) =& -\frac{4\gamma\sigma^2 e^{rT}}{r^2}K_0\big(\frac{2}{r}\sqrt{2\gamma\sigma^2 e^{rT}}e^{-rt/2}\big)G^{2,0}_{1,3}\left(-\frac{8}{r^2}\gamma\sigma^2 e^{r(T-t)}\middle|\begin{matrix}1\\0,\ 0,\ 0\end{matrix}\right)\\ &+\frac{2\gamma\sigma^2 e^{rT}}{r^2}I_0\big(\frac{2}{r}\sqrt{2\gamma\sigma^2 e^{rT}}e^{-rt/2}\big)G^{3,0}_{1,3}\left(\frac{8}{r^2}\gamma\sigma^2 e^{r(T-t)}\middle|\begin{matrix}1\\0,\ 0,\ 0\end{matrix}\right).\end{aligned} \tag{66}$$

Then we have due to (55), (56), (60), (61), (64)

$$\begin{aligned}
\xi &= \alpha_1\varphi_1(t) + \alpha_2\varphi_2(t) + K\psi(t),\\
\alpha &= -2\int(\alpha_1\varphi_1(t) + \alpha_2\varphi_2(t) + K\psi(t) - Ke^{rt})dt,\\
D(t) &= \frac{(\alpha_1\varphi_1'(t) + \alpha_2\varphi_2'(t) + K\psi'(t))e^{r(t-T)}}{\gamma\sigma^2}\\
&\quad - 2\int(\alpha_1\varphi_1(t) + \alpha_2\varphi_2(t) + K\psi(t) - Ke^{rt})dt,\\
J(t) &= J_0e^{rt} - \frac{\mu}{\gamma\sigma^2}(\alpha_1\varphi_1(t) + \alpha_2\varphi_2(t) + K\psi(t))e^{r(t-T)},
\end{aligned}$$

consequently

$$\begin{aligned}
\eta =& \frac{(\alpha_1\varphi_1'(t) + \alpha_2\varphi_2'(t) + K\psi'(t))e^{r(t-T)}}{\gamma\sigma^2}S\\
&- 2S\int(\alpha_1\varphi_1(t) + \alpha_2\varphi_2(t) + K\psi(t) - Ke^{rt})dt\\
&+ (\alpha_1\varphi_1(t) + \alpha_2\varphi_2(t) + K\psi(t) - Ke^{rt})q\\
&+ J_0e^{rt} - \frac{\mu}{\gamma\sigma^2}(\alpha_1\varphi_1(t) + \alpha_2\varphi_2(t) + K\psi(t))e^{r(t-T)}.
\end{aligned}$$

2.5 Group Classification Theorem

Thus, the results of this section can be formulated as the next theorem on the group classification.

Theorem 2 1. *The principal Lie algebra of the equation*

$$\theta_t = r\theta + (\mu - rS)q - \mu\theta_S - \frac{1}{2}\sigma^2\theta_{SS} - \frac{1}{2}\gamma\sigma^2 e^{r(T-t)}(\theta_S - q)^2 + F(\theta_q),$$

where $F'' \neq 0$, *is generated by the operators*

$$X_1 = e^{rt}\partial_\theta, \quad X_2 = \partial_q + S\partial_\theta.$$

2. *The algebra Lie of the equation*

$$\theta_t = r\theta + (\mu - rS)q - \mu\theta_S - \frac{1}{2}\sigma^2\theta_{SS} - \frac{1}{2}\gamma\sigma^2 e^{r(T-t)}(\theta_S - q)^2 + e^{\nu\theta_q}, \ \nu \in \mathbb{R},$$

is generated by the operators

$$X_1 = e^{rt}\partial_\theta, \quad X_2 = \partial_q + S\partial_\theta, \quad X_3 = \nu\partial_t + r\partial_S + r\nu q\partial_q + (r\nu\theta + rq)\partial_\theta.$$

3. *The algebra Lie of the equation*

$$\theta_t = r\theta + (\mu - rS)q - \mu\theta_S - \frac{1}{2}\sigma^2\theta_{SS} - \frac{1}{2}\gamma\sigma^2 e^{r(T-t)}(\theta_S - q)^2 + \theta_q^2$$

is generated by the operators

$$\begin{aligned}
X_1 &= e^{rt}\partial_\theta, \quad X_2 = \partial_q + S\partial_\theta, \\
X_3 &= \varphi_1(t)\partial_S - 2\int \varphi_1(t)dt\partial_q \\
&\quad + \Big[-\frac{\mu e^{r(t-T)}\varphi_1(t)}{\gamma\sigma^2} + \varphi_1(t)q + \Big(\frac{e^{r(t-T)}\varphi_1'(t)}{\gamma\sigma^2} - 2\int \varphi_1(t)dt\Big)S\Big]\partial_\theta, \\
X_4 &= \varphi_2(t)\partial_S - 2\int \varphi_2(t)dt\partial_q \\
&\quad + \Big[-\frac{\mu e^{r(t-T)}\varphi_2(t)}{\gamma\sigma^2} + \varphi_2(t)q + \Big(\frac{e^{r(t-T)}\varphi_2'(t)}{\gamma\sigma^2} - 2\int \varphi_2(t)dt\Big)S\Big]\partial_\theta, \\
X_5 &= \psi(t)\partial_S - 2\int (\psi(t) - e^{rt})dt\partial_q + \Big[-\frac{\mu e^{r(t-T)}\psi(t)}{\gamma\sigma^2} + (\psi(t) - e^{rt})q \\
&\quad + \Big(\frac{e^{r(t-T)}\psi'(t)}{\gamma\sigma^2} - 2\int (\psi(t) - e^{rt})dt\Big)S\Big]\partial_\theta,
\end{aligned}$$

where φ_1, φ_2 *and* ψ *have form* (65).

3 Invariant Submodels and Invariant Solutions

Now we will use the obtained symmetries for the search of non-equivalent invariant solutions and invariant submodels of the equations.

3.1 General Case

The Lie algebra L_2 with the basis $X_1 = e^{rt}\partial_\theta$, $X_2 = \partial_q + S\partial_\theta$ has no nonzero structural constants. Therefore, the optimal system of one-dimensional subalgebras for L_2 has the form $\Theta_1 = \{\langle X_1\rangle, \langle bX_1 + X_2\rangle, b \in \mathbb{R}\}$.

It is obvious that the subalgebra $\langle X_1\rangle$ and all algebra $\langle X_1, X_2\rangle$ have not invariant submodels, the submodel

$$\varphi_t = r\varphi - \mu\varphi_S - \frac{1}{2}\sigma^2\varphi_{SS} - \frac{1}{2}\gamma\sigma^2 e^{r(T-t)}\varphi_S^2 + F(S), \;\; \theta = (be^{rt} + S)q + \varphi(t, S)$$

is invariant for the subalgebra $\langle bX_1 + X_2\rangle$.

3.2 Invariant Solutions at $F = e^{\nu\theta_q}$

Consider the Lie algebra L_3 with the basis

$$X_1 = e^{rt}\partial_\theta, \quad X_2 = \partial_q + S\partial_\theta, \quad X_3 = \nu\partial_t + r\partial_S + r\nu q\partial_q + (r\nu\theta + rq)\partial_\theta. \quad (67)$$

It has nonzero structural constants $c_{23}^2 = r\nu, c_{32}^2 = -r\nu$. So, the inner automorphisms of L_3 are $E_1 : \bar{e}_2 = e_2 + a_1 e_3$ and $E_2 : \bar{e}_2 = e^{a_2}e_2$. Morover, we have the inner automorphism $E_3 : \bar{e}_1 = -e_1$. We will find the basis operators of one-dimensional subalgebras in the form $X = e_1X_1 + e_2X_2 + e_3X_3$.

1. If $e_3 \neq 0$, then by E_1, we obtain $e_2 = 0$ and $(e_1, e_2, e_3) = (b, 0, 1), X = bX_1 + X_3, b \in \mathbb{R}$.

2. Let $e_3 = 0$, $e_1 \neq 0$, $e_2 \neq 0$, then by the automorphism E_3, we can make the coefficients e_1 and e_2 positive, and by E_2 get $(1, 1, 0)$, or $X = X_1 + X_2$. Besides, at $e_3 = 0$, we have the cases $X = X_1$ and $X = X_2$.

Lemma 1 *The optimal system of one-dimensional subalgebras for Lie algebra L_3 with basis* (67) *has the form* $\Theta_1 = \{\langle X_1\rangle, \langle X_2\rangle, \langle X_1 + X_2\rangle, \langle bX_1 + X_3\rangle, b \in \mathbb{R}\}$.

The subalgebras $\langle X_2\rangle$, $\langle X_1 + X_2\rangle$ are the partial cases at $b = 0$ and $b = 1$ of the considered above subalgebra.

Consider the subalgebra $\langle bX_1 + X_3\rangle$, where $bX_1 + X_3 = \nu\partial_t + r\partial_S + r\nu q\partial_q + (r\nu\theta + rq + be^{rt})\partial_\theta$. Its invariants are $rt - \nu S$, qe^{-rt}, $\theta e^{-rt} - \nu^{-1}(b + re^{-rt}q)t$. Hence, we will find a solution in the form

$$\theta = \nu^{-1}(be^{rt} + rq)t + e^{rt}\varphi(rt - \nu S, qe^{-rt}).$$

Denote $u = rt - \nu S$, $w = qe^{-rt}$, then we have the invariant submodel

$$\frac{1}{2}\sigma^2\nu^2\varphi_{uu} + \left(r - \mu\nu + \frac{1}{2}\gamma\sigma^2\nu e^{rT}\right)\varphi_u - rw\varphi_w - e^{\nu\varphi_w}$$
$$+ \nu^{-1}r(1-u)w + \left(\frac{1}{2}\gamma\sigma^2 e^{rT} - \mu\right)w + \nu^{-1}b = 0.$$

Now we will find the optimal system of two-dimensional subalgebras.

1. For the basis vector X_1 of the one-dimensional subalgebra $\langle X_1\rangle$, the second vector will be searched in the form $c_2X_2 + c_3X_3$, then we have the equality $[X_1, c_2X_2 + c_3X_3] = 0$, where $[\cdot, \cdot]$ is the commutator. Hence, we have the subalgebra $\langle X_1, c_2X_2 + c_3X_3\rangle$ for arbitrary $c_2, c_3 \in \mathbb{R}$. Using E_1, we will get the subalgebras $\langle X_1, X_2\rangle$, $\langle X_1, X_3\rangle$.

2. We have $[X_2, c_1X_1 + c_3X_3] = c_3r\nu X_2$, therefore, we get the subalgebra $\langle X_2, c_1X_1 + c_3X_3\rangle$ for every $c_1, c_3 \in \mathbb{R}$. Consider the cases $c_3 = 0$ and $c_3 \neq 0$, and obtain the subalgebras $\langle X_2, X_1\rangle$ (it was obtained before), $\langle X_2, bX_1 + X_3\rangle, b \in \mathbb{R}$.

3. Calculate the commutator $[X_1 + X_2, c_2X_2 + c_3X_3] = c_3r\nu X_2 = \alpha_1(X_1 + X_2) + \alpha_2(c_2X_2 + c_3X_3)$. Consequently, $\alpha_1 = 0, \alpha_2c_3 = 0, \alpha_2c_2 = c_3r\nu$. If $\alpha_2 = 0$, then $c_3 = 0$ and the subalgebra coincides with $\langle X_1, X_2\rangle$. If $\alpha_2 \neq 0$, then $c_2 = c_3 = 0$.

4. Let $[bX_1 + X_3, c_1X_1 + c_2X_2] = -c_2 r\nu X_2 = \alpha_1(bX_1 + X_3) + \alpha_2(c_1X_1 + c_2X_2)$. So, it will be $c_1 = 0$ or $c_2 = 0$ and there are no new subalgebras.

Lemma 2 *The optimal system of two-dimensional subalgebras for Lie algebra* L_3 *with basis* (67) *has the form* $\Theta_2 = \{\langle X_1, X_2\rangle, \langle X_1, X_3\rangle, \langle X_2, bX_1 + X_3\rangle, b \in \mathbb{R}\}$.

The subalgebras $\langle X_1, X_2\rangle$, $\langle X_1, X_3\rangle$ have no invariant submodels. Consider the subalgebra $\langle X_2, bX_1 + X_3\rangle$. The operator X_2 has invariants $J(t, S, \theta - Sq)$, denote $z = \theta - Sq$ and act on such function by the operator $bX_1 + X_3$:

$$\nu J_t + r J_S + (be^{rt} + r\nu z)J_z = 0.$$

Therefore, $J_1 = rt - \nu S$, $J_2 = e^{-rt}(\theta - Sq) - bt/\nu$ are invariants of the two-dimensional algebra. Hence, invariants solutions will be searched in the form

$$\theta = Sq + \frac{b}{\nu}te^{rt} + e^{rt}\varphi(rt - \nu S).$$

Consequently,

$$\theta_t = \frac{b}{\nu}e^{rt} + \frac{br}{\nu}te^{rt} + re^{rt}\varphi + re^{rt}\varphi', \quad \theta_q = S, \quad \theta_S = q - \nu e^{rt}\varphi', \quad \theta_{SS} = \nu^2 e^{rt}\varphi''.$$

Substitite them into the equation and obtain

$$\sigma^2\nu^2\varphi'' + 2(r - \mu\nu)\varphi' + \gamma\sigma^2\nu^2 e^{rT}\varphi'^2 - 2e^{-z} + 2\frac{b}{\nu} = 0, \tag{68}$$

where $\varphi = \varphi(z)$, $z = rt - \nu S$. Make the change of variables

$$\varphi = \frac{\ln|y|}{\gamma e^{rT}} + \frac{\mu\nu - r}{\sigma^2\nu^2\gamma e^{rT}}z, \quad \varphi' = \frac{y'}{\gamma e^{rT}y} + \frac{\mu\nu - r}{\sigma^2\nu^2\gamma e^{rT}}, \quad \varphi'' = \frac{y''}{\gamma e^{rT}y} - \frac{y'^2}{\gamma e^{rT}y^2}.$$

Then (68) has the form

$$y'' - (e^{-z}\frac{2\gamma e^{rT}}{\sigma^2\nu^2} - 2\frac{b\gamma e^{rT}}{\sigma^2\nu^3} + \frac{(\mu\nu - r)^2}{\sigma^4\nu^4})y = 0,$$

After the change of the independent variable

$$u = 2e^{-z/2}\frac{\sqrt{2\gamma e^{rT}}}{\sigma\nu}$$

we have

$$u^2 y_{uu} + u y_u - (u^2 - \frac{8b\gamma e^{rT}}{\sigma^2 \nu^3} + \frac{4(\mu\nu - r)^2}{\sigma^4 \nu^4})y = 0,$$

It is the modified Bessel equation of the order

$$p = \sqrt{-\frac{8b\gamma e^{rT}}{\sigma^2 \nu^3} + \frac{4(\mu\nu - r)^2}{\sigma^4 \nu^4}}.$$

Its solution has the form

$$y = \alpha_1 I_p(u) + \alpha_2 K_p(u),$$

where I_p is the modified Bessel function of the first kind, and K_p is the modified Bessel function of the second kind. So

$$\begin{aligned}\theta =& Sq + \frac{b}{\nu} t e^{rt} + \frac{(\mu\nu - r)(rt - \nu S)e^{r(t-T)}}{\gamma\sigma^2\nu^2}\\ &+ \ln\left|\alpha_1 I_p(\frac{2\sqrt{2\gamma e^{rT}}}{\sigma\nu} e^{\frac{\nu S - rt}{2}}) + \alpha_2 K_p(\frac{2\sqrt{2\gamma e^{rT}}}{\sigma\nu} e^{\frac{\nu S - rt}{2}})\right|.\end{aligned}$$

Act by the group, which is generated X_1 and obtain a more general solution that is invariant with respect to the entire algebra L_3

$$\begin{aligned}\theta =& Sq + \frac{b}{\nu} t e^{rt} + \frac{(\mu\nu - r)(rt - \nu S)e^{r(t-T)}}{\gamma\sigma^2\nu^2}\\ &+ \ln\left|\alpha_1 I_p(\frac{2\sqrt{2\gamma e^{rT}}}{\sigma\nu} e^{\frac{\nu S - rt}{2}}) + \alpha_2 K_p(\frac{2\sqrt{2\gamma e^{rT}}}{\sigma\nu} e^{\frac{\nu S - rt}{2}})\right| + a e^{rt}.\end{aligned}$$

3.3 Invariant Submodels for $F = \theta_q^2$

Consider the Lie algebra L_5 with the basis

$$\begin{aligned}&X_1 = e^{rt}\partial_\theta, \quad X_2 = \partial_q + S\partial_\theta,\\ &X_3 = \varphi_1(t)\partial_S - 2\int \varphi_1(t)dt\,\partial_q\\ &\quad + [-\frac{\mu e^{r(t-T)}\varphi_1(t)}{\gamma\sigma^2} + \varphi_1(t)q + (\frac{e^{r(t-T)}\varphi_1'(t)}{\gamma\sigma^2} - 2\int \varphi_1(t)dt)S]\partial_\theta,\\ &X_4 = \varphi_2(t)\partial_S - 2\int \varphi_2(t)dt\,\partial_q\end{aligned}$$

$$+\big[-\frac{\mu e^{r(t-T)}\varphi_2(t)}{\gamma\sigma^2}+\varphi_2(t)q+\big(\frac{e^{r(t-T)}\varphi_2'(t)}{\gamma\sigma^2}-2\int\varphi_2(t)dt\big)S\big]\partial_\theta,$$

$$X_5=\psi(t)\partial_S-2\int(\psi(t)-e^{rt})dt\partial_q+\big[-\frac{\mu\psi(t)e^{r(t-T)}}{\gamma\sigma^2}+(\psi(t)-e^{rt})q$$

$$+\big(\frac{\psi' e^{r(t-T)}}{\gamma\sigma^2}-2\int(\psi(t)-e^{rt})dt\big)S\big]\partial_\theta,$$

where φ_1, φ_2 and ψ have form (66). It is the Lie algebra of

$$\theta_t=r\theta+(\mu-rS)q-\mu\theta_S-\frac{1}{2}\sigma^2\theta_{SS}-\frac{1}{2}\gamma\sigma^2e^{r(T-t)}(\theta_S-q)^2+\theta_q^2. \tag{69}$$

Let us calculate the commutators of the basis elements. We have

$$[X_1,X_2]=[X_1,X_3]=[X_1,X_4]=[X_1,X_5]=[X_2,X_3]=[X_2,X_4]=0,$$

$$[X_2,X_5]=-X_1,\quad [X_3,X_4]=\frac{e^{r(t-T)}}{\gamma\sigma^2}(\varphi_1\varphi_2'-\varphi_2\varphi_1')\partial_\theta=\frac{re^{-rT}}{2\gamma\sigma^2}X_1,$$

since

$$\varphi_1'(t)\varphi_2(t)-\varphi_1(t)\varphi_2'(t)=-\frac{r}{2}x(I_0'(x)K_0(x)-I_0(x)K_0'(x))=\frac{r}{2}xW(x)=-\frac{r}{2},$$

where $x=\frac{2}{r}\sqrt{2\gamma\sigma^2e^{rT}}e^{-rt/2}$, the Wronskian $W(x)=I_0(x)K_0'(x)-I_0'(x)K_0(x)$ of the modified Bessel functions $I_0(x)$ and $K_0(x)$ is equal to $-1/x$. Analogously

$$[X_3,X_5]=\frac{e^{r(t-T)}}{\gamma\sigma^2}(\varphi_1\psi'-\psi\varphi_1')\partial_\theta+2e^{rt}\int\varphi_1ds\partial_\theta$$

$$=\frac{4e^{rt}}{r}(\varphi_1'\varphi_2-\varphi_1\varphi_2')\int\varphi_1(t)dt+2e^{rt}\int\varphi_1ds\partial_\theta=0,$$

$$[X_4,X_5]=\frac{e^{r(t-T)}}{\gamma\sigma^2}(\varphi_2\psi'-\psi\varphi_2')\partial_\theta+2e^{rt}\int\varphi_2ds\partial_\theta=0.$$

So, we have the nonzero structural constants

$$c_{34}^1=\frac{re^{-rT}}{2\gamma\sigma^2},\quad c_{43}^1=-\frac{re^{-rT}}{2\gamma\sigma^2},\quad c_{25}^1=-1,\quad c_{52}^1=1$$

and the inner automorphisms

$$E_3:\bar e_1=e_1+\frac{re^{-rT}}{2\gamma\sigma^2}e_4a_3,\quad E_4:\bar e_1=e_1-\frac{re^{-rT}}{2\gamma\sigma^2}e_3a_4,$$

$$E_2:\bar e_1=e_1-e_5a_2,\quad E_5:\bar e_1=e_1+e_2a_5.$$

Using these automorphisms, it is easy to obtain the next assertion.

Lemma 3 *An optimal system of one-dimensional subalgebras for* L_5 *is* $\Theta_1 = \{\langle X_1\rangle, \langle X_2\rangle, \langle bX_2 + X_3\rangle, \langle bX_2 + cX_3 + X_4\rangle, \langle bX_2 + cX_3 + dX_4 + X_5\rangle, b, c, d \in \mathbb{R}\}$.

The operator X_2 has the invariant submodel

$$U_t = -\frac{\sigma^2}{2}U_{SS} - \frac{\gamma\sigma^2}{2}e^{r(T-t)}U_S^2 - \mu U_S + rU + S^2, \quad \theta = U(t, S) + Sq.$$

Consider the operator $X = bX_2 + cX_3 + dX_4 + KX_5$. Denote

$$\Phi(t) = c\varphi_1(t) + d\varphi_2(t) + K\psi(t), \quad A(t) = b + 2\int (Ke^{rt} - \Phi(t))dt, \quad C(t) = \frac{e^{r(t-T)}}{\gamma\sigma^2},$$

then the operator

$$X = \Phi(t)\partial_S + A(t)\partial_q + (-\mu C(t)\Phi(t) + (\Phi(t) - Ke^{rt})q + (A(t) + C(t)\Phi'(t))S)\partial_\theta$$

has the invariants $J_1 = t$, $J_2 = u := A(t)S - \Phi(t)q$,

$$J_3 = \theta + \mu C(t)S + \frac{Ke^{rt} - \Phi(t)}{\Phi(t)}Sq + \frac{A(t)(\Phi(t) - Ke^{rt})}{2\Phi(t)^2}S^2 - \frac{A(t) + C(t)\Phi'(t)}{2\Phi(t)}S^2.$$

So, we will search the invariant solution in the form

$$\begin{aligned}\theta =& U(t, A(t)S - \Phi(t)q) - \mu C(t)S + \frac{\Phi(t) - Ke^{rt}}{\Phi(t)}Sq + \frac{A(t)(Ke^{rt} - \Phi(t))}{2\Phi(t)^2}S^2 \\ &- \frac{A(t) + C(t)\Phi'(t)}{2\Phi(t)}S^2.\end{aligned}$$

Substitute it in (69) and obtain the invariant submodel

$$\begin{aligned}U_t =& -\frac{\sigma^2 A(t)^2}{2}U_{uu} + \left(\Phi(t)^2 - \frac{\gamma\sigma^2 e^{r(T-t)}A(t)^2}{2}\right)U_u^2 \\ &- \left(\frac{\Phi'(t)}{\Phi(t)} + \frac{\gamma\sigma^2 Ke^{rT}A(t)}{\Phi(t)^2}\right)uU_u + rU - \frac{\gamma\sigma^2 K^2 e^{r(t+T)}}{2\Phi(t)^4}u^2 \\ &- \frac{\sigma^2 Ke^{rt}A(t)}{2\Phi(t)^2} - \frac{e^{r(t-T)}\Phi'(t)}{2\gamma\Phi(t)} + \frac{\mu^2 e^{r(t-T)}}{2\gamma\sigma^2}.\end{aligned} \tag{70}$$

If $c = 1$, $d = K = 0$, then $\Phi(t) \equiv \varphi_1(t)$, $A = b - 2\int \varphi_1(t)dt := A_1$, and we have the invariant submodel

$$U_t = -\frac{\sigma^2 A_1(t)^2}{2}U_{uu} + \big(\varphi_1(t)^2 - \frac{\gamma\sigma^2 e^{r(T-t)}A_1(t)^2}{2}\big)U_u^2$$
$$-\frac{\varphi_1'(t)}{\varphi_1(t)}uU_u + rU - \frac{e^{r(t-T)}\varphi_1'(t)}{2\gamma\varphi_1(t)} + \frac{\mu^2 e^{r(t-T)}}{2\gamma\sigma^2}.$$

For $d = 1$, $K = 0$, we have $\Phi(t) \equiv c\varphi_1(t) + \varphi_2$, $A = b - 2\int(c\varphi_1(t) + \varphi_2(t))dt :=$ A_2, and due to (70), we have the invariant submodel for $bX_2 + cX_3 + X_4$

$$U_t = -\frac{\sigma^2 A_2(t)^2}{2}U_{uu} + \big((c\varphi_1(t) + \varphi_2(t))^2 - \frac{\gamma\sigma^2 e^{r(T-t)}A_2(t)^2}{2}\big)U_u^2$$
$$-\frac{c\varphi_1'(t) + \varphi_2'(t)}{c\varphi_1(t) + \varphi_2(t)}uU_u + rU - \frac{e^{r(t-T)}(c\varphi_1'(t) + \varphi_2'(t))}{2\gamma(c\varphi_1(t) + \varphi_2(t))} + \frac{\mu^2 e^{r(t-T)}}{2\gamma\sigma^2}.$$

And at $K = 1$ $\Phi(t) \equiv c\varphi_1(t) + d\varphi_2 + \psi := \Phi_3(t)$, $A = b - 2\int(c\varphi_1(t) + d\varphi_2(t) +$ $\psi(t))dt := A_3$, and the invariant submodel for $bX_2 + cX_3 + dX_4 + X_5$ is

$$U_t = -\frac{\sigma^2 A_3(t)^2}{2}U_{uu} + \big(\Phi_3(t)^2 - \frac{\gamma\sigma^2 e^{r(T-t)}A_3(t)^2}{2}\big)U_u^2$$
$$-\big(\frac{\Phi_3'(t)}{\Phi_3(t)} + \frac{\gamma\sigma^2 e^{rT}A_3(t)}{\Phi_3(t)^2}\big)uU_u + rU - \frac{\gamma\sigma^2 e^{r(t+T)}}{2\Phi_3(t)^4}u^2$$
$$-\frac{\sigma^2 e^{rt}A_3(t)}{2\Phi_3(t)^2} - \frac{e^{r(t-T)}\Phi_3'(t)}{2\gamma\Phi_3(t)} + \frac{\mu^2 e^{r(t-T)}}{2\gamma\sigma^2}$$

4 The Case of a Linear F

Now let $F'' = 0$, therefore, $F = c\theta_q + d$. Taking into account the equivalence transforms after Theorem 1, it is sufficient to consider the case $F = c\theta_q$.

So, consider the equation

$$\theta_t = r\theta + (\mu - rS)q - \mu\theta_S - \frac{1}{2}\sigma^2\theta_{SS} - \frac{1}{2}\gamma\sigma^2 e^{r(T-t)}(\theta_S - q)^2 + c\theta_q. \quad (71)$$

Take the change of variables $\theta = Sq + e^{rt}\varphi$, then

$$\theta_t = re^{rt}\varphi + e^{rt}\varphi_t, \quad \theta_q = S + e^{rt}\varphi_q, \quad \theta_S = q + e^{rt}\varphi_S, \quad \theta_{SS} = e^{rt}\varphi_{SS}.$$

Substitute them into (71) and obtain

$$\varphi_t = -\mu\varphi_S - \frac{1}{2}\sigma^2\varphi_{SS} - \frac{1}{2}\gamma\sigma^2 e^{rT}\varphi_S^2 + cSe^{-rt} + c\varphi_q.$$

After the substitution $\varphi = \ln y/(\gamma e^{rT})$, we have

$$y_t = -\frac{1}{2}\sigma^2 y_{SS} - \mu y_S + c y_q + cS\gamma e^{r(T-t)} y.$$

Now use the change of variables $v = q + ct$, then

$$y_t = -\frac{1}{2}\sigma^2 y_{SS} - \mu y_S + cS\gamma e^{r(T-t)} y.$$

Set

$$y(t, S, v) = z(t, S, v) \exp\Big(- \frac{c\gamma e^{r(T-t)}}{r}\big(S + \frac{\mu}{r}\big) + \frac{c^2\gamma^2\sigma^2}{4r^3} e^{2r(T-t)}\Big),$$

then

$$z_t = -\frac{1}{2}\sigma^2 z_{SS} + \big(\frac{c\gamma\sigma^2}{r} e^{r(T-t)} - \mu\big) z_S.$$

After the next change of the variables

$$u = -t, \quad w = \frac{\sqrt{2}}{\sigma}\Big(S - \mu t - \frac{c\gamma\sigma^2}{r^2} e^{r(T-t)}\Big)$$

we obtain the equation

$$\zeta_u(u, v, w) = \zeta_{ww}(u, v, w), \tag{72}$$

where

$$\zeta(u, w, v) = z\Big(- u, \frac{\sigma}{\sqrt{2}} w - \mu u + \frac{c\gamma\sigma^2}{r^2} e^{r(T+u)}, v\Big).$$

Thus, we can use the known symmetries of the heat equation for the group analysis of (72), if to take into account the dependence of the unknown function on three variables, including the additional variable v, which is absent in Eq. (72) in an explicit form.

As a result, the replacement of variables has the form

$$t = -u, \quad q = v + cu, \quad S = \frac{\sigma}{\sqrt{2}} w - \mu u + \frac{c\gamma\sigma^2}{r^2} e^{r(T+u)},$$

$$\theta = \frac{e^{-r(T+u)}}{\gamma} \ln z + \big(v + cu - \frac{c}{r}\big)\big(\frac{\sigma}{\sqrt{2}} w - \mu u + \frac{c\gamma\sigma^2}{r^2} e^{r(T+u)}\big)$$
$$- \frac{c\mu}{r^2} + \frac{c^2\gamma\sigma^2}{4r^3} e^{r(T+u)};$$

and the inverse change of variables is

$$u = -t, \quad v = ct + q, \quad w = \frac{\sqrt{2}}{\sigma}\Big(S - \mu t - \frac{c\gamma\sigma^2}{r^2} e^{r(T-t)}\Big),$$

$$z = \exp\left\{\gamma e^{r(T-t)}\left[\theta + (\frac{c}{r} - q)S + \frac{c\mu}{r^2}\right] - \frac{c^2\gamma^2\sigma^2}{4r^3}e^{2r(T-t)}\right\}.$$

Therefore, we have

$$\partial_u = -\partial_t + c\partial_q + (\frac{c\gamma\sigma^2}{r}e^{r(T-t)} - \mu)\partial_S + \left[-r\theta + \frac{2c\mu}{r} - \frac{c^2\gamma\sigma^2}{2r^2}e^{r(T-t)} + q\big(rS + \frac{c\gamma\sigma^2}{r}e^{r(T-t)} - \mu\big)\right]\partial_\theta,$$

$$\partial_v = \partial_q + S\partial_\theta, \quad \partial_w = \frac{\sigma}{\sqrt{2}}\partial_S + \frac{\sigma}{\sqrt{2}}(q - \frac{c}{r})\partial_\theta,$$

$$\partial_z = \frac{e^{r(t-T)}}{\gamma}\exp\left\{-\gamma e^{r(T-t)}\left[\theta + \left(\frac{c}{r} - q\right)S - \frac{c\mu}{r^2}\right] + \frac{c^2\gamma^2\sigma^2}{4r^3}e^{2r(T-t)}\right\}\partial_\theta.$$

Using the known symmetries of heat equation, we obtain the symmetries of Eq. (72)

$$X_1 = c_1(v)\partial_u, \quad X_2 = c_2(v)\partial_w, \quad X_3 = c_3(v)\left(2u\partial_u + w\partial_w\right),$$

$$X_4 = c_4(v)z\partial_z, \quad X_5 = c_5(v)\left(2u\partial_w - wz\partial_z\right),$$

$$X_6 = c_6(v)\left[u^2\partial_u + uw\partial_w - (\frac{u}{2} + \frac{w^2}{4})z\partial_z\right],$$

$$X_Z = Z(u, w, v)\partial_z, \quad X_V = V(v)\partial_v,$$

where c_i, $i = 1, 2, \ldots, 6$, are functions depending on the implicit variable v, $Z = Z(u, v, w)$ is an arbitrary solution of (72) and $V = V(v)$ is an arbitrary function. The symmetry X_V arises due to the presence of the implicit variable v. Using the system of one-dimensional subalgebras of the Lie algebra L_6 of the heat equation [37], we can obtain non-equivalent invariant solutions for Eq. (71), for example,

$$z_1 = A(v)w + B(v), \quad z_2 = e^{au}(A(v)e^{\sqrt{a}w} + B(v)e^{-\sqrt{a}}),$$

$$z_3 = e^{\frac{2}{3}u^3 - uw}(A(v)\mathrm{Ai}(u^2 - w) + B(v)\mathrm{Bi}(u^2 - w)),$$

where A and B are arbitrary function on v; Ai, Bi are the Airy functions of the 1st and the 2nd kind, respectively.

Returning to the original variables, we get the symmetries of Eq. (71)

$$Y_1 = c_1(ct + q)\left\{-\partial_t + c\partial_q + (\frac{c\gamma\sigma^2}{r}e^{r(T-t)} - \mu)\partial_S\right.$$

$$+\left[-r\theta+\frac{2c\mu}{r}-\frac{c^2\gamma\sigma^2}{2r^2}e^{r(T-t)}+q\big(rS+\frac{c\gamma\sigma^2}{r}e^{r(T-t)}-\mu\big)\right]\partial_\theta\Big\},$$

$$Y_2=c_2(ct+q)\left[\partial_S+(q-\frac{c}{r})\partial_\theta\right],$$

$$Y_3=c_3(ct+q)\Big\{2t\partial_t-2ct\partial_q+\big(S+\mu t-(1+2rt)\frac{c\gamma\sigma^2}{r^2}e^{r(T-t)}\big)\partial_S$$

$$+\Big[2rt\theta-\frac{3c\mu t}{r}+(1+rt)\frac{c^2\gamma\sigma^2}{r^3}e^{r(T-t)}-\frac{c}{r}S$$

$$+(1-2rt)qS-(1+2rt)\frac{c\gamma\sigma^2q}{r^2}e^{r(T-t)}+\mu tq\Big]\partial_\theta\Big\},$$

$$Y_4=c_4(ct+q)e^{r(t-T)}\partial_\theta,$$

$$Y_5=c_5(ct+q)\left\{\sigma t\partial_S+\big(\sigma tq+\frac{e^{r(t-T)}(S-\mu t)}{\gamma\sigma}-\frac{c\sigma}{r^2}(1+rt)\big)\partial_\theta\right\},$$

$$Y_6=c_6(ct+q)\Big\{-t^2\partial_t+ct^2\partial_q+\left[(1+rt)\frac{c\gamma\sigma^2t}{r^2}e^{r(T-t)}-tS\right]\partial_S$$

$$+\Big[-rt^2\theta+\frac{c\mu t^2}{r}-(2+rt)\frac{c^2\gamma\sigma^2t}{2r^3}e^{r(T-t)}+\frac{ctS}{r}+(1+rt)tSq$$

$$+(1+rt)\frac{c\gamma\sigma^2tq}{r^2}e^{r(T-t)}+\frac{te^{r(t-T)}}{2\gamma}-\frac{\left(S-\mu t-\frac{c\gamma\sigma^2}{r^2}e^{r(T-t)}\right)^2e^{r(t-T)}}{2\gamma\sigma^2}\Big]\partial_\theta\Big\},$$

$$Y_\Theta=\Theta(t,S,q)\frac{e^{r(t-T)}}{\gamma}\exp\left\{-\gamma e^{r(T-t)}\left[\theta+(\frac{c}{r}-q)S-\frac{c\mu}{r^2}\right]+\frac{c^2\gamma^2\sigma^2}{4r^3}e^{2r(T-t)}\right\}\partial_\theta,$$

$$Y_V=V(ct+q)\left(\partial_q+S\partial_\theta\right),$$

where $\Theta(t,S,q)=\zeta\left(-t,\frac{\sqrt{2}}{\sigma}\left[S-\mu t-\frac{c\gamma\sigma^2}{r^2}e^{r(T-t)}\right],ct+q\right)$ is an arbitrary solution of Eq. (71), and invariant solutions

$$\theta_1=\frac{e^{r(t-T)}}{\gamma}\ln\left[\frac{\sqrt{2}A}{\sigma}\big(S-\mu t-\frac{c\gamma\sigma^2}{r^2}e^{r(T-t)}\big)+B\right]$$

$$+\frac{c^2\gamma\sigma^2}{4r^3}e^{r(T-t)}+\left(q-\frac{c}{r}\right)S-\frac{c\mu}{r^2},$$

$$\theta_2 = \frac{e^{r(t-T)}}{\gamma}\left\{-at + \ln\left[Ae^{\frac{\sqrt{2a}}{\sigma}\left(S-\mu t-\frac{c\gamma\sigma^2}{r^2}e^{r(T-t)}\right)} + Be^{\frac{-\sqrt{2a}}{\sigma}\left(S-\mu t-\frac{c\gamma\sigma^2}{r^2}e^{r(T-t)}\right)}\right]\right\}$$
$$+ \frac{c^2\gamma\sigma^2}{4r^3}e^{r(T-t)} + (q - \frac{c}{r})S - \frac{c\mu}{r^2},$$

$$\theta_3 = \frac{e^{r(t-T)}}{\gamma}[\frac{\sqrt{2}t}{\sigma}(S - \mu t - \frac{c\gamma\sigma^2}{r^2}e^{r(T-t)}) - \frac{2}{3}t^3]$$
$$+ \frac{e^{r(t-T)}}{\gamma}\ln\left\{A(ct+q)\mathrm{Ai}\left[t^2 - \frac{\sqrt{2}}{\sigma}(S - \mu t - \frac{c\gamma\sigma^2}{r^2}e^{r(T-t)})\right]\right.$$
$$\left.+B(ct+q)\mathrm{Bi}\left[t^2 - \frac{\sqrt{2}}{\sigma}(S - \mu t - \frac{c\gamma\sigma^2}{r^2}e^{r(T-t)})\right]\right\}$$
$$+ \frac{c^2\gamma\sigma^2}{4r^3}e^{r(T-t)} + \left(q - \frac{c}{r}\right)S - \frac{c\mu}{r^2},$$

where A, B are arbitrary functions on $ct + q$. Acting by the entire finite-dimensional group of the heat equation, we can obtain more general multi-parameter invariant with respect to L_6 solutions.

A detailed study of the Lie algebra $\langle X_1, X_2, \ldots, X_6, X_Z, X_V\rangle$ will yield other invariant solutions of (71) by the same way.

5 Conclusion

The group classification is obtained for a class of the Guéant and Pu models. Three specifications of the free element of the execution costs function represent the three various classes of equations with non-equivalent Lie algebras. The algebras are used for the search of invariant solutions and submodels of the Guéant and Pu models.

In the case of the linear execution costs function, the equation is reduced to the heat equation with one implicit variable. Using the symmetries of such equation, we obtain the algebra Lie and some invariant solutions of the Guéant and Pu equation.

Acknowledgements The reported study was funded by the Russian Foundation for Basic Research, project number 19-01-00244; and by Act 211 of Government of the Russian Federation, contract 02.A03.21.0011.

References

1. Bachelier L (1900) Théorie de la spaculation. Annales de l'Ecole Normale Supérieure 17:21–86
2. Samuelson PA (1965) Rational theory of warrant pricing. Ind Manag Rev 6:13–31
3. Black F (1976) The pricing of commodity contracts. J Financ Econ 3:167–179

4. Black F, Scholes M (1973) The pricing of options and corporate liabilities. J Polit Econ 81:637–659
5. Merton RC (1973) Theory of rational option pricing. Bell J Econ Manag Sci 4:141–183
6. Kyle AS (1985) Continuous auctions and insider trading. Econometrica 53:1315–1335
7. Bank P, Baum D (2004) Hedging and portfolio optimization in financial markets with a large trader. Math Financ 14:1–18
8. Çetin U, Jarrow R, Protter P (2004) Liquidity risk and arbitrage pricing theory. Financ Stoch 8:311–341
9. Çetin U, Rogers LC (2007) Modelling liquidity effects in discrete time. Math Financ 17:15–29
10. Grossman S (1998) An Analysis of the implications for stock and futures price volatility of program trading and dynamic hedging strategies. J Bus 61(3):275–298
11. Platen E, Schweizer M (1998) On feedback effects from hedging derivatives. Math Financ 8:67–84
12. Sircar R, Papanicolaou G (1998) Generalized Black – Scholes models accounting for increased market volatility from hedging strategies. Appl Math Financ 5(1):45–82
13. Schönbucher P, Wilmott P (2000) The feedback-effect of hedging in illiquid markets. SIAM J Appl Math 61:232–272
14. Magill MJP, Constantinides GM (1976) Portfolio selection with transactions costs. J Econ Theory 13:245–263
15. Leland HE (1985) Option pricing and replication with transactions costs. J Financ 40:1283–1301
16. Barles G, Soner HM (1998) Option pricing with transaction costs and a nonlinear Black – Scholes equation. Finance Stochast 2:369–397
17. Cvitanić J, Karatzas I (1996) Hedging and portfolio optimization under transaction costs: a martingale approach. Math Financ 6:133–165
18. Rogers LC, Singh LS (2010) The cost of illiquidity and its effects on hedging. Math Financ 20(4):597–615
19. Li TM, Almgren R (2013) Closed-form solutions for option hedging with market impact. SSRN Electron J. 13 September 2013. http://dx.doi.org/10.2139/ssrn.2318897. Accessed 04 Oct 2020
20. Gazizov RK, Ibragimov NH (1998) Lie symmetry analysis of differential equations in finance. Nonlinear Dyn 17:387–407
21. Bordag LA (2008) On option-valuation in illiquid markets: invariant solutions to a nonlinear model. In: Sarychev A, Shiryaev A, Guerra M, Grossinho MR (eds) Mathematical control theory and finance. Springer, Berlin, pp 71–94
22. Bordag LA (2015) Geometrical properties of differential equations: applications of the lie group analysis in financial mathematics. World Scientific Publishing Company, Singapore
23. Bordag LA, Chmakova AY (2007) Explicit solutions for a nonlinear model of financial derivatives. Int J Theor Appl Fin 10(1):1–21
24. Bordag LA, Frey R (2008) Pricing options in illiquid markets: symmetry reductions and exact solutions. In: Ehrhardt M (ed) Nonlinear models in mathematical finance: research trends in option pricing. Nova Science Publishers, New York, pp 83–109
25. Bordag LA, Mikaelyan A (2011) Models of self-financing hedging strategies in illiquid markets: symmetry reductions and exact solution. J Lett Math Phys 96(1–3):191–207
26. Dyshaev MM (2016) Group analysis of a nonlinear generalization for Black – Scholes equation. Chelyabinsk Phys Math J 1(3):7–14
27. Dyshaev MM, Fedorov VE (2016) Symmetry analysis and exact solutions for a nonlinear model of the financial markets theory. Math Notes NEFU 23(1):28–45
28. Fedorov VE, Dyshaev MM (2016) Group classification for a general nonlinear model of option pricing. Ural Math J 2(2):37–44
29. Dyshaev MM, Fedorov VE (2017) Symmetries and exact solutions of a nonlinear pricing options equation. Ufa Math J 9:29–40
30. Fedorov VE, Dyshaev MM (2018) Invariant solutions for nonlinear models in illiquid markets. Math Methods Appl Sci 41(18):8963–8972

31. Dyshaev MM, Fedorov VE (2019) Comparing of some sensitivities (Greeks) for nonlinear models of option pricing with market illiquidity. Math Notes NEFU 26(2):94–108
32. Fedorov VE, Dyshaev MM (2021) Group classification for a class of non-linear models of the RAPM type. Commun Nonlinear Sci Numer Simul 92:105471
33. Guéant O, Pu J (2015) Option pricing and hedging with execution costs and market impact. arXiv:1311.4342
34. Guéant O (2016) The financial mathematics of market liquidity: from optimal execution to market making. Chapman and Hall/CRC, London
35. Ovsiannikov LV (1982) Group analysis of differential equations. Academic, New York
36. Bateman H, Erdelyi A (1953) Higher transcendental functions, vol 2. McGraw-Hill Book Company, New York
37. Olver PJ (1986) Applications of lie groups to differential equations. Springer, New York

On Involutive Systems of Partial Differential Equations

A. A. Talyshev

Abstract This chapter presents a criterion of involutive systems of partial differential equations. The criterion is based on the concept of formal extended Pfaffian systems with fixed independent variables which introduced in this chapter. The system is involutive if and only if the formal extended of the system coincides with the usual extended. This criterion was proved that the order of nontrivial contact transformations allowed by the involutive system of partial differential equations cannot exceed the order of this system. This criterion can also be useful for constructing computer algorithms for reducing a system of differential equations to an involutive form.

1 Introduction

Compatibility theory of overdetermined systems and, in particular, the notion of involutive systems of differential equations has many applications in theoretical and applied researches. For example, in the method of differential constraints, when constructing partially invariant solutions with respect to Lie groups and invariant solutions with respect to Lie–Bäcklund algebras [6, 7].

The method of differential constraints is to build such differential equations (constraints), joining which to a given system gives an integrable or even involutive system. The resulting overridden the system, as a rule, is easier to integrate, due to reducing arbitrariness in building a solution. Method differential constraints was proposed in the paper [18].

The concept of involutive systems of differential equations was introduced by Cartan [1], studying systems with partial derivatives in the form of Pfaffian systems. The presentation of Cartan's method is also contained in the papers [2, 13].

A. A. Talyshev (✉)
Novosibirsk State University, Pirogova st. 2, Novosibirsk, Russia
e-mail: tal@academ.org

A. C. J. Luo and R. K. Gazizov (eds.), *Symmetries and Applications of Differential Equations*, Nonlinear Physical Science,
https://doi.org/10.1007/978-981-16-4683-6_7

Modern consistency theory on fiber manifolds developed in papers [3–5, 9, 10, 12, 14]. The advantage of this theory is geometric, invariant under of changes of variables, the presentation, although for practical goals naturally have to use coordinate approach.

Condition of invariance or partial invariance solutions with respect to the Lie group can be written as addition to the system additional equations. In this sense, the method of differential constraints can be considered more versatile, but the construction invariant and partially invariant solutions by passing to the space of invariants much easier.

To build solutions that are invariant or partially Lie–Bäcklund groups invariant, until such a convenient theory has been developed as in the case of Lie groups. This is, in particular, related to difficulties in determining the universal invariant Lie–Bäcklund algebras. But the invariance condition solutions with respect to the Lie–Bäcklund group immediately leads, as in the case of Lie groups, to the system with differential constraints. In the paper [11] it is shown under certain conditions for an admissible Lie–Bäcklund group existence of invariant regarding her decisions.

In the papers [8, 15, 16] the existence of nontrivial tangential transformations, transformations which preserve the tangent structure on solutions of differential equations.

This chapter presents a criterion of involutive systems of partial differential equations. In the paper [17] this criterion was proved that the order of nontrivial contact transformations allowed by the involutive system of partial differential equations cannot exceed the order of this system.

2 Preliminary Information

2.1 *The Systems of Partial Differential Equations*

The term manifold will mean a connected finite-dimensional manifold of class C^∞. Everything the mappings under consideration, unless otherwise stated, will be assumed class C^∞. Restriction mapping to submanifold, when it does not cause confusion, will be denoted by the same symbol as the mapping itself. If a $\mathfrak{N}$ is a manifold, then by $T(\mathfrak{N})$, $T^*(\mathfrak{N})$, $\Lambda^p(\mathfrak{N})$ will denote the tangent, cotangent, and external forms of degree p of the vector bundle of $\mathfrak{N}$. Manifold chart $\mathfrak{N}$ will be denoted by (V, ψ), where V is the region in $\mathfrak{N}$ and ψ is a homeomorphism of V to R^n.

The triple $(\mathfrak{M}, \mathfrak{N}, \rho)$ is called fiber manifold if $\mathfrak{M}, \mathfrak{N}$ are manifolds and $\rho : \mathfrak{M} \to \mathfrak{N}$ is surjective mapping of constant rank.

If $m + n$ and n are the dimensions of the manifolds $\mathfrak{M}$ and $\mathfrak{N}$, respectively, then for each point $\omega \in \mathfrak{M}$ there is a chart (U, φ) of the manifolds $\mathfrak{M}$ and chart (V, ψ) of the manifolds $\mathfrak{N}$ such that $\omega \in U$, $\rho(U) \subset V$ and $\pi \circ \varphi = \psi \circ \rho$, where π is canonical projection R^{n+m} onto R^n.

The chart (U, φ) is called a fiber chart over the chart (V, ψ). The coordinates of the points of the manifold $\mathfrak{M}$ with respect to the fiber chart will be denoted by (x, y), $x \in R^n$, $y \in R^m$, and the coordinates of the point $\rho(\varphi^{-1}(x, y))$ relative to the chart (V, ψ) through (x).

The mapping f of the open set $V \subset \mathfrak{N}$ to $\mathfrak{M}$ is called a cross-section of the fiber manifold $(\mathfrak{M}, \mathfrak{N}, \rho)$ if the mapping $\rho \circ f$ is the identity on V. V is called the domain of the cross-section f. Coordinate representation cross-section f in the fiber chart has the form $x \to (x, \tilde{f}(x))$, where $\tilde{f} = \varphi \circ f \circ \psi^{-1}$.

Let $k \geqslant 0$ be an integer. Two cross-sections f and g, whose domains contain the point ω are called k-equivalent at the point ω if all partial derivatives at the point $\psi(\omega)$ of the mappings $\tilde{f}$ and $\tilde{g}$ match up to order k inclusive. k-equivalence is an equivalence relation and is invariant with respect to the choice of the fiber chart. A class of this equivalence is called a k-jet at the point ω. k-jet containing the section f is denoted by $j^k_\omega(f)$.

Set of all k-jets $J^k(\mathfrak{M}, \mathfrak{N}, \rho)$ has the structure manifolds of dimension $n + m(k + n)!/(k!n!)$. If a (U, φ) is a fiber chart over (V, ψ), then in $J^k(\mathfrak{M}, \mathfrak{N}, \rho)$ is introduced as follows associated chart:

$$\Phi(j^k_\omega(f)) = \left(x_0, \tilde{f}(x_0), \frac{\partial \tilde{f}}{\partial x_1}(x_0), \ldots, \frac{\partial^k \tilde{f}}{\partial x_n^k}(x_0)\right),$$

where $x_0 = \psi(\omega)$. Associated chart is a fiber chart over the chart (V, ψ). Coordinates points of the manifold $J^k(\mathfrak{M}, \mathfrak{N}, \rho)$ in associated chart will be denoted by x, y, p and variables p will be numbered with multi-indexes α (p_α, $0 < |\alpha| \leqslant k$).

If through $J^0(\mathfrak{M}, \mathfrak{N}, \rho)$, $J^{-1}(\mathfrak{M}, \mathfrak{N}, \rho)$ denote respectively $\mathfrak{M}$, $\mathfrak{N}$ and maps $\rho^k_l : J^k(\mathfrak{M}, \mathfrak{N}, \rho) \to J^l(\mathfrak{M}, \mathfrak{N}, \rho)$ is defined as follows formulas:

$$\rho^k_l(j^k_\omega(f)) = j^l_\omega(f), \quad 1 \leqslant l < k,$$

$$\rho^k_0(j^k_\omega(f)) = f(\omega), \quad \rho^k_{-1}(j^k_\omega(f)) = \omega,$$

then $(J^k(\mathfrak{M}, \mathfrak{N}, \rho), J^l(\mathfrak{M}, \mathfrak{N}, \rho), \rho^k_l)$ are fiber manifolds for $-1 \leqslant l < k$.

For each cross-section f of the fiber manifold $(\mathfrak{M}, \mathfrak{N}, \rho)$ mapping $\omega \to j^k_\omega(f)$ defines a section of the fiber manifold $(J^k(\mathfrak{M}, \mathfrak{N}, \rho), \mathfrak{N}, \rho^k_{-1})$ which will be denoted by $j^k(f)$.

The mapping $j^{k+1}_\omega(f) \to j^1_\omega(j^k(f))$ defines an embedding of the manifold $J^{k+1}(\mathfrak{M}, \mathfrak{N}, \rho)$ in $J^1(J^k(\mathfrak{M}, \mathfrak{N}, \rho), \mathfrak{N}, \rho^k_{-1})$.

The system of partial differential equations of the order k is called the submanifold $E \subset J^k(\mathfrak{M}, \mathfrak{N}, \rho)$ such that

1. $(E, \mathfrak{N}, \rho^k_{-1})$ is fiber manifold.
2. $J^1(E \cap J^l(\mathfrak{M}, \mathfrak{N}, \rho), \mathfrak{N}, \rho^l_{-1}) \cap J^{l+1}(\mathfrak{M}, \mathfrak{N}, \rho) \supset E \cap J^{l+1}(\mathfrak{M}, \mathfrak{N}, \rho)$, $l = 1, 2, \ldots, k-1$.

The solution of the system E is the cross-section f of the fiber manifold $(\mathfrak{M}, \mathfrak{N}, \rho)$ such that $j^k_\omega(f) \in E$ for each point ω from the region definition of the section f.

Locally, this definition coincides with the classical definition systems of differential equations. Condition (2) imposes the restriction only on the form of notation of the classical system. It condition means that together with any algebraic consequence $\Phi = 0$ of order $l < k$ by algebraic consequences of the system are equations whose left-hand sides are equal to all possible derivatives of Φ with respect to independent variables to order $k - l$ inclusive.

Prolongation of the order l of the system of differential equations $E \subset J^k(\mathfrak{M}, \mathfrak{N}, \rho)$ called the manifold $P^l(E)$ defined by the formula:

$$P^l(E) = J^l(E, \mathfrak{N}, \rho^k_{-1}) \cap J^{k+l}(\mathfrak{M}, \mathfrak{N}, \rho).$$

System continuation E may not be system differential equations, since $P^l(E)$ can not be a fiber manifold over $\mathfrak{N}$ with the projection ρ^{k+l}_{-1}.

3 Systems of External Differential Equations

System external differential equations on a manifold $\mathfrak{M}$ is called locally finitely generated ideal Σ of the outer form algebra $\Lambda(\mathfrak{M})$. Local finite generation means that for each points $\omega \in \mathfrak{M}$ there is such an open the set $U \ni \omega$ and a finite number of forms on U, which generate the constraint Σ on U.

The manifold $F \subset \mathfrak{M}$ is called integral manifold of the system Σ, if the restriction of any forms from Σ to F is the zero form. An integral point of the Σ system is a point in which all forms of degree zero from Σ become to zero.

$\overline{\Sigma}$ will denote the closure of the ideal Σ relative to external derivation. On any integral manifold of the system Σ any form from ideal $\overline{\Sigma}$ vanishes.

A system of external differential equations with prescribed independent variables are a pair $(\Sigma, (\mathfrak{M}, \mathfrak{N}, \rho))$, where $(\mathfrak{M}, \mathfrak{N}, \rho)$ is a fiber manifold and Σ is system of external differential equations to $\mathfrak{M}$.

The image of a cross section of the fiber manifold $(\mathfrak{M}, \mathfrak{N}, \rho)$, which is the integral manifold of the system Σ is called the integral manifold of the system $(\Sigma, (\mathfrak{M}, \mathfrak{N}, \rho))$.

For each $k > 0$ there is a system $(\Sigma^k, (J^k(\mathfrak{M}, \mathfrak{N}, \rho), \mathfrak{N}, \rho^k_{-1}))$, whose set of integral manifolds coincides with the set of images of cross sections of the form $j^k(f)$, where f is cross-section $(\mathfrak{M}, \mathfrak{N}, \rho)$. This system is generated by first-order forms that in the associated fiber chart can be written as

$$\begin{aligned} & dy - \sum_{i=1}^{n} p_{\gamma_i} dx_i, \\ & dp_\alpha - \sum_{i=1}^{n} p_{\alpha+\gamma_i} dx_i, \quad 0 < |\alpha| < k, \end{aligned} \tag{1}$$

where $|\gamma_i| = 1$, and the component with number i is equal to one.

If in a fiber chart we write down the condition that the image cross-sections f of the fiber manifold $(\mathfrak{M}, \mathfrak{N}, \rho)$ is an integral manifold of the system $(\overline{\Sigma}, (\mathfrak{M}, \mathfrak{N}, \rho))$, then this condition will be the classical system differential equations for the mapping $\tilde{f}$. If this system is consistent in some region $U \in \mathfrak{M}$, then it defines the submanifold $E_0 \subset J^1(U, \rho(U), \rho)$.

Sequence

$$E_i = J^1(E_{i-1} \cap U, \rho(U), \rho) \cap E_{i-1}, \ i = 1, 2, \ldots$$

stabilizes on some l, i.e., $E_l = E_{l+1}$ and if $(E_l, \rho(U), \rho)$ is a fiber manifold, then E_l will be denoted by $E(\overline{\Sigma})$. Thus, $E(\overline{\Sigma})$ in this case is system of differential equations equivalent to the system $(\overline{\Sigma}, (U, \rho(U), \rho))$.

Let $E \subset J^k(\mathfrak{M}, \mathfrak{N}, \rho)$ be the system differential equations. Limiting the Σ^k system to the manifold E leads to a system of exterior differential equations $(\Sigma(E), (E, \mathfrak{N}, \rho^k_{-1}))$, equivalent to the system E.

The use of the term equivalence here and above is justified in that for any solution of the system E the image of this solution is an integral manifold of the system $(\Sigma(E), (E, \mathfrak{N}, \rho^k_{-1}))$ and, conversely, for of each integral manifold of the system $(\Sigma(E), (E, \mathfrak{N}, \rho^k_{-1}))$ the cross-section defining this manifold is a solution to the system E.

The first prolongation of the system of external differential equations $(\Sigma, (\mathfrak{M}, \mathfrak{N}, \rho))$ is called system limitation $(\Sigma^1, (J^1(\mathfrak{M}, \mathfrak{N}, \rho), \mathfrak{N}, \rho^1_{-1}))$ on manifold $E(\overline{\Sigma}) \subset J^1(\mathfrak{M}, \mathfrak{N}, \rho)$, i.d. system $(\Sigma(E(\overline{\Sigma})), (E(\overline{\Sigma}), \mathfrak{N}, \rho^1_{-1}))$. Thus, the first prolongation of the system of external differential equations will be a Pfaffian system, i.d. the ideal that is generated by the forms of the first order.

4 Involutive Systems

Let be G and F are finite-dimensional vector spaces, whose dimensions are equal to m and n, respectively.

Prolongation of the space $A \subset G \otimes S^k F$ is called the space $P(A)$, where

$$P(A) = (A \otimes F) \cap (G \otimes S^{k+1} F).$$

Let be

$$\tau_i = \min \dim(G \otimes S^k(F_i) \cap A), \ i = 0, \ldots, n-1,$$

where the minimum is taken over all subspaces $F_i \subset F$ dimension $n - i$. For any A

$$\dim P(A) \leqslant \tau_0 + \tau_1 + \cdots + \tau_{n-1}.$$

A space A is called involutive if

$$\dim P(A) = \tau_0 + \tau_1 + \cdots + \tau_{n-1}. \tag{2}$$

Let $T_z(J^l)$ be the layer over the point $z \in J^l(\mathfrak{M}, \mathfrak{N}, \rho)$ in the tangent vector bundle $T(J^l(\mathfrak{M}, \mathfrak{N}, \rho))$. $Q_z(J^l)$ will denote the kernel of the mapping

$$d\rho^l_{l-1} : T_z(J^l) \to T_{z'}(J^{l-1}),$$

where $z' = \rho^l_{l-1}(z)$. $Q_z(J^l)$ is isomorphic to the space $T_y(\mathfrak{M}_\omega) \otimes S^l(T^*_\omega\mathfrak{N})$, where $y = \rho^l_0(z), \omega = \rho^l_{-1}(z)$ and $\mathfrak{M}_\omega = \rho^{-1}(\omega)$ is layer over point $\omega \in \mathfrak{N}$ in fiber manifold $(\mathfrak{M}, \mathfrak{N}, \rho)$.

System of differential equations $E \subset J^k(\mathfrak{M}, \mathfrak{N}, \rho)$ is called involutive at the point $z \in E$ if the subspace $C_z(E) = T_z(E) \cap Q_z(J^k)$ is involutive in $T_y(\mathfrak{M}_\omega) \otimes S^k(T^*_\omega\mathfrak{N})$, where $y = \rho^k_0(z)$, $\omega = \rho^k_{-1}(z)$ and there exists a neighborhood U of z in $J^k(\mathfrak{M}, \mathfrak{N}, \rho)$ that

$$\rho^{k+1}_k(P^1(E)) \cap U \supset E \cap U. \tag{3}$$

If the system E is involutive at the point $z \in E$, then it is involutive at each point of some neighborhood of the point z.

System of external differential equations $(\Sigma, (\mathfrak{M}, \mathfrak{N}, \rho))$ is called involutive if there exists and is involutive an equivalent system of differential equations $E(\overline{\Sigma})$.

In what follows, we will often consider Pfaffian systems, therefore, the involutive condition for them is useful to write in coordinate form.

Locally the forms generating the Pfaffian system $(\Sigma, (\mathfrak{M}, \mathfrak{N}, \rho))$ can always be written as

$$dy - \varphi(x, y, y_1)dx. \tag{4}$$

Here the dimension $\mathfrak{M}$ is $n + m + m_1$, the dimension $\mathfrak{N}$ equals n and $y \in R^m$, $y_1 \in R^{m_1}$. Then the system $S(\overline{\Sigma})$ in coordinates is

$$\begin{gathered} u_x = \varphi(x, u, u_1), \\ \varphi^i_{x_j} + \varphi^i_y\varphi^j + \varphi^i_{y_1}u_{1x_j} = \varphi^j_{x_i} + \varphi^j_y\varphi^i + \varphi^j_{y_1}u_{1x_i}, \\ i, j = 1, \ldots, n, \end{gathered} \tag{5}$$

where $x \to (x, u(x), u_1(x))$ is the coordinate representation of sections of a fiber manifold $(\mathfrak{M}, \mathfrak{N}, \rho)$.

Field components

$$L = \eta \cdot \partial_p + \zeta \cdot \partial_{p_1}$$

at every point z such that $L(z) \in C_z$ satisfy the equations

$$\begin{gathered} \eta = 0, \\ \varphi^i_{y_1}\zeta_j = \varphi^j_{y_1}\zeta_i, \quad i, j = 1, \ldots, n, \end{gathered} \tag{6}$$

where x, y, y_1, p, p_1 coordinates of points in the associated map. Whence it follows that the dimension $P(C_z)$ is $m_1 n - N$, where N is the rank of the system (6), and

$$\tau_{n-i} = m_1 - \rho_{i-1}, \quad i = 1, \ldots, n,$$

where $\rho_0 = 0$

$$\rho_i = \max_{c \in L(R^n)} \operatorname{rank} \begin{pmatrix} \varphi^{\alpha}_{y_1} c_{\alpha 1} \\ \ldots \\ \varphi^{\alpha}_{y_1} c_{\alpha i} \end{pmatrix}. \tag{7}$$

Thus, the condition (2) is rewritten here as

$$N = \rho_1 + \cdots + \rho_{n-1}, \tag{8}$$

and the condition (3) means that in some neighborhood the rank of the linear system (5) with respect to u_{1x} is equal N.

5 Group Analysis of Differential Systems and External Differential Equations

Classical group analysis of differential equations studies groups transformations of the space of dependent and independent variables, which translate the solutions of the system back into solutions. Group analysis relies on geometric interpretation differential equations and their solutions as submanifolds in the corresponding jet bundle. This approach combined with the locality of the considered transformations reduces the task to the study of certain classes of vector fields on $J^k(\mathfrak{M}, \mathfrak{N}, \rho)$ concerning systems of differential equations (submanifolds in $J^k(\mathfrak{M}, \mathfrak{N}, \rho)$).

From the point of view of differential equations, represent only tangent transformations of the manifold are of interest $J^k(\mathfrak{M}, \mathfrak{N}, \rho)$, i.e., those who translate images of cross sections of the form $j^k(f)$ again into the images of such cross sections.

If the dimension of the manifold $\mathfrak{M}$ is greater than $n + 1$, then tangent transformations of manifold $J^k(\mathfrak{M}, \mathfrak{N}, \rho)$ are always prolongation of point transformations, i.e., transformations manifold $\mathfrak{M}$. In the case when $m = 1$ tangents transforming manifold $J^k(\mathfrak{M}, \mathfrak{N}, \rho)$ are prolongations tangent transformations of the manifold $J^1(\mathfrak{M}, \mathfrak{N}, \rho)$ or dotted transformations.

The set of tangent transformations of a manifold $J^k(\mathfrak{M}, \mathfrak{N}, \rho)$ forms a group and the system $(\Sigma^k, (J^k(\mathfrak{M}, \mathfrak{N}, \rho), \eta, \rho_1^k)$ is invariant under the action of this group.

Further, we consider local continuous Lie groups tangent transformations and the Lie algebra of vector fields, corresponding to these groups.

Lie group of tangent transformations of a manifold $J^k(\mathfrak{M}, \mathfrak{N}, \rho)$ is called main group for the system $E \subset J^k(\mathfrak{M}, \mathfrak{N}, \rho)$, if the manifold E is invariant under actions of this group. The main group is denoted by GE. The corresponding Lie algebra of vector fields on $J^k(\mathfrak{M}, \mathfrak{N}, \rho)$ is denoted by LE. Vector fields from LE are tangent to the manifold E. The coordinate representation of this fact gives a linear system of differential equations for the components of vector fields. This system is called the determinative system for LE.

Lie algebra of vector fields on $J^\infty(\mathfrak{M}, \mathfrak{N}, \rho)$ under whose action the system Σ^∞ is invariant is called the Lie–Bäcklund algebra [6, 7] and is denoted by $L\Sigma^\infty$.

Action of the vector field L on the form ω is defined by the following formulas

$$\begin{aligned}(L)\omega &= d(L \lrcorner \omega) + L \lrcorner d\omega,\\ (L)d\omega &= d(L \lrcorner d\omega) = d((L)\omega),\end{aligned}$$

where the symbol $\lrcorner$ denotes the inner product, which is completely determined by the following formulas:

$$\begin{aligned}(\partial x_i) \lrcorner dx^j &= \delta^{ij},\\ L \lrcorner (\omega_1 \wedge \omega_2) &= (L \lrcorner \omega_1) \wedge \omega_2 + (-1)^{i_1} \omega_1 \wedge (L \lrcorner \omega_2),\\ L \lrcorner (\omega + \omega') &= L \lrcorner \omega + L \lrcorner \omega',\\ (L_1 + L_2) \lrcorner \omega &= L_1 \lrcorner \omega + L_2 \lrcorner \omega.\end{aligned}$$

Here i_1 is the order of the form ω_1.

Let be

$$L = \xi \cdot \partial_x + \sum_{|\alpha| \geqslant 0} \zeta^\alpha \cdot \partial y_\alpha,$$

where each of the components ξ, ζ^α depends on a finite number of coordinates. If $L \in L\Sigma^\infty$, then

$$\Sigma^\infty \ni (L)\omega_\alpha = d(\zeta^\alpha - y_{\alpha+\gamma_\beta}\xi^\beta) - \zeta^{\alpha+\gamma_\beta} dx_\beta + \xi^\beta dy_{\alpha+\gamma_\beta}.$$

Whence it follows that

$$(D_\beta\tilde{\zeta}^\alpha - \tilde{\zeta}^{\alpha+\gamma_\beta})dx_\beta = 0$$

and thus

$$D_j\tilde{\zeta}^\alpha - \tilde{\zeta}^{\alpha+\gamma_j} = 0, \quad j = 1, \dots, n, \ |\alpha| \geqslant 0, \tag{9}$$

where

$$\begin{aligned}\tilde{\zeta}^\alpha &= \zeta^\alpha - y_{\alpha+\gamma_\beta}\xi^\beta,\\ D_j &= \partial_{x_j} + \sum_{|\alpha| \geqslant 0} y_{\alpha+\gamma_j}\partial_{y_\alpha}, \quad j = 1, \dots, n.\end{aligned}$$

Let $\tilde{L} = L - \xi^\beta D_\beta$, then the condition (9) is rewritten as

$$\left[\tilde{L}, D_j\right] = 0, \quad j = 1, \dots, n. \tag{10}$$

Equations (9) or (10) are called determinative equations of the algebra $L\Sigma^\infty$. From Eq. (9) it follows that every vector field from $L\Sigma^\infty$ is completely defined by the components ξ and ζ. For all $j = 1, \dots, n$ fields $D_j \in L\Sigma^\infty$ and generate ideal $J\Sigma^\infty$ of $L\Sigma^\infty$. Usually instead of the algebra $L\Sigma^\infty$ consider factorization of the algebra $L\Sigma^\infty / J\Sigma^\infty$. This is justified, in particular, the fact that for each cross section f of the fiber manifold $(\mathfrak{M}, \mathfrak{N}, \rho)$ ideal $J\Sigma^\infty$ touches the image of the cross section $j^\infty(f)$.

6 Some Statements on Pfaffian Systems with Prescribed Independent Variables

To the integration of Pfaffian systems with prescribed independent variables reduce the integration of systems differential equations and the problem of constructing integral manifolds of a given dimension of systems of external differential equations.

The definitions and statements in this chapter are used in the following section and, in addition, are of independent interest, in particular, simplify specific calculations associated with the study Pfaffian systems.

6.1 Canonical Form

Lemma 1 *By transforming the variables x, y, y_1 the form system (4) can be reduced to the canonical form, in which*

$$\varphi^{i_0(l)k_0(l)} \equiv y_1^l, \quad l = 1, \dots, \rho_n,$$

where

$$(i_0(l), k_0(l)) = \begin{cases} (1, m - \rho_1 + l), \ 0 < l \leqslant \rho_1, \\ (2, m - \rho_2 + l), \ \rho_1 < l \leqslant \rho_2, \\ \dots \\ (n, m - \rho_n + l), \ \rho_{n-1} < l \leqslant \rho_n. \end{cases} \tag{11}$$

Proof Due to the definition of the values $\rho_1, \dots, \rho_n$ by a linear change of variables x, we can achieve that

$$\operatorname{rank} \begin{pmatrix} \varphi^1_{y_1} \\ \cdots \\ \varphi^i_{y_1} \end{pmatrix} = \rho_i, \quad i = 1, \dots, n.$$

Let the variables x be chosen in the indicated way, then there is such a mapping $k_0' : (1, \dots, \rho_n) \to (1, \dots, m)$ and such functions $\theta^1(x, y, y_1), \dots, \theta^{m_1 - \rho_n}(x,$

y, y_1), that the lines $\varphi_{y_1}^{i_0(l)k_0'(l)}$, $l = 1, \ldots, \rho_n$, $\theta_{y_1}^1, \ldots, \theta_{y_1}^{m_1-\rho_n}$ linear independent. If the mapping k_0' is such that for each $i = 1, \ldots, n-1$

$$k_0'(\rho_i + 1, \ldots, \rho_{i+1}) \subset k_0'(\rho_{i-1} + 1, \ldots, \rho_i), \tag{12}$$

then the transformation of variables

$$\begin{aligned} y_1^{1l} &= \varphi^{i_0(l)k_0'(l)}, \; l = 1, \ldots, \rho_n, \\ y_1^{1l} &= \theta^{l-\rho_n}, \quad\; l = \rho_n + 1, \ldots, m_1, \end{aligned} \tag{13}$$

and the corresponding renumbering of the y variables results in system (4) to the desired canonical form. It remains to prove that one can always choose k_0' satisfying the condition (12). Let for some k_0' the transformation (13), i.e., the system satisfies the condition:

$$\varphi^{i_0(l)k_0'(l)} \equiv y_1^l, \quad l = 1, \ldots, \rho_n. \tag{14}$$

If for some j_0 and $l_0 \in (\rho_{j_0} + 1, \ldots, \rho_{j_0+1})$ $k_0'(l_0) \notin k_0'(\rho_{j_0-1} + 1, \ldots, \rho_{j_0})$, then there is a $p_0 \in k_0'(\rho_{j_0-1} + 1, \ldots, \rho_{j_0})$, that $\varphi_{y_1^{l_0}}^{j_0+1p_0} \neq 0$. Indeed, if $\varphi_{y_1^{l_0}}^{j_0+1p} = 0$ for all $p \in k_0'(\rho_{j_0-1} + 1, \ldots, \rho_{j_0})$, then there are numbers c_1 and c_2 such that

$$\operatorname{rank} \begin{pmatrix} \varphi_{y_1}^1 \\ \ldots \\ \varphi_{y_1}^{j_0-1} \\ c_1\varphi_{y_1}^{j_0} + c_2\varphi_{y_1}^{j_0+1} \end{pmatrix} \geqslant \rho_{j_0} + 1,$$

but this contradicts the definition of the quantities $\rho_1, \ldots, \rho_n$. In this way, $\varphi_{y_1^{l_0}}^{j_0+1p_0} \neq 0$ for some $p_0 \in k_0'(\rho_{j_0-1} + 1, \ldots, \rho_{j_0})$ and as the new value $k_0'(l_0)$ take p_0. □

Remark 1 Coefficients of the system (4) in canonically form independent of the variables $y_1^{\rho_n+1}, \ldots, y_1^{m_1}$, so it makes sense confine ourselves to considering systems with $m_1 = \rho_n$.

Further on the system (4) it will always be assumed to be recorded in canonical form and that $m_1 = \rho_n$.

6.2 *Formal Prolongation*

The definition of a formal prolongation of the Pfaffian system is given with prescribed independent variables, which in the case an involutive system coincides with the usual prolongation.

In this section and in what follows, the following denotes:

$$\overset{0}{\rho}_i = \rho_i, \quad i = 0, 1, \ldots, n,$$

$$\overset{k}{\rho}_0 = 0, \ \overset{k}{\rho}_i = i \overset{k-1}{\rho}_n - \overset{k-1}{\rho}_0 - \cdots - \overset{k-1}{\rho}_{i-1}, \quad i = 1, \ldots, n,$$

$$m_k = \overset{k-1}{\rho}_n, \ k > 0,$$

$$(i_p(l), k_p(l)) = \begin{cases} (1, m_p - \overset{p}{\rho}_1 + l), \ 0 < l \leqslant \overset{p}{\rho}_1, \\ (2, m_p - \overset{p}{\rho}_2 + l), \ \overset{p}{\rho}_1 < l \leqslant \overset{p}{\rho}_2, \\ \ldots \\ (n, m_p - \overset{p}{\rho}_n + l), \ \overset{p}{\rho}_{n-1} < l \leqslant \overset{p}{\rho}_n, \end{cases} \tag{15}$$

$$X = R^n, \ Y_p = R^{m_p}, \ Z_p = X \times Y_0 \times \cdots \times Z \times Y_p, \ p = 0, 1, 2, \ldots.$$

Variables from Y_p are denoted by y_p.

Definition 1 Formal prolongation Ω_k of order k of a system of forms (4) is called the ideal generated by forms

$$\overset{p}{\omega} = dy_p - \overset{p}{\varphi}(x, y_0, \ldots, y_{p+1})dx, \quad p = 0, 1, \ldots, k,$$

where the mappings $\overset{p}{\varphi}$ are defined sequentially in $p = 0, 1, \ldots, k$. Formulas

$$\overset{p}{\varphi}{}^{i_p(l)k_p(l)} \equiv y^l_{p+1}, \quad l = 1, \ldots, m_{p+1}$$

define the part of components of the map $\overset{p}{\varphi}$. Other components, since $i_{p-1}(l) < i$ for $l \leqslant \overset{p-1}{\rho}_{i-1}$, are determined sequentially by $i = 2, \ldots, n$ from the formulas:

$$\overset{p}{\varphi}{}^{il} = \overset{p}{D}_{i_{p-1}(l)} \overset{p-1}{\varphi}{}^{ik_{p-1}(l)}, \quad 0 < l \leqslant \overset{p-1}{\rho}_{i-1}, \tag{16}$$

where

$$\overset{p}{D}_j = \partial_{x_j} + \overset{0}{\varphi}{}^j \cdot \partial_{y_0} + \cdots + \overset{p}{\varphi}{}^j \cdot \partial_{y_p}, \quad j = 1, \ldots, n. \tag{17}$$

Remark 2 In the same way, you can build a formal prolongation for a system that satisfies condition (14) with mapping k'_0 not satisfying condition (12). In this case, the first and subsequent formal prolongations as well as in the case when the original the system is presented in the canonical form will have canonical form.

Along with the final prolongations, we will be considered infinite prolongations of Pfaffian systems. Operators $\overset{\infty}{D}_j$ will be denoted just D_j. Operators D_j will be operated only on functions of a finite number of variables, therefore the infinite sum in the definition of the operators D_j it is enough to understand it as a convenient formal notation.

Theorem 1 *The following statements are equivalent*

1. *The Ω_0 system is involutive.*
2. $D_i \overset{0}{\varphi}{}^j - D_j \overset{0}{\varphi}{}^i = 0, \quad i, j = 1, \ldots, n.$
3. $[D_i, D_j] = 0, \quad i, j = 1, \ldots, n.$
4. $d\Omega_k \subset \Omega_{k+1}, \quad k = 0, 1, 2, \ldots.$
5. *For each $k > 0$ the system Ω_k is an prolongation of the order k (informal) system Ω_0.*

Proof $1 \Rightarrow 2$.

System Eq. (5) with $i < j$ and $l > m_0 - \overset{0}{\rho}_i$ are linearly independent and have the form:

$$u^{l'}_{1x_j} = \overset{0}{\varphi}{}^{jl}_{x_i} + \overset{0}{\varphi}{}^{jl}_{y} \overset{0}{\varphi}{}^{i} + \overset{0}{\varphi}{}^{jl}_{y_1} u_{1x_i}, \quad l' = l + \overset{0}{\rho}_i - m_0. \tag{18}$$

Since the system Ω_0 is involutive, equality (8), which means that all the equations of the system (5) are expressed through equations of the system (18). Therefore the formulas:

$$u_{1x_j} = \overset{1}{\varphi}{}^{j}, \quad j = 1, \ldots, n, \tag{19}$$

determining the general solution of the system (18), vanish all equations of the system (5). Substitution of values u_{1x_j} from (19) to the system (5) leads to the desired equality

$$D_i \overset{0}{\varphi}{}^j - D_j \overset{0}{\varphi}{}^i = 0, \quad i, j = 1, \ldots, n.$$

$2 \Rightarrow 3$.

$$[D_i, D_j] = \sum_{p=0}^{\infty} (D_i \overset{p}{\varphi}{}^j - D_j \overset{p}{\varphi}{}^i) \cdot \partial y_p.$$

Therefore, it is required to prove that

$$\overset{p}{\Phi}{}^l_{ij} = D_i \overset{p}{\varphi}{}^{jl} - D_j \overset{p}{\varphi}{}^{il},$$

$$i, j = 1, \ldots, n, \quad p = 1, 2, \ldots, \quad l = 1, \ldots, m_p,$$

given that $\overset{0}{\Phi}{}^{l}_{ij} = 0, i, j = 1, \ldots, n, l = 1, \ldots, m_p$. $\overset{p}{\Phi}{}^{l}_{ij}$=0 at $i < j$ and $l > \overset{p-1}{\rho}{}_{i-1}$ strength of determination mappings $\overset{p}{\varphi}{}^{j}$. Let $i < j$ $l \leqslant \overset{p-1}{\rho}{}_{i-1}$ then

$$\begin{aligned}\overset{p}{\Phi}{}^{l}_{ij} &= D_i \overset{p}{\varphi}{}^{jl} - D_j \overset{p}{\varphi}{}^{il} = D_i D_{i_{p-1}(l)} \overset{p-1}{\varphi}{}^{jk_{p-1}(l)} - D_j D_{i_{p-1}(l)} \overset{p-1}{\varphi}{}^{ik_{p-1}(l)} = \\ &= D_{i_{p-1}(l)}\big(D_i \overset{p-1}{\varphi}{}^{jk_{p-1}(l)} - D_j \overset{p-1}{\varphi}{}^{ik_{p-1}(l)}\big) + \big[D_i, D_{i_{p-1}(l)}\big] \overset{p-1}{\varphi}{}^{jk_{p-1}(l)} - \\ &- \big[D_j, D_{i_{p-1}(l)}\big] \overset{p-1}{\varphi}{}^{ik_{p-1}(l)} = D_{i_{p-1}(l)} \overset{p-1}{\Phi}{}^{k_{p-1}(l)}_{ij} + \overset{p-1}{\varphi}{}^{jk_{p-1}(l)}_{y^{\beta}_{\alpha}} \overset{\alpha}{\Phi}{}^{\beta}_{ii_{p-1}(l)} - \\ &- \overset{p-1}{\varphi}{}^{ik_{p-1}(l)}_{y^{\beta}_{\alpha}} \overset{\alpha}{\Phi}{}^{\beta}_{ji_{p-1}(l)},\end{aligned} \tag{20}$$

i.d. $\overset{p}{\Phi}{}^{l}_{ij}$ at $i < j$ and $l \leqslant \overset{p-1}{\rho}{}_{i-1}$ expressed through $\overset{p}{\Phi}{}^{l}_{i'j'} = 0$ with $i' < i$ and $j' \leqslant j$. Because $\overset{p}{\Phi}{}^{l}_{1j} = 0$ for all j, l, then with repeated application of the formula (20) $\overset{p}{\Phi}{}^{l}_{ij}$ with $i < j$ and $l \leqslant \overset{p-1}{\rho}{}_{i-1}$ expressed through $\overset{0}{\Phi}, \ldots \overset{p-1}{\Phi}$. Thus, the application of mathematical induction on p completes the proof.

$3 \Rightarrow 4$.

$$\begin{aligned}d \overset{p}{\omega}{}^{l} &= -d \overset{p}{\varphi}{}^{\alpha l} \wedge dx_\alpha = -\big(\overset{p}{\varphi}{}^{\alpha l}_{x_\beta} dx_\beta + \overset{p}{\varphi}{}^{\alpha l}_{y_\gamma} dy_\gamma \big) dx_\alpha = \\ &= -\sum_{\gamma=1}^{p+1} \sum_{nu=1}^{m_\gamma} \big(\overset{\gamma}{\omega}{}^{\nu} \wedge \sum_{\alpha=1}^{n} \overset{p}{\varphi} y_\gamma{}^{\nu} dx_\alpha \big) - D_\beta \overset{p}{\varphi}{}^{\alpha l} dx_\beta \wedge dx_\alpha,\end{aligned}$$

where the last term is equal to zero due to condition 3.

$4 \Rightarrow 5$.

Statement 5 is a reformulation of Statement 4 and therefore, naturally, follows from it.

$5 \Rightarrow 1$.

Statement 5 means that the formulas (18) give a general solution of the system (5), i.d. the rank of the system (5) satisfies the condition (8), and therefore the system Ω_0 is involutive. □

Corollary 1 *If the system Ω_0 is involutive, then for each $k > 0$ the system Ω_k also involutive.*

The proof is similar to the proof of the transition $5 \Rightarrow 1$ of the previous theorem.

6.3 Group Analysis

The Lie algebra of vector fields on Z_{k+1} with respect to which the invariant ideal Ω_k will be denoted by the symbol $L\Omega_k$.

In this section it will be shown that in the case of an involutive system $L\Omega_k$ for $k < \infty$ is an extension of $L\Omega_0$, and $L\Omega_0$ is a subalgebra of $L\Omega_\infty$. The algebra $L\Omega_\infty$ following [6] will be called the Lie–Bäcklund algebra.

So let $L \in L\Omega_k$ and

$$L = \xi \cdot \partial_x + \zeta \cdot \partial_y + \cdots + \zeta^{k+1} \cdot \partial_{y_{k+1}},$$

then

$$\begin{aligned}&(L) \overset{p}{\omega} = d(L \lrcorner \overset{p}{\omega}) + L \lrcorner d \overset{p}{\omega} = d(\zeta^p - \overset{p}{\varphi} \xi) - \\ &(\overset{p}{\varphi}_{x_\alpha} \xi^\alpha + \overset{p}{\varphi}_{y_0} \zeta^0 + \cdots + \overset{p}{\varphi}_{y_{k+1}} \zeta^{k+1}) dx + d \overset{p}{\varphi} \xi, \\ &p = 0, 1, \ldots, k.\end{aligned} \tag{21}$$

Since $(L) \overset{p}{\omega} \in \Omega_k$, if and only if $(L) \overset{p}{\omega} |_{\Omega_k = 0} = 0$, then from (21) for $k < \infty$ follows

$$\begin{aligned}&\overset{k}{D}_j (\zeta^p - \overset{p}{\varphi} \xi) - L \overset{p}{\varphi}{}^j + \overset{k}{D}_j \overset{p}{\varphi} \xi = 0, \quad j = 1, \ldots, n, \\ &(\zeta^p - \overset{p}{\varphi} \xi)_{y_{k+1}} + \overset{p}{\varphi}_{y_{k+1}} \xi = 0, \\ &p = 0, 1, \ldots, k\end{aligned} \tag{22}$$

and for $k = \infty$

$$\begin{aligned}&D_j(\zeta^p - \overset{p}{\varphi} \xi) - L \overset{p}{\varphi}{}^j + D_j \overset{p}{\varphi} \xi = 0, \quad j = 0, \ldots, n, \\ &j = 1, \ldots, n, \quad p = 0, 1, 2. \ldots\end{aligned} \tag{23}$$

With the notation

$$\begin{aligned}&\tilde{\zeta}^p = \zeta^p - \overset{p}{\varphi} \xi, \quad p = 0, 1, 2, \ldots, \\ &\tilde{L}^k = \tilde{\zeta}^0 \cdot \partial_{y_0} + \cdots + \tilde{\zeta}^k \cdot \partial_{y_k}, \\ &\tilde{L} = \tilde{\zeta}^0 \cdot \partial_{y_0} + \cdots = L - \xi^\alpha D_\alpha\end{aligned}$$

the system (22) is written as

$$\begin{aligned}&\overset{k}{D}_j \tilde{\zeta}^p - \tilde{L}^k \overset{p}{\varphi}{}^j - \overset{p}{\varphi}{}^j_{y_{k+1}} + \xi^\alpha (\overset{k}{D}_j \overset{p}{\varphi}{}^\alpha - \overset{k}{D}_\alpha \overset{p}{\varphi}{}^j), \quad j = 1, \ldots, n, \\ &\tilde{\zeta}^p_{y_{k+1}} + \overset{p}{\varphi}_{y_{k+1}} \xi = 0, \\ &p = 0, 1, \ldots, k\end{aligned} \tag{24}$$

and the system (23) is written as

$$D_j\tilde{\xi}^p - \tilde{L}\overset{p}{\varphi}{}^j + \xi^\alpha(D_j\overset{p}{\varphi}{}^\alpha - D_\alpha\overset{p}{\varphi}{}^j) = 0, \qquad j = 1, \ldots, n, \quad p = 0, 1, 2 \ldots. \tag{25}$$

Theorem 2 *If the system Ω_0 is involutive, then the algebra $L\Omega_1$ is a prolongation of the algebra $L\Omega_0$.*

Proof Let the field $L \in L\Omega_1$, then its coordinates satisfy the system (24) with $k = 1$, which needs to be rewritten in more detail here.

$$\overset{0}{D}_i\,\tilde{\xi}^0 + \tilde{\zeta}^0_{y_1}\,\overset{1}{\varphi}{}^i - \overset{0}{\tilde{L}}\overset{0}{\varphi}{}^i - \overset{0}{\varphi}{}^i_{y_1}\zeta^1 + \xi^\alpha\big(\overset{0}{D}_i\overset{0}{\varphi}{}^\alpha + \overset{0}{\varphi}{}^i_{y_1}\,\overset{1}{\varphi}{}^i - \overset{0}{D}_\alpha\overset{0}{\varphi}{}^i\big) = 0, \qquad i = 1, \ldots, n, \tag{26}$$

$$F_i^l = \overset{0}{D}_i\,\tilde{\xi}^{1l} - \overset{1}{\tilde{L}}\overset{1}{\varphi}{}^{il} - \overset{1}{\varphi}{}^{il}_{y_2}\zeta^2 + \xi^\alpha\big(\overset{1}{D}_i\overset{1}{\varphi}{}^{\alpha l} - \overset{1}{D}_\alpha\overset{1}{\varphi}{}^{il}\big) = 0, \qquad l = 1, \ldots, m_1, \quad i = 1, \ldots, n, \tag{27}$$

$$\tilde{\zeta}^0_{y_2} = 0, \tag{28}$$

$$\tilde{\zeta}^1_{y_2} + \xi^\alpha\,\overset{1}{\varphi}{}^\alpha_{y_2} = 0. \tag{29}$$

Equation (26) can also be written as

$$D_i^1\tilde{\xi}^0 - \overset{1}{\tilde{L}}\overset{0}{\varphi}{}^i = 0, \quad i = 1, \ldots, n.$$

Differentiating Eq. (26) with respect to y_2 gives

$$\tilde{\zeta}^0_{y_2}\,\overset{1}{\varphi}{}^i_{y_2} - \overset{0}{\varphi}{}^i_{y_1}\tilde{\zeta}^1_{y_2} = 0, \quad i = 1, \ldots, n,$$

whence given (29) it follows

$$\tilde{\zeta}^0_{y_1}\,\overset{1}{\varphi}{}^i_{y_2} - \overset{0}{\varphi}{}^i_{y_1}\,\overset{1}{\varphi}{}^\alpha_{y_2} = 0, \quad i = 1, \ldots, n. \tag{30}$$

Formula differentiation

$$\overset{1}{D}_j\overset{0}{\varphi}{}^i = \overset{1}{D}_i\overset{0}{\varphi}{}^j, \quad i, j = 1, \ldots, n,$$

by y_2 gives

$$\overset{0}{\varphi}{}^i_{y_1}\,\overset{1}{\varphi}{}^j_{y_2} = \overset{0}{\varphi}{}^j_{y_1}\,\overset{1}{\varphi}{}^i_{y_2}, \quad i, j = 1, \ldots, n,$$

therefore from (30) it follows

$$(\tilde{\zeta}^0_{y_1} + \xi^\alpha \overset{0}{\varphi}{}^\alpha_{y_1})\, \overset{1}{\varphi}{}^i_{y_2} = 0, \quad i = 1, \ldots, n.$$

In particular, $\operatorname{rank}(\overset{1}{\varphi}{}^1_{y_2}) = m_1$ therefore

$$\tilde{\zeta}^0_{y_1} + \xi^\alpha \overset{0}{\varphi}{}^\alpha_{y_1} = 0. \tag{31}$$

From (26) taking into account (31) it follows

$$\overset{0}{D}_i\, \tilde{\zeta}^0 - \overset{0}{\tilde{L}}\overset{0}{\varphi}{}^i - \overset{0}{\varphi}{}^i_{y_1} \zeta^1 + \xi^\alpha (\overset{0}{D}_i \overset{0}{\varphi}{}^\alpha - \overset{0}{D}_\alpha \overset{0}{\varphi}{}^i) = 0, \quad i = 1, \ldots, n. \tag{32}$$

Equations (31), (32) form the determinative system for algebra $L\Omega_0$. Thus, to prove the assertion of the theorem it remains to show that Eqs. (27), (28) do not follows additional equations for the coordinates ξ, ζ, ζ^1.

From (26) it follows

$$\tilde{\zeta}^{1l} = D_{i_0(l)} \tilde{\zeta}^{0k_0(l)}, \quad l = 1, \ldots, m_1. \tag{33}$$

Substitution of values $\tilde{\zeta}^1$ from (33) and $\overset{1}{\varphi}$ from (16) to (29) results in equality:

$$\left(D_{i_0(l)} \tilde{\zeta}^{k_0(l)}\right)_{y_2} + \xi^\alpha \left(D_{i_0(l)}\, \overset{0}{\varphi}{}^{\alpha k_0(l)}\right)_{y_2} = (\tilde{\zeta}^{0k_0(l)}_{y_1} + \xi^\alpha\, \overset{0}{\varphi}{}^{\alpha k_0(l)}_{y_1})\, \overset{1}{\varphi}{}^{i_0(l)} = 0.$$

Thus Eq. (29) is satisfied by virtue of Eqs. (31), (32).

Substitution of values $\tilde{\zeta}^1$ from (33) and $\overset{1}{\varphi}$ from (16) to (27) gives

$$\begin{aligned}
F_i^l =& \overset{1}{D}_i \overset{1}{D}_{i_0(l)}\, \tilde{\zeta}^{0k_0(l)} - \overset{1}{\tilde{L}} \overset{1}{D}_{i_0(l)} \overset{0}{\varphi}{}^{ik_0(l)} - (\overset{1}{D}_{i_0(l)} \overset{0}{\varphi}{}^{ik_0(l)})_{y_2} \zeta^2 + \xi^\alpha (\overset{1}{D}_i \overset{1}{D}_{i_0(l)} \overset{0}{\varphi}{}^{\alpha k_0(l)} - \overset{1}{D}_\alpha \overset{1}{D}_{i_0(l)} \overset{0}{\varphi}{}^{ik_0(l)}) \\
=& \overset{1}{D}_{i_0(l)} (\overset{1}{D}_i\, \tilde{\zeta}^{0k_0(l)} - \overset{1}{\tilde{L}} \overset{0}{\varphi}{}^{ik_0(l)}) + [\overset{1}{D}_i, \overset{1}{D}_{i_0(l)}] \tilde{\zeta}^{0k_0(l)} - [\overset{1}{\tilde{L}}, \overset{1}{D}_{i_0(l)}]\, \overset{0}{\varphi}{}^{ik_0(l)} - \overset{0}{\varphi}{}^{ik_0(l)}_{y_1}\, \overset{1}{\varphi}{}^{i_0(l)}_{y_2} \zeta^2 \\
&+ \xi^\alpha (\overset{1}{D}_{i_0(l)} (\overset{1}{D}_i \overset{0}{\varphi}{}^{\alpha k_0(l)} - \overset{1}{D}_\alpha \overset{0}{\varphi}{}^{ik_0(l)}) + [\overset{1}{D}_i, \overset{1}{D}_{i_0(l)}]\, \overset{0}{\varphi}{}^{\alpha k_0(l)} - [\overset{1}{D}_\alpha, \overset{1}{D}_{i_0(l)}]\, \overset{0}{\varphi}{}^{ik_0(l)}) \\
=& \tilde{\zeta}^{0k_0(l)}_{y_1} (\overset{1}{D}_i \overset{1}{\varphi}{}^{i_0(l)} - \overset{1}{D}_{i_0(l)} \overset{1}{\varphi}{}^i) - \overset{0}{\varphi}{}^{ik_0(l)}_{y_1} (\overset{1}{\tilde{L}} \overset{1}{\varphi}{}^{i_0(l)} - \overset{1}{D}_{i_0(l)}\, \tilde{\zeta}^1) - \overset{0}{\varphi}{}^{ik_0(l)}_{y_1}\, \overset{1}{\varphi}{}^{i_0(l)}_{y_2} \zeta^2 \\
&+ \xi^\alpha (\overset{0}{\varphi}{}^{\alpha k_0(l)}_{y_1} (\overset{1}{D}_i \overset{1}{\varphi}{}^{i_0(l)} - \overset{1}{D}_{i_0(l)} \overset{0}{\varphi}{}^i) - \overset{0}{\varphi}{}^{ik_0(l)}_{y_1} (\overset{1}{D}_\alpha \overset{1}{\varphi}{}^{i_0(l)} - \overset{1}{D}_{i_0(l)} \overset{0}{\varphi}{}^\alpha)) \\
=& (\tilde{\zeta}^{0k_0(l)}_{y_1} + \xi^\alpha\, \overset{0}{\varphi}{}^{\alpha k_0(l)}_{y_1}) (\overset{1}{D}_i \overset{1}{\varphi}{}^{i_0(l)} - \overset{1}{D}_{i_0(l)} \overset{0}{\varphi}{}^i) + \overset{0}{\varphi}{}^{ik_0(l)}_{y_1} (\overset{1}{D}_{i_0(l)}\, \tilde{\zeta}^1 - \overset{1}{\tilde{L}} \overset{0}{\varphi}{}^{i_0(l)} - \overset{1}{\varphi}{}^{i_0(l)}_{y_2} \zeta^2) \\
&+ \xi^\alpha (\overset{1}{D}_{i_0(l)} \overset{1}{\varphi}{}^\alpha - \overset{1}{D}_\alpha \overset{0}{\varphi}{}^{i_0(l)}).
\end{aligned}$$

Whence follows:

$$F_i^l = \overset{0}{\varphi}{}^{ik_0(l)}_{y_1^\beta} F^\beta_{i_0(l)}, \quad l = 1, \ldots, m_1, \quad i = 1, \ldots, n. \tag{34}$$

The part of Eq. (27) defines

$$\zeta^{2l} = D_{i_1(l)} \tilde{\zeta}^{1k_1(l)} + \xi^\alpha D_{i_1(l)} \overset{1}{\varphi}{}^{\alpha k_1(l)}. \tag{35}$$

The rest of Eq. (27) is satisfied due to (31), (32), and (35). Indeed, for each $i = 1, \ldots, n$ the equation $F_i^l = 0$ with $l > \rho_{i-1}^0$ belongs to the part that defines ζ^2 and F_i^l with $l \leqslant \rho_{i-1}^0$ are expressed by the formula (34) through $F_{i'}^j$ with $i' < i$ (for F_i^l with $l \leqslant \rho_{i-1}^0$ due to (15) $i_0(l) < i$). Since $\overset{0}{\rho}_0 = 0$, then by successive application formulas (34) each $F_i^l, l = 1, \ldots, m_1$, $i = 1, \ldots, n$ can be expressed in terms of those that define ζ^2. □

Corollary 2 *If Ω_0 is involutive, then for all $k < \infty$ the algebra $L\Omega_k$ is an prolongation of the order k of the algebra $L\Omega_0$.*

Proof Since for each $k < \infty$ the system Ω_k has the same form as the system Ω_0 and is involutive due to system Ω_0 is involutive, then the application of mathematical induction using the previous theorem proves the assertion. □

If the system Ω_0 is involutive, then Eq. (25) is rewritten as

$$D_i \tilde{\zeta}^p - \tilde{L} \overset{p}{\varphi}{}^i, \quad i = 1, \ldots, n, \quad p = 0, 1, 2, \ldots, \tag{36}$$

or

$$\left[D_i, \tilde{L}\right] = 0, \quad i = 1, \ldots, n. \tag{37}$$

From Eq. (36) follows:

$$\tilde{\zeta}^{p+1l} = D_{i_p(l)} \tilde{\zeta}^{pk_p(l)}, \quad l = 1, \ldots, m_{p+1}, \quad p = 0, 1, 2, \ldots. \tag{38}$$

Thus, all coordinates of the operator $\tilde{L}$ are expressed through $\tilde{\zeta}^0$.

Theorem 3 **2.** *Equation (36) are satisfied due to Eq. (38) and equations*

$$D_i \tilde{\zeta}^0 - \tilde{L} \overset{0}{\varphi}{}^i = 0, \quad i = 1, \ldots, n.$$

Proof Let

$$\overset{pl}{F}_i \equiv D_i \tilde{\zeta}^{pl} - \tilde{L} \overset{pil}{\varphi}, \quad i = 1, \ldots, n, \quad l = 1, \ldots, m_p, \quad p = 0, 1, 2, \ldots.$$

Then substitution of values $\tilde{\zeta}^p$ from (38) and $\overset{p}{\varphi}$ from (16) to (36) gives

$$\begin{aligned}\overset{pl}{F}_i &= D_i D_{i_{p-1}(l)} \tilde{\zeta}^{p-1\,k_{p-1}(l)} - \tilde{L} D_{i_{p-1}(l)} \overset{p-1}{\varphi}{}^{ik_{p-1}(l)} \\ &= D_{i_{p-1}(l)} \big(D_i \tilde{\zeta}^{p-1\,k_{p-1}(l)} - \tilde{L} \overset{p-1}{\varphi}{}^{ik_{p-1}(l)} \big) \\ &\quad - \big[\tilde{L}, D_{i_{p-1}(l)} \big] \overset{p-1}{\varphi}{}^{ik_{p-1}(l)} \\ &= D_{i_{p-1}(l)} \overset{p-1}{F}{}^{k_{p-1}(l)}_{i} - \overset{p-1}{\varphi}{}^{ik_{p-1}(l)}_{y^\beta_\alpha} \overset{\alpha\,\beta}{F}_{i_{p-1}(l)}.\end{aligned} \tag{39}$$

Thus, repeated application of the formula (39) and using the method of mathematical induction proves required approval. □

From the formulas (37) it follows that the vector fields D_i, $i = 1, \dots, n$ belong to the algebra $L\Omega$ and generate the ideal of this algebra, which will be denoted by $J\Omega$.

7 Involutive Pfaffian Systems with Finite Relations

A Pfaffian system with finite relations is involutive if involutive the restriction of the Pfaffian system to the manifold defined by these relations. However, for applications it is useful to have a criterion for the involutivity of such systems in the form of some conditions on finite relations.

In this section, we consider the problem of constructing for a given the involutive Pfaffian system of finite relations, together with which it forms again an involutive system. Such a problem arises, for example, in connection with the implementation of the method of differential constraints.

The system is considered:

$$\begin{aligned}&\Omega = dy - \varphi(x, y, y_1) = 0, \\ &x \in R^n, \quad y \in R^m, \quad y \in R^{m_1},\end{aligned} \tag{40}$$

$$\Phi(x, y, y_1) = 0, \quad \Phi : z_1 \to R^s, \tag{41}$$

where the system (40) is involutive system and

$$\operatorname{rank}\left(\frac{\partial \Phi}{\partial y_1}\right) = s.$$

The last condition allows, by solving Eq. (41) with respect to some of the variables y_1, write (41) as

$$\bar{y}_1 = \psi(x, y, \tilde{y}_1), \tag{42}$$

where $\bar{y}_1$ and $\tilde{y}_1$ together coincide with a set of variables y_1.

Let $\widetilde{\Omega}$ denote the system restriction Ω on the manifold defined by Eq. (42), obtained by excluding variables $\bar{y}_1$ from Ω.

Without loss of generality, we can assume that the systems $\Omega = 0$ and $\widetilde{\Omega} = 0$ simultaneously satisfy the condition (11) and the system $\Omega = 0$ is written in canonical form. Indeed, by linear change of variables x it is possible to ensure that the condition (11) will be satisfied simultaneously for both systems $\Omega = 0$ and $\widetilde{\Omega} = 0$. Further, after transforming the variables $y_1 \to y_1'$ and the renumbering of variables y_1' described in the proof of the Lemma 1 relations (42) can be rewritten as

$$\bar{y}_1' = \psi'(x, y, \tilde{y}_1'),$$

where $\bar{y}_1'$ and $\tilde{y}_1'$ together coincide with the set variables y_1' and $\det(\partial \bar{y}_1'/\partial y_1) \neq 0$.

So let the systems $\Omega = 0$ and $\widetilde{\Omega} = 0$ simultaneously satisfy the condition (11) and the numbers defined relations (7), for these systems are designated accordingly via $\rho_1, \ldots, \rho_n$ and $\tilde{\rho}_1, \ldots, \tilde{\rho}_n$.

Equation (42) can be split into n groups

$$f^1 = 0,$$
$$\ldots\ldots$$
$$f^n = 0$$

in such a way that the ith group contains those equations from (42), whose variable numbers $\bar{y}_1$ are greater than ρ_{i-1} and is less than or equal to ρ_i.

Definition 2 Finite relations

$$\begin{aligned} &D_{j_1} \cdots D_{j_l} f^i = 0, \quad i = 1, \ldots, n, \\ &l = 1, \ldots, k, \quad j_1, \ldots, j_l \leqslant i \end{aligned} \tag{43}$$

are called the formal prolongation of the order k of the finite relations (41) relative to system (40).

The left-hand sides of Eq. (43) will be denoted by the symbol Φ_k. Formal prolongation of the order k of the system $\widetilde{\Omega} = 0$ coincides with the restriction of the formal prolongation of the order k of the system $\Omega = 0$ onto the manifold defined by the equations $\Phi_k = 0$.

Theorem 4 *The system (40), (41) is involutive system if and only if the vector fields generating the ideal $J\Omega$ tangent to the manifold, defined by equations $\Phi_\infty = 0$.*

Proof Necessity. If the system (40), (41) is involutive, then formal prolongation $\widetilde{\Omega}_k$ for every $k > 0$ is a prolongation of the order k of the $\widetilde{\Omega}$ system. Therefore, the equations $d\Phi_\infty = 0$ should not give new relations for the variables $\widetilde{Z}_\infty$, but

$$d\Phi_\infty = D_i \Phi_\infty dx_i.$$

So $D_i\Phi_\infty, i = 1, \dots, n$ must vanish by virtue of the equalities $\Phi_\infty = 0$.

Adequacy. Since the system (40) is involutive, then

$$D_i\varphi_j - D_j\varphi_i = 0, \quad i, j = 1, \dots, n,$$

and therefore

$$D_i\varphi_j - D_j\varphi_i|_{\Phi_\infty=0} = 0, \quad i, j = 1, \dots, n.$$

Since $D_1, \dots, D_n$ touch the manifold $\Phi_\infty = 0$, then

$$D_i\varphi_j|_{\Phi_\infty=0} = (D_i|_{\Phi_\infty=0})(\varphi_j|_{\Phi_\infty=0}) = \widetilde{D}_i\tilde{\varphi}_i, \quad i, j = 1, \dots, n$$

and thus,

$$\widetilde{D}_i\tilde{\varphi}_j - \widetilde{D}_j\tilde{\varphi}_i = 0, \quad i, j = 1, \dots, n,$$

whence by Theorem 1 it follows that system $\widetilde{\Omega} = 0$ is involutive. □

Thus, the criterion for the involutive of the system (40), (41) is the condition

$$D_i\Phi_\infty|_{\Phi_\infty=0} = 0, \quad i = 1, \dots, n,$$

which can be rewritten as

$$D_i\Phi_k|_{\Phi_{k+1}=0} = 0, \quad i = 1, \dots, n, \quad k = 0, 1, 2, \dots. \tag{44}$$

From Theorem 1 it follows that the condition (44) is satisfied if and only if

$$D_i\Phi_0|_{\Phi_1=0} = 0, \quad i = 1, \dots, n. \tag{45}$$

Since for any $f : Z_k \to R$ the expression $D_i f$ linearly depends in variables y_{k+1}, then there are linear operators $A^i_{\alpha\beta}$ (matrix) that the condition (45) can be written in the form

$$\begin{aligned} &D_i f^l = \textstyle\sum_{\beta=1}^{n} \sum_{\alpha=1}^{\beta} A^i_{\alpha\beta} D_\alpha f^\beta \Big|_{f^1=0,\dots,f^n=0} = 0, \\ &l = 1, \dots, n, \quad i = l+1, \dots, n. \end{aligned} \tag{46}$$

In the case of a scalar finite relation, further simplification of the criterion of involutivity.

Theorem 5 *System (40) with finite relations*

$$\Phi(x, y, y_1) = 0, \quad \Phi : Z_1 \to R$$

is involutive system if and only if there exist vector fields from $J\Omega$ such that the manifold defined by the equation $\Phi = 0$ is an invariant manifold under the action of these vector fields.

Proof Under the conditions of the theorem, the matrices $A^i_{\alpha\beta}$ in (46) will be scalar functions, so the sought vector fields will be

$$D_i = \sum_{\alpha=1}^{\beta} A^i_\alpha D_\alpha, \quad i = \beta + 1, \dots, n, \tag{47}$$

where β is the number of the first of the numbers $\rho_1, \dots, \rho_n$ which will decrease by one when joining the system (40) finite relation $\Phi = 0$. □

Investigation of the system (40) with relations

$$\Phi(x, y, y_1, \dots, y_k) = 0$$

reduces to studying the system $\Omega_k = 0$ with these relations, and the latter has the same form as the system (40), (41). Therefore, all the statements in this section are also true for this case.

Conditions (46) for Φ depending on $(x, y, y_1, \dots, y_k)$ split in variables y_{k+1} and in the case of scalar Φ mean that some collection of vector fields on Z_k touches the manifold defined by the equation $\Phi = 0$.

Thus, this manifold is an invariant manifold of the group Lie algebra corresponding to the Lie algebra generated by these vector fields. Therefore, the methods of constructing invariant manifolds can be used here classical group analysis. In particular, when constructing non-singular invariant manifolds, it suffices to find all the invariants of this group, which is an easier task.

The universal invariant of this group coincides with the set of invariants (depending on variables $x, y, y_1, \dots, y_k$) vector fields (47).

Further, if there is a vector field L on Z_∞ commuting with fields (47), then the set of invariants of operators (47) is closed under the action of the field L.

The $L\psi$ function for $\psi : Z_k \to R$ is usually mapping $Z_{k'} \to R$ with $k' > k$, therefore, the indicated fact allows, for example, in the method of differential constraints to construct constraints higher order from bonds of lower order.

Example 1 A system of partial differential equations with two independent variables

$$W_t + A(t, x, W)W_x = f(t, x, W),$$
$$W : R^2 \to R^m, \quad A : R^2 \times R^m \to \mathfrak{L}(R^m, R^m)$$

is equivalent to the Pfaffian system

$$\Omega = dy_0 - y_1 dx - (f - Ay_1)dt = 0$$

with prescribed independent variables (x, t). The continuation of the order k of this system is the system

$$dy_l - y_{l+1}dx - D^l_x(f - Ay_1)dt = 0, \quad l = 0, 1, \dots, k,$$

where

$$D_x = \partial_x + y_1 \cdot \partial_{y_0} + y_2 \cdot \partial_{y_1} + \cdots$$

The ideal $J\Omega$ of the Lie–Bäcklund algebra $L\Omega$ is generated here vector fields

$$D_x, \quad D_t = \partial_t + (f - Ay_1) \cdot \partial_{y_0} + D_x(f - Ay_1) \cdot \partial_{y_1} + \cdots$$

The operator (47) here has the form $D_t + \lambda D_x$, and its invariants $\psi : Z_k \to R$ are determined from the equation

$$\begin{aligned} (D_t + \lambda D_x)\psi = \psi_t + \lambda\psi_x + \psi_{y_0}(f - Ay_1 + \lambda y_1) + \cdots \\ + \psi_{y_k}(D_x^k(f - Ay_1) + \lambda y_{k+1}) = 0 \end{aligned} \tag{48}$$

whence, in particular, it follows

$$\psi_{y_k}(\lambda E - A) = 0.$$

Therefore, for the existence of nontrivial solutions, it is necessary so that λ is an eigenvalue of the matrix A.

Example 2 The system of equations of one-dimensional gas dynamics

$$\begin{pmatrix} \rho \\ u \\ p \end{pmatrix}_t + \begin{pmatrix} u & \rho & 0 \\ 0 & u & 1/\rho \\ 0 & \gamma p & u \end{pmatrix} \begin{pmatrix} \rho \\ u \\ p \end{pmatrix}_x + \frac{1}{x} \begin{pmatrix} \nu\rho u \\ 0 \\ \nu\gamma p u \end{pmatrix} = 0$$

is equivalent to the Pfaffian system

$$\begin{aligned} d\rho - \rho_1 dx + \big(\frac{\nu\rho u}{x} + u\rho_1 + \rho u_1\big)dt &= 0, \\ du - u_1 dx + \big(uu_1 + \frac{1}{\rho} p_1\big)dt &= 0, \\ dp - p_1 dx + \big(\gamma p u_1 + u p_1 + \frac{\nu\gamma p u}{x}\big)dt &= 0 \end{aligned} \tag{49}$$

with prescribed independent variables (x, t).

The corresponding matrix here has eigenvalues $u, u \pm \sqrt{\gamma p \rho^{-1}}$. Equation (48) for $\lambda = u \pm \sqrt{\gamma p \rho^{-1}}$ and $\psi : Z_k \to R$ for $k = 0, 1, 2, 3$ has no nontrivial solutions. For $\lambda = u$ Eq. (48) has solutions $\psi : Z_k \to R$ for every $k \geqslant 0$. If $k = 0$, then $\psi_0 = p\rho^{-\gamma}$.

The operator $L = \rho^{-1} x^{-\nu} D_x$ commutes with the operator $D_t + u D_x$ and therefore $\psi_k = L^k(p\rho^{-\gamma})$ for every $k \geqslant 0$ will be a solution to Eq. (48). Whence it follows that the system (49) with finite relations

$$\Phi(\psi_0, \psi_1, \ldots, \psi_k) = 0$$

for every $k \geqslant 0$ forms an involutive system.

Example 3 System (49) with finite relations $2p\rho^{-1} = g$ and $\gamma = 2,\ \nu = 0$ is reduced to the system

$$d\rho - \rho_1 dx + (u\rho_1 + \rho u_1)dt = 0,$$
$$du - u_1 dx + (uu_1 + g\rho_1)dt = 0,$$

which is equivalent to the system of equations of motion of 'shallow water' (ρ is depth, u is speed). The solutions of Eq. (48) here are the functions

$$u \pm 2\sqrt{g\rho}, \quad g\rho_1 \pm \sqrt{g\rho}\, u_1.$$

Example 4 Equation

$$u_{tt} - u_{xx} = a(u)$$

is equivalent to the Pfaffian system

$$\begin{aligned} du - pdx - qdt &= 0,\\ dp - p_1 dx - q_1 dt &= 0,\\ dq - q_1 dx - (p_1 + a)dt &= 0 \end{aligned} \tag{50}$$

with prescribed independent variables (x, t).

The operator (47) here has the form $D_t + \lambda D_x$, where

$$D_x = \partial_x + p\partial_u + p_1\partial_p + q_1\partial_q + p_2\partial_{p_1} + q_2\partial_{q_1} + \cdots,$$
$$D_t = \partial_t + q\partial_u + q_1\partial_p + (p_1 + a)\partial_q + q_2\partial_{p_1} + D_x(p_1 + a)\partial_{q_1} + \cdots$$

In order for Eq. (48) to have nontrivial solutions here, it is necessary that $\lambda = \varepsilon$, where $\varepsilon^2 = 1$.

Let $\psi : Z_2 \to$, then Eq. (48) here has the form

$$\begin{aligned} &\psi_t + \varepsilon\psi_x + (p + \varepsilon q)\psi_u + (p_1 + \varepsilon q_1)\psi_p + (q_1 + \varepsilon p_1 + a)\psi_q +\\ &+ (p_2 + \varepsilon q_2)\psi_{p_1} + (q_2 + \varepsilon p_2 + a + a'p)\psi_{q_1} = 0, \end{aligned} \tag{51}$$

whence it follows that

$$\begin{aligned} L_0\psi &= (\partial_t + \varepsilon\partial_x + (p + \varepsilon q)\partial_u + (p_1 + \varepsilon q_1)\partial_p + (q_1 + \varepsilon p_1 + a)\partial_q +\\ &\quad + a'p\partial_{q_1})\psi = 0\\ L_1\psi &= (\partial_{p_1} + \varepsilon\partial_{q_1})\psi = 0. \end{aligned}$$

Further

$$\begin{aligned}
[L_1, L_0] &= 2\varepsilon(\partial_p + \varepsilon\partial_q) = 2\varepsilon L_2, \\
[L_2, L_0] &= 2\varepsilon\partial_u + a'\partial_{q_1} = L_3, \\
[L_3, L_0] &= a'\partial_p + (\varepsilon a' + 2\varepsilon a')\partial_q + 2\varepsilon a'' p\partial_{q_1} - (q + \varepsilon p)a''\partial_{q_1} \\
&= a'L_2 + 2\varepsilon a'\partial_q + a''(\varepsilon p - q)\partial_{q_1} = a'L_2 + L_4, \\
[L_3, L_4] &= a''\partial_q + 2\varepsilon a'''(\varepsilon p - q)\partial_{q_1} = L_5.
\end{aligned}$$

Whence it follows that for the existence of a nontrivial solution it is necessary so that

$$\operatorname{rank}\begin{pmatrix} L_4 \\ L_5 \end{pmatrix} = 1,$$

i.d.

$$\det\begin{pmatrix} 2\varepsilon a' & a''(\varepsilon p - q) \\ 4a'' & 2\varepsilon a'''(\varepsilon p - q) \end{pmatrix} = 0.$$

Thus, the function a must satisfy the equation

$$a'a''' - (a'')^2 = 0,$$

integration of which gives

$$a = \frac{c_2}{c_1}e^{c_1 u} + c_3, \tag{52}$$

where c_1, c_2, c_3 are constants.

If a satisfies (52) then

$$\begin{aligned}
&[L_1, L_2] = [L_1, L_3] = [L_1, L_4] = [L_1, L_0'] = [L_2, L_3] = 0, \\
&[L_2, L_4] = [L_2, L_0'] = [L_3, L_0'] = 0, \quad [L_3, L_4] = 2\varepsilon c_1 L_4, \\
&[L_4, L_0'] = c_1 c_2 c_3 e^{c_1 u}\partial_{q_1},
\end{aligned}$$

where

$$L_0' = L_0 - (q_1 + \varepsilon p_1)L_2 - \frac{1}{2}\varepsilon(q + \varepsilon p)L_3 - \frac{\varepsilon}{2c_1}L_4 = \partial_t + \varepsilon\partial_x + c_3\partial_q.$$

Thus, Eq. (51) has a nontrivial solution if and only if $a = c$ or $a = ce^u$ (constant c_1 eliminated by stretching u).

Let $a = ce^u$, then

$$\psi = \theta\big(q_1 - \varepsilon p_1 + \frac{\varepsilon}{4}(q - \varepsilon p)^2 - \frac{c}{2\varepsilon}e^u,\ x - \varepsilon t\big),$$

where θ is an arbitrary function of two arguments.

Since D_x commutes with $D_t + \varepsilon D_x$, the system (50) with finite relations

$$\Phi(\psi, D_x\psi, \ldots, D_x^k\psi) = 0$$

is involutive system for each $k \geqslant 0$.

Example 5 Pfaffian system

$$\begin{aligned}
&d\rho - \rho_x dx - \rho_y dy - \rho_z dz + (u\rho_x + v\rho_y + w\rho_z + \rho(u_x + v_y + w_z))dt = 0,\\
&du - u_x dx - u_y dy - u_z dz + (uu_x + vu_y + wu_z + \frac{1}{\rho} p_x)dt = 0,\\
&dv - v_x dx - v_y dy - v_z dz + (uv_x + vv_y + wv_z + \frac{1}{\rho} p_y)dt = 0,\\
&dw - w_x dx - w_y dy - w_z dz + (uw_x + vw_y + ww_z + \frac{1}{\rho} p_z)dt = 0,\\
&dp - p_x dx - p_y dy - p_z dz + (up_x + vp_y + wp_z + \gamma p(u_x + v_y + w_z))dt = 0
\end{aligned} \tag{53}$$

with prescribed independent variables (x, y, z, t) corresponds to the system of equations of gas dynamics. (Here $\rho_x, \ldots, p_z$ denotes parametric variables y_1, rather than derivatives of functions, as in the rest of the work.)

The operator (47) here looks like:

$$L = D_t + \lambda_1 D_x + \lambda_2 D_y + \lambda_3 D_z.$$

Let $\psi : Z_0 \to R$ then it follows from the equation $L\psi = 0$ that

$$\begin{aligned}
&(\lambda_1 - u)\psi_\rho = 0, \quad (\lambda_2 - v)\psi_\rho = 0, \quad (\lambda_3 - w)\psi_\rho = 0,\\
&(\lambda_1 - u)\psi_u - \rho\psi_\rho - \gamma p\psi_p = 0, \quad (\lambda_2 - v)\psi_u = 0, \quad (\lambda_3 - w)\psi_u = 0,\\
&(\lambda_1 - u)\psi_v = 0, \quad (\lambda_2 - v)\psi_v - \rho\psi_\rho - \gamma p\psi_p = 0, \quad (\lambda_3 - w)\psi_v = 0,\\
&(\lambda_1 - u)\psi_w = 0, \quad (\lambda_2 - v)\psi_w = 0, \quad (\lambda_3 - w)\psi_w - \rho\psi_\rho - \gamma p\psi_p = 0,\\
&(\lambda_1 - u)\psi_p - \rho^{-1}\psi_u = 0, \quad (\lambda_2 - v)\psi_p - \rho^{-1}\psi_v = 0,\\
&(\lambda_3 - w)\psi_p - \rho^{-1}\psi_w = 0, \quad \psi_t + \lambda_1\psi_x + \lambda_2\psi_y + \lambda_3\psi_z = 0,
\end{aligned}$$

whence it follows that$\lambda_1 = u$, $\lambda_2 = v$, $\lambda_3 = w$, $\psi = \theta(p\rho^{-\gamma})$.

Further operator

$$M = (w_y - v_z)D_x + (u_z - w_x)D_y + (v_x - u_y)D_z$$

satisfies the relation

$$\begin{aligned}
[L, M] &= -(u_x + v_y + w_z)M\\
&+ \frac{1}{2}((\rho_y p_z - \rho_z p_y)D_x - (\rho_x p_z - \rho_z p_x)D_y + (\rho_x p_y - \rho_y p_x)D_z),
\end{aligned}$$

whence it follows that

$$[L, M](p\rho^{-\gamma}) = -(u_x + v_y + w_z)M(p\rho^{-\gamma}),$$

so

$$L(M(p\rho^{-\gamma}))\big|_{M(p\rho^{-\gamma})=0} = ML(p\rho^{-\gamma}) - (u_x + v_y + w_z)M(p\rho^{-\gamma})\big|_{M(p\rho^{-\gamma})=0} = 0,$$

i.d. system (53) with finite relations

$$\Phi(p\rho^{-\gamma}, M(p\rho^{-\gamma})) = 0$$

is involutive system.

8 Tangent Transformations

Tangent transformations are prolongations of point transformations in the case of many dependent variables, or prolongations of tangent transformations of the first order in the case of one dependent variable. However, when studying differential equations, it is sufficient require transformations to preserve the tangent structure only on solutions of these equations. With this approach, it is convenient to consider instead of systems differential equations are equivalent to them Pfaffian systems.

This section describes some of the classes of systems differential equations that admit nontrivial, thus defined tangent transformations.

8.1 Definition, General Statements

We consider an involutive system in partial derivatives $E \subset J^k(\mathfrak{M}, \mathfrak{N}, \rho)$. The continuation of the order l is here denoted by E^l.

System Σ^k in local coordinates $(x, \overset{0}{y}, \overset{1}{y}, \ldots, \overset{k}{y})$, $\overset{i}{y} \in R^m \otimes S^i R^n$, $i = 1, \ldots, k$ written in the form

$$d\overset{i}{y} - \overset{i+1}{y}dx = 0, \quad i = 0, 1, \ldots, k-1. \tag{54}$$

$$d\overset{i}{y} - \overset{i+1}{y}dx = 0, \quad i = 0, 1, \ldots, k-1.$$

Let the manifold E in these local coordinates is defined by equations

$$\Phi(x, \overset{0}{y}, \overset{1}{y}, \ldots, \overset{k}{y}) = 0. \tag{55}$$

Lie algebra of vector fields with respect to which the invariant system (54), (55) will be denoted LE and the transformation group GE defined by this algebra, will be called the group of tangent transformations of order k of system E.

Transformations of the GE group preserve the tangent structure of order k on the manifold defined by the relations (55), i.e., on the manifold to which the any integral manifold of the system (54), (55), and through each point of which, if Φ is R-analytic, due to the involutive system E, passes at least one n-dimensional integral manifold with independent variables x of the system (54), (55). Thus, the manifold defined by the relations (55) is a minimal manifold containing 'graphs' in the space Z_k of all solutions of the system E. It is in this sense that it will be said that the group GE is the group of tangent transformations of order k on the solutions of the system E and the restriction of the group GE on the manifold defined by the relations (55) will be called the restriction of the group GE to the set solutions of the system E.

It will be shown below that for every $l > 0$ the group GE^l is an prolongation of the group GE and therefore it makes no sense to talk about tangent transformations more higher order than the order of the system. In fact and group GE for most equations of the mathematical physics is a prolongation of the group of point transformations.

Equation (55) is solvable with respect to different variables $\overset{i}{y}$, $i = 0, \ldots, k$, and so that on the left side there is the maximum possible number of variables $\overset{k}{y}$. Thus, if the variables $\overset{i}{y}$, $i = 0, \ldots, k-1$ and $\overset{k}{y}$ in the right-hand sides, denote $\bar{y}$ and $\bar{y}_1$, respectively, and the remaining in the right-hand sides, respectively, y and y_1, then Eq. (55) takes the form:

$$\begin{aligned} \bar{y} &= f(x, y), \\ \bar{y}_1 &= f^1(x, y, y_1). \end{aligned} \tag{56}$$

Equation (54) after substitution of values in them $\bar{y}$ and $\bar{y}_1$ take the form:

$$\overset{0}{\omega} = dy - \overset{0}{\varphi}(x, y, y_1)dx = 0, \tag{57}$$

since the left-hand sides of the equations $d\bar{y} - \cdots = 0$ in this case, since the system E is involutive system, become identically equal to zero.

A similarly constructed Pfaffian system equivalent to system E^1, contains Eqs. (56), (57) and additional equations that can be written as

$$\bar{y}_2 = f^2(x, y, y_1, y_2), \tag{58}$$

$$\overset{1}{\omega} = dy_1 - \overset{1}{\varphi}(x, y, y_1, y_2)dx = 0, \tag{59}$$

where $\bar{y}_2$ and y_2 together coincide with the set variables $\overset{k+1}{y}$. Mappings $\overset{1}{\varphi} = (\overset{1}{\varphi}{}^1, \ldots, \overset{1}{\varphi}{}^n)$ satisfy the identities:

$$c \overset{0}{\varphi}{}^i_{x_j} + \overset{0}{\varphi}{}^i_y \overset{0}{\varphi}{}^j + \overset{0}{\varphi}{}^i_{y_1} \overset{1}{\varphi}{}^j = \overset{0}{\varphi}{}^j_{x_i} + \overset{0}{\varphi}{}^j_y \overset{0}{\varphi}{}^i + \overset{0}{\varphi}{}^j_{y_1} \overset{1}{\varphi}{}^i,$$
$$i, j = 1, \ldots, n,$$

and the system (59) is an prolongation of the Pfaffian system (57).

Infinitesimal operator coordinates

$$L = \xi \cdot \partial_x + \zeta \cdot \partial_y + \zeta^1 \cdot \partial_{y_1} + \bar{\zeta} \cdot \partial_{\bar{y}} + \bar{\zeta}^1 \cdot \partial_{\bar{y}_1}$$

for $L \in LE$ on the manifold defined by the relations (56), satisfy the equations:

$$\bar{\zeta} = f_x \xi + f_y \zeta, \quad \bar{\zeta}^1 = f^1_x \xi + f^1_y \zeta + f^1_{y_1} \zeta^1,$$
$$\tilde{\zeta}_{y_1} + \tilde{\zeta}_{\bar{y}_1} f^1_{y_1} + \overset{0}{\varphi}{}^\alpha_{y_1} \xi^\alpha = 0,$$
$$\tilde{\zeta}_{x_i} + \tilde{\zeta}_y \overset{0}{\varphi}{}^i + \tilde{\zeta}_{\bar{y}} (f_{x_i} + f_y \overset{0}{\varphi}{}^i) + \tilde{\zeta}_{\bar{y}_1} (f^1_{x_i} + f^1_y \overset{0}{\varphi}{}^i) - \overset{0}{\varphi}{}^i_{x_\alpha} \xi^\alpha$$
$$- \overset{0}{\varphi}{}^i_y \zeta - \overset{0}{\varphi}{}^i_{y_1} \zeta^1 + \xi^\alpha (\overset{0}{\varphi}{}^\alpha_{x_i} + \overset{0}{\varphi}{}^\alpha_y \overset{0}{\varphi}{}^i) = 0, \quad i = 1, \ldots, n,$$

where $\tilde{\zeta} = \zeta - \overset{0}{\varphi}{}^\alpha \xi^\alpha$. Coordinates of the restriction of the operator L

$$\hat{L} = \hat{\xi} \cdot \partial_x + \hat{\zeta} \cdot \partial_y + \hat{\zeta}^1 \cdot \partial_{y_1}$$

on the manifold defined by the relations (56) satisfy equations:

$$\tilde{\hat{\zeta}}_{y_1} + \overset{0}{\varphi}{}^\alpha_{y_1} \hat{\xi}^\alpha = 0,$$

$$\tilde{\hat{\zeta}}_{x_i} + \tilde{\hat{\zeta}}_y \overset{0}{\varphi}{}^i - \overset{0}{\varphi}{}^i_{x_\alpha} \hat{\xi}^\alpha - \overset{0}{\varphi}{}^i_y \hat{\zeta} - \overset{0}{\varphi}{}^i_{y_1} \hat{\zeta}^1 + \hat{\xi}^\alpha (\overset{0}{\varphi}{}^\alpha_{x_i} + \overset{0}{\varphi}{}^\alpha_y \overset{0}{\varphi}{}^i) = 0, \quad i = 1, \ldots, n,$$

but these equations are the defining equations of the algebra $L(\overset{0}{\omega})$ vector fields, relative to actions which the Pfaffian system (57) is invariant. In this way, the restriction of the algebra LE to the set of solutions of the system E, that is, on the manifold defined by the relations (56) coincides with the algebra $L(\overset{0}{\omega})$. Similarly, the restriction of the algebra LE^1 onto the manifold defined by the relations (56), (58), coincides with the algebra $L(\overset{0}{\omega}, \overset{1}{\omega})$. The same statement is true for the algebra LE^l for $l > 1$, but these equations are the defining equations of the algebra $L(\overset{0}{\omega})$ vector fields, relative to actions which the Pfaffian system (57) is invariant. In this way, the restriction of the algebra LE to the set of solutions of the system E, that is, on the manifold defined by the relations (56) coincides with the algebra $L(\overset{0}{\omega})$. Similarly, the restriction of the algebra LE^1 onto the manifold defined by the relations (56), (58), coincides with the algebra $L(\overset{0}{\omega}, \overset{1}{\omega})$. The same statement is true for the algebra LE^l for $l > 1$.

Theorem 6 *For every $l > 0$ the restriction of LE^l to the set solutions of the system E^l coincides with the continuation of the order l restrictions of the algebra LE to the set of solutions of the system E.*

The validity of the statement of the theorem follows from Theorem 2.

Remark 3 For non-involutive systems, the statement of the theorem is false.

References

1. Cartan E (1946) Les systèmes différential extérior et leurs applications scietifiques. Herman
2. Finikov SP (1948) Cartan external forms method. M.-L.: GITL (in Russian)
3. Goldshmidt H (1967) Existens theorems for analitic linear partial differential equations. Annalis of Mathimatics 86(2):246–270
4. Goldshmidt H (1967) Integrability criteria for systems of nonlinear partial differential equations. J Diff Geom 1(3):269–307
5. Guillemin V, Kuranishi M (1968) Some algebraic results conserning involutive subspaces. Amer J Math 90(4):1307–1320
6. Ibragimov NH (1979) On the theory of Lie-Bäcklund transformation groups. Math Coll 109(2):229–253 (in Russian)
7. Ibragimov NH, Anderson RL (1977) Lie-Bäcklund tangent transformations. J Math Anal Appl 59(1):145–162
8. Krendelev SF, Talyshev AA (1979) Partial systems admitting nontrivial tangent transformations. Dyn Solid Med 40:134–140 (in Russian)
9. Kuranishi M (1957) On E. Cartan's prolongations theorem of exterior differential systems. Amer J Math 79(1):1–47
10. Kuranishi M (1967) Lectures of involutive systems of partial differential equations. San Paulo
11. Meleshko SV (1983) Differential constraints and one-parameter Lie-Bäcklund transformation groups. DAN USSR 271(1):42–46 (in Russian)
12. Pommaret JF (1978) Systems of partial differential equations and Lie pseudogroups. Gordon and Breach Science Publisher
13. Rashevskii PK (1947) Geometry theory of partial differential equation. M.: Gostexizdat (in Russian)
14. Spencer DC (1969) Overdeterminated systems of linear partial differential equations. Bull Amer Math Soc 75(2):179–239
15. Talyshev AA (1980) Tangent transformations equations of one-dimensional gas dynamics with one differential constraints. Dyn Solid Med (44):121–126 (in Russian)
16. Talyshev AA (1980) Differential constraints method and invariant solutions of strictly hyperbolic systems with two independent variables. Numer Methods Cont Mech 11(6):130–134 (in Russian)
17. Talyshev AA (1982) On high order tangent transforms of systems of partial differential equation. Dyn Solid Med 54:142–152 (in Russian)
18. Yanenko NN (1964) Compatibility theory and the method of integrating nonlinear partial differential equations. Tr. IV All-Union Mathematical Congress, L.: Nauka 2:247–259

Group Analysis of Some Camassa–Holm-Type Equations

Igor Leite Freire and Júlio César Santos Sampaio

Abstract In this chapter, we consider symmetries and conservation laws for some shallow water models including the Camassa–Holm equation.

1 Introduction

Consider the equation

$$\begin{aligned} &u_t - u_{txx} + 3uu_x + \lambda(u - u_{xx}) \\ &= 2u_x u_{xx} + uu_{xxx} + \alpha u_x + \beta u^2 u_x + \gamma u^3 u_x + \Gamma u_{xxx}, \end{aligned} \tag{1}$$

where α, β, γ, Γ and λ are constants. This equation, introduced in [15], encloses several equations coming from hydrodynamics, such as

- The Camassa–Holm (CH) equation [3], whenever $\beta = \gamma = \Gamma = \lambda = 0$, and the weakly dissipative CH equation [30, 31], if $\beta = \gamma = \Gamma = 0$ and $\lambda > 0$;
- The Dullin–Gottwald–Holm (DGH) equation [13], when $\beta = \gamma = \lambda = 0$. In case $\beta = \gamma = 0$, but $\lambda > 0$, we have the weakly dissipative DGH equation [23, 24];
- The rotation-CH equation [4, 16, 17, 25], provided that $\lambda = 0$ and

$$u_t - u_{txx} + 3uu_x - \frac{\beta_0}{\beta}u_{xxx} + \frac{\omega_1}{\alpha^2}u^2 u_x + \frac{\omega_2}{\alpha^3}u^3 u_x = 2u_x u_{xx} + uu_{xxx},$$

I. L. Freire (✉)
Centro de Matemática, Computação e Cognição, Universidade Federal do ABC, Avenida dos Estados, 5001, Bairro Bangu, Santo André, SP 09210-580, Brazil
e-mail: igor.freire@ufabc.edu.br

J. C. S. Sampaio
Universidade Tecnológica Federal do Paraná, Curitiba, PR, Brazil

A. C. J. Luo and R. K. Gazizov (eds.), *Symmetries and Applications of Differential Equations*, Nonlinear Physical Science,
https://doi.org/10.1007/978-981-16-4683-6_8

where

$$c=\sqrt{1+\Omega^2}-\Omega,\quad \alpha=\frac{c^2}{1+c^2},\quad \beta_0=\frac{c(c^4+6c^2-1)}{6(c^2+1)^2},$$

$$\beta=\frac{3c^4+8c^2-1}{6(c^2+1)^2},\quad \omega_1=-\frac{3c(c^2-1)(c^2-2)}{2(1+c^2)^3},\quad \omega_2=\frac{(c^2-1)^2(c^2-2)(8c^2-1)}{2(1+c^2)^5},$$

and Ω is a constant related to the speed of Earth's rotation.

All of the aforementioned models come from the study of shallow water models and they have been subject of intense studies in view of their rich mathematical and physical features, see [9] and references therein for a wide discussion about these properties.

Our main interest in Eq. (1) is its study from the point of view of Lie symmetries and related topics, such as conservation laws and invariants obtained from them. To celebrate Nail's memory, in order to establish conservation laws, we use the approach he developed around 15 years ago [18–20].

The problem considered in these notes was chosen influenced by some works by Nail:

- firstly, the work by Ibragimov, Khamitova and Valent [21], where they investigated a Camassa–Holm-type equation. Such equation can be recovered from (1) taking $\lambda=\beta=\gamma=\Gamma=0$ and replacing u_{txx} by ϵu_{txx}, where ϵ is a constant;
- secondly, a joint work of Nail, the first author of the present work, and Bozhkov [2], in which they showed that the Novikov equation, discovered some years earlier, is quasi self-adjoint.

We give now a picture of these notes: the next section is concerned with Lie symmetries of Eq. (1), while in Sect. 3, we revisit Ibragimov theorem and related topics to establish conservation laws for Eq. (1). In Sect. 4, we discuss some consequences of the conservation laws we found using Ibragimov theorem, and explore some consequences of the solutions of (1).

2 Lie Symmetries

Let us revisit some concepts about Lie symmetries. We begin with assuming that we have n independent variables $x:=(x^1,\ldots,x^n)$ and m dependent variables $u:=(u^1,\ldots,u^m)$. By $u_{(k)}$, we denote the set of kth order derivatives of u. Throughout this section, we use the summation over repeated indices.

A function depending on (x,u) and derivatives of u up to a finite, but arbitrary, order is called differential function, while the collection of all differential functions is denoted by $\mathcal{A}$. Note that we can sum differential functions and multiply members of $\mathcal{A}$ by real numbers, which endows an algebraic structure in that set. Moreover, given a positive integer ℓ, we can define the product space

$$\mathcal{A}^{\ell} := \underbrace{\mathcal{A} \times \cdots \times \mathcal{A}}_{\ell \text{ copies}}.$$

Given $f \in \mathcal{A}$, we define the order of f, $\text{ord}(f)$ as being the highest derivative appearing on f. In case f does not depend on any derivatives of u, then we say that its order is 0. More generally, if $F = (f_1, \dots, f_\ell) \in \mathcal{A}^\ell$, we can define the order of F as $\text{ord}(F) = \max\{\text{ord}(f_1), \dots, \text{ord}(f_\ell)\}$.

Example 1 Let $f_1 := u_t - u_{txx} + 3uu_x + \lambda(u - u_{xx}) - 2u_x u_{xx} - uu_{xxx} - \alpha u_x - \beta u^2 u_x - \gamma u^3 u_x - \Gamma u_{xxx}$. Then $\text{ord}(f_1) = 3$, since we have explicit dependence on u_{xxx}. We might also invoke the dependence on u_{txx} to get the same result. Moreover, note that f_1 can be seen as a linear combination of the differential functions of order

- 0, given by u;
- 1, given by u_t, u_x, uu_x, u^2u_x and u^3u_x;
- 2, given by u_{xx} and $u_x u_{xx}$ and
- 3, given by u_{txx} and u_{xxx}.

Example 2 Let us consider f_1 from Example 1 and $f_2 = v_t - v_{txx} + 3uv_x - \lambda(v - v_{xx}) - u_{xx}v_x - u_x v_{xx} - (\alpha + \beta u^2 + \gamma u^3)v_x - uv_{xxx} - \Gamma v_{xxx}$. Then $f := (f_1, f_2) \in \mathcal{A}^2$ and it is easy to see that $\text{ord}(f) = 3$.

Similarly as smooth functions, in which we have the derivatives mapping a function into another (possibly different) function, on $\mathcal{A}$, we can define some operators, that can be seen as generalisations of the ordinary derivatives. More precisely, we have the total derivative operators

$$D_i = \frac{\partial}{\partial x^i} + u_i^\alpha \frac{\partial}{\partial u^\alpha} + u_{ij}^\alpha \frac{\partial}{\partial u_j^\alpha} + u_{ijk}^\alpha \frac{\partial}{\partial u_{jk}^\alpha} + \cdots, \quad 1 \le i \le n, \quad 1 \le \alpha \le m.$$

To our purposes, the total derivatives operators we are interested on are the one with respect to t

$$D_t = \frac{\partial}{\partial t} + u_t^\alpha \frac{\partial}{\partial u^\alpha} + u_{tt}^\alpha \frac{\partial}{\partial u_t^\alpha} + u_{tx}^\alpha \frac{\partial}{\partial u_x^\alpha} + \cdots, \quad 1 \le \alpha \le m, \tag{2}$$

and the total derivative with respect to x

$$D_x = \frac{\partial}{\partial x} + u_x^\alpha \frac{\partial}{\partial u^\alpha} + u_{xt}^\alpha \frac{\partial}{\partial u_t^\alpha} + u_{xx}^\alpha \frac{\partial}{\partial u_x^\alpha} + \cdots, \quad 1 \le \alpha \le m. \tag{3}$$

Another important operators defined on $\mathcal{A}$ are the Euler–Lagrange operators

$$\mathcal{E}_{u^\alpha} = \frac{\partial}{\partial u^\alpha} - D_i \frac{\partial}{\partial u_i^\alpha} + D_i D_j \frac{\partial}{\partial u_{ij}^\alpha} + \cdots, \quad 1 \le \alpha \le m. \tag{4}$$

Example 3 Consider the function $\mathcal{L} : \mathcal{A}^2 \to \mathcal{A}$, given by

$$\mathcal{L} = v\Big(u_t - \frac{u_{txx} + u_{xtx} + u_{xxt}}{3} + 3uu_x + \lambda(u - u_{xx})$$
$$-2u_x u_{xx} - uu_{xxx} - \alpha u_x - \beta u^2 u_x - \gamma u^3 u_x - \Gamma u_{xxx}\Big).$$

Then, we have

$$\begin{aligned}\mathcal{E}_v(\mathcal{L}) = {} & u_t - u_{txx} + 3uu_x + \lambda(u - u_{xx}) - 2u_x u_{xx} - uu_{xxx} \\ & - (\alpha + \beta u^2 + \gamma u^3)u_x - \Gamma u_{xxx}\end{aligned}$$

and

$$\begin{aligned}\mathcal{E}_u(\mathcal{L}) = {} & v_t - v_{txx} + 3uv_x - \lambda(v - v_{xx}) - u_{xx}v_x - u_x v_{xx} \\ & - (\alpha + \beta u^2 + \gamma u^3)v_x - uv_{xxx} - \Gamma v_{xxx}.\end{aligned}$$

Assume that in the space (x, u) we have a transformation, depending on a parameter ϵ, which we assume to be analytic in this variable. Also, at $\epsilon = 0$ we assume that the transformation is the identity. Let $(x, u) \mapsto (\overline{x}(x, u, \epsilon), \overline{u}(x, u, \epsilon))$ be a one-parameter group of transformations that at $\epsilon = 0$ corresponds to the identity. Making a Maclaurin expansion, we have

$$\begin{aligned}\overline{x}^i &= x^i + \epsilon\xi^i(x, u) + O(\epsilon^2), \\ \overline{u}^\alpha &= u^\alpha + \epsilon\zeta^\alpha(x, u) + O(\epsilon^2).\end{aligned} \tag{5}$$

The expansion above enables us to define the operator

$$X = \xi^i(x, u)\frac{\partial}{\partial x^i} + \zeta^\alpha(x, u)\frac{\partial}{\partial u^\alpha} \tag{6}$$

which is the generator of the one-parameter group of transformations.

Example 4 Let us consider the transformation $(t, x, u) \mapsto (\overline{t}(t, x, u, \epsilon), \overline{x}(t, x, u, \epsilon), \overline{u}(t, x, u, \epsilon))$ in $\mathbb{R}^3$. Supposing that such transformation is analytic with respect to the parameter ϵ, we have

$$\begin{aligned}\overline{t}(t, x, u, \epsilon) &= t + \epsilon\tau(t, x, u) + O(\epsilon^2), \\ \overline{x}(t, x, u, \epsilon) &= x + \epsilon\xi(t, x, u) + O(\epsilon^2), \\ \overline{u}(t, x, u, \epsilon) &= u + \epsilon\eta(t, x, u) + O(\epsilon^2).\end{aligned} \tag{7}$$

The coefficients τ, ξ, η, which depend only on (t, x, u), define the infinitesimal generator

$$X = \tau(t, x, u)\frac{\partial}{\partial t} + \xi(t, x, u)\frac{\partial}{\partial x} + \eta(t, x, u)\frac{\partial}{\partial u}, \tag{8}$$

of the group of transformations.

Note that if we know the transformation, then the Maclaurin expansion gives the generator of the transformation. On the opposite side, if we have the latter, than the former can be recovered from the exponential $e^{\epsilon X}(x, u) := (e^{\epsilon X}x^1, \ldots, e^{\epsilon X}x^n, e^{\epsilon X}u^1, \ldots, e^{\epsilon X}u^m)$.

Example 5 Consider the transformation $(t, x, u) \mapsto (t + \epsilon, x, u) =: (\overline{t}, \overline{x}, \overline{u})$ in $\mathbb{R}^3$. Then we have $\tau = 1, \xi = \eta = 0$ and (8) gives $X = \partial_t$.

So far we have seen transformations, but we have not connected them with Lie symmetries of differential equations. It is time to overcome this gap. First we note that a Lie symmetry is a transformation. But what kind of transformation? The transformations we call Lie symmetries of a differential equation are those transformations mapping solutions of a given equation or system into another solution of the same equation or system.

Very often we know the equation, but not necessarily its symmetries. They are found in the following way:

- Let $F = 0$ be a system of differential equations, where $F \in \mathcal{A}^\ell$.
- Let us assume that any Lie symmetry of the equation has an expansion like (5), and therefore, defines (and is defined by) the generator (6).
- Let $k = \text{ord}(F)$ and, from (6), we construct the operator

$$X^{(k)} := X + \zeta_i^\alpha \frac{\partial}{\partial u_i^\alpha} + \zeta_{ij}^\alpha \frac{\partial}{\partial u_{ij}^\alpha} + \cdots \zeta_{i_1 \ldots i_p}^\alpha \frac{\partial}{\partial u_{i_1 \ldots i_p}^\alpha}, \tag{9}$$

  and

$$\begin{aligned} \zeta_i^\alpha &= D_i \zeta^\alpha - (D_i \xi^j) u_{ij}, \\ \zeta_{i_1 \ldots i_p}^\alpha &= D_{i_p} \zeta_{i_1 \ldots i_{p-1}}^\alpha - (D_{i_p} \xi^j) u_{i_1 \ldots i_{p-1} j}^\alpha. \end{aligned}$$

- We impose that

$$X^{(k)} F = 0 \quad \text{when} \quad F = 0. \tag{10}$$

  Condition (10) is called invariance condition and it leads us to obtain a system of linear partial differential equations for the coefficient functions of the generator (6). Solving such system (which is always possible) and substituting the solution into (6), we obtain a linear combination of generators, which gives us a basis to the generators of the Lie symmetries of the equation.

Let us concretely apply the process mentioned above to find the symmetries of the Eq. (1).

Let us assume that (8) is a generator of a one-parameter group of transformation (7) and that such transformation is a symmetry of (1). Since the equation is of third order, we must then find the prolongation of the generator (8). Taking the structure of the equation into account, we have

$$X^{(3)} = X + \zeta^{t}\frac{\partial}{\partial u_t} + \zeta^{x}\frac{\partial}{\partial u_x} + \zeta^{xx}\frac{\partial}{\partial u_{xx}} + \zeta^{xxx}\frac{\partial}{\partial u_{xxx}} + \zeta^{txx}\frac{\partial}{\partial u_{txx}}, \tag{11}$$

where

$$\begin{aligned}
\zeta^{t} &= D_t(\eta) - (D_t\tau)u_t - (D_t\xi)u_x, \quad \zeta^{x} = D_x(\eta) - (D_x\tau)u_t - (D_x\xi)u_x,\\
\zeta^{xx} &= D_x(\zeta^{x}) - (D_x\tau)u_{xt} - (D_x\xi)u_{xx},\\
\zeta^{xxx} &= D_x(\zeta^{xx}) - (D_x\tau)u_{xxt} - (D_x\xi)u_{xxx},\\
\zeta^{txx} &= D_t(\zeta^{xx}) - (D_t\tau)u_{ttx} - (D_t\xi)u_{txx}.
\end{aligned}$$

Note that (1) does not have dependence on u_{tx}, u_{tt}, u_{ttx} and u_{ttt} and this is reflected in (11), where we omitted these corresponding components because they will not have any contribution in (10). The invariance condition is

$$X^{(3)}(u_t - u_{txx} + \lambda(u - u_{xx}) - uu_{xxx} - 2u_xu_{xx} + 3u^2u_x) = 0, \tag{12}$$

on the solutions of (1). This last sentence means that we calculate the left side of (12), substitute $u_{txx} = u_t + \lambda(u - u_{xx}) - uu_{xxx} - 2u_xu_{xx} + 3u^2u_x$ and equates the result to 0. This will give us a polynomial identity in the derivatives of u, which gives us the following set of determining equations:

$$\begin{aligned}
&\tau_x = 0,\ \eta_{uu} = 0,\ \tau_u = 0,\ \xi_u = 0,\ 2\eta_{xu} - \xi_{xx} = 0,\ \xi_t + (\Gamma + u)\eta_u - \eta(x,t,u) = 0,\\
&(\Gamma + u)(\tau_t + \eta_u - \xi_x) = 0, \quad \xi_t - (\Gamma + u)(\tau_t + \eta_{xxu} - 3\xi_x) - \eta(x,t,u) = 0,\\
&\lambda\eta(x,t,u) - \lambda\xi_t - (\Gamma + u)\big(\eta_{,tu} - 2\xi_{xt} + 3\Gamma\eta_{xu} + 3u\eta_{xu} - 3u\xi_{xx} - 3\Gamma\xi_{xx}\\
&\qquad\qquad + 2\eta_x + \lambda\xi_x\big) = 0,\\
&\eta(x,t,u)(\alpha + 3\Gamma - 2\gamma u^3 - u^2(\beta + 3\gamma\Gamma) - 2\beta\Gamma u) - \xi_t(\alpha + \Gamma + \gamma u^3 + \beta u^2 - 2u)\\
&-(\Gamma + u)[2\eta_{,xtu} - \xi_{xxt} + 2\xi_x(\alpha + \gamma u^3 + \beta u^2 - 3u) + 3\Gamma\eta_{xxu} + 2\lambda\eta_{xu}\\
&+3u\eta_{xxu} - u\xi_{xxx} - \Gamma\xi_{xxx} + 2\eta_{xx} - \lambda\xi_{xx}] = 0,\\
&\Gamma\lambda\eta(x,t,u) + \Gamma\eta_t + u\eta_t + \lambda u\xi_t - u\eta_{xxt} - \Gamma\eta_{xxt} - \gamma u^4\eta_x - \beta u^3\eta_x - \gamma\Gamma u^3\eta_x\\
&-\beta\Gamma u^2\eta_x + 3u^2\eta_x - u^2\eta_{xxx} + 3\lambda u^2\xi_x - \lambda u(\Gamma + u)\eta_u - \alpha u\eta_x + 3\Gamma u\eta_x - 2\Gamma u\eta_{xxx}\\
&+3\Gamma\lambda u\xi_x - \lambda u\eta_{xx} - \alpha\Gamma\eta_x - \Gamma^2\eta_{xxx} - \Gamma\lambda\eta_{xx} = 0.
\end{aligned} \tag{13}$$

We would like to point out some observations.

Remark 1 A couple of lines above we mentioned that we should substitute the relation $u_{txx} = u_t + \lambda(u - u_{xx}) - uu_{xxx} - 2u_x u_{xx} + 3u^2 u_x$ into the condition (12). It is not mandatory such a choice. Actually, we could use $u_t = u_{txx} - \lambda(u - u_{xx}) + uu_{xxx} + 2u_x u_{xx} - 3u^2 u_x$.

Remark 2 Currently, we have at our disposal several packages for finding the determining equations and also solve them, which is the same to say that they find the symmetries. Particularly, we used the ones developed by Dimas [11, 12] to find the determining equations and also to find the Lie symmetries.

The solution of (14) proves the following result:

Theorem 1 *Let* (2) *be a Lie point symmetry generator of the Eq. (1). Then X is spanned by the generators*

$$X_1 = \frac{\partial}{\partial t} \quad \text{and} \quad X_2 = \frac{\partial}{\partial x}. \tag{14}$$

For some specific choices of the parameters α, β, γ and Γ, we have additional generators. They are:

1. *If $\beta = \gamma = \lambda = \Gamma = 0$ and $\alpha \neq 0$, then*

$$X_3 = 2t\frac{\partial}{\partial t} + \alpha t\frac{\partial}{\partial x} + (\alpha - 2u)\frac{\partial}{\partial u}; \tag{15}$$

2. *If $\beta = \gamma = 0$, $\Gamma = -\alpha$ and $\lambda \neq 0$, then*

$$X_4 = e^{\lambda t}\frac{\partial}{\partial t} + \Gamma e^{\lambda t}\frac{\partial}{\partial x} - \lambda e^{\lambda t} u\frac{\partial}{\partial u}; \tag{16}$$

3. *If $\beta = \gamma = \lambda = 0$, then*

$$X_5 = 2t\frac{\partial}{\partial t} + (\alpha + 3\Gamma)t\frac{\partial}{\partial x} + (\alpha + \Gamma - 2u)\frac{\partial}{\partial u}. \tag{17}$$

Theorem 1 encloses some equations that have already been previously considered in other papers, see [1, 5, 29] and references therein.

We note that the arbitrary case, that is, no restrictions on the parameters, gives the Lie symmetries for the rotating Camassa–Holm equation, derived in [4, 17], see also [9, 16, 25].

It is worth mentioning that the symmetries considered in this section are Lie point symmetries, that is, symmetries coming from transformations of the form (5). There are other symmetries whose coefficients of the generator (6), instead of depending only on (x, u), also depends on the derivatives of u. Transformations depending of (x, u) and the first-order derivatives of u are often called *tangent transformations*, while those depending on higher order derivatives are referred as *generalised symmetries*. For further details, see [26, Chap. 5].

3 Ibragimov Theorem and Conservation Laws

Let us consider a system of differential equations, with independent variables $x \in \mathbb{R}^n$,

$$F_\alpha = 0, \quad \alpha = 1, \dots, m, \tag{18}$$

for a certain positive integer m, where $F_\alpha \in \mathcal{A}$. Let $v^\beta = v^\beta(x)$ be new functions. We can define the *formal Lagrangian*

$$\mathcal{L} := v^\alpha F_\alpha. \tag{19}$$

The formal Lagrangian enables us to embed the system (18), which is not necessarily variational (i.e., a system coming from the Euler–Lagrange equations), into the following set Euler–Lagrange equations

$$\begin{cases} \mathcal{E}_{v^\alpha}(\mathcal{L}) \; = \; F_\alpha = 0, \\ \\ \mathcal{E}_{u^\alpha}(\mathcal{L}) =: F_\alpha^* = 0 \quad \alpha = 1, \dots, m. \end{cases} \tag{20}$$

The set of equations $F_\alpha^* = 0$, $\alpha = 1, \dots m$, is called adjoint system to the system (18).

Our main focus is on single differential equation and an example may be salutary at this stage.

Example 6 Let us consider the Eq. (1). The corresponding formal Lagrangian is given by

$$\begin{aligned} \mathcal{L} = v\big(u_t - \frac{u_{txx} + u_{xtx} + u_{xxt}}{3} + 3uu_x + \lambda(u - u_{xx}) \\ -2u_x u_{xx} - uu_{xxx} - \alpha u_x - \beta u^2 u_x - \gamma u^3 u_x - \Gamma u_{xxx}\big). \end{aligned} \tag{21}$$

Let us find

$$F^* = \mathcal{E}_{u^\alpha}(\mathcal{L})$$

where $\mathcal{L}$ is given by (21). We have

$$
\begin{aligned}
F^* &= \frac{\partial \mathcal{L}}{\partial u} - D_t \frac{\partial \mathcal{L}}{\partial u_t} - D_x \frac{\partial \mathcal{L}}{\partial u_x} + D_x^2 \frac{\partial \mathcal{L}}{\partial u_{xx}} - D_t D_x^2 \frac{\partial \mathcal{L}}{\partial u_{txx}} - D_x^3 \frac{\partial \mathcal{L}}{\partial u_{xxx}} \\
&= \lambda(v - v_{xx}) - v_t + (\alpha - 3u + \beta u^2 + \gamma u^3)v_x + v_{txx} + uv_{xxx} \\
&\quad + u_{xx}v_x + u_x v_{xx} + \Gamma v_{xxx}.
\end{aligned}
\tag{22}
$$

Therefore, the adjoint equation to (1) is

$$
v_t - v_{txx} - \lambda(v - v_{xx}) - (\alpha - 3u + \beta u^2 + \gamma u^3)v_x - uv_{xxx} - u_{xx}v_x - u_x v_{xx} - \Gamma v_{xxx} = 0. \tag{23}
$$

Now we revisit the machinery developed by Ibragimov with respect to conservation laws. In [18] Ibragimov proved the following Noether-type theorem:

Theorem 2 (Ibragimov theorem, [18, Theorem 3.3]) *Let*

$$
X = \xi^i \frac{\partial}{\partial x^i} + \eta^\alpha \frac{\partial}{\partial u^\alpha}
$$

be any symmetry of Eq. (18). Then the system (20) *has the conservation law* $D_i C^i = 0$, *where*

$$
\begin{aligned}
C^i &= \xi^i \mathcal{L} + W^\alpha \left[\frac{\partial \mathcal{L}}{\partial u_i^\alpha} - D_j\left(\frac{\partial \mathcal{L}}{\partial u_{ij}^\alpha}\right) + D_j D_k \frac{\partial \mathcal{L}}{\partial u_{ijk}^\alpha} - \cdots\right] \\
&\quad + D_j(W^\alpha)\left[\frac{\partial \mathcal{L}}{\partial u_{ij}^\alpha} - D_k\left(\frac{\partial \mathcal{L}}{\partial u_{ijk}^\alpha}\right) + \cdots\right] \\
&\quad + D_j D_k(W^\alpha)\left[\frac{\partial \mathcal{L}}{\partial u_{ijk}^\alpha} - \cdots\right] + \cdots
\end{aligned}
\tag{24}
$$

and $W^\alpha = \eta^\alpha - \xi^i u_i^\alpha$.

Let us consider Eq. (1). Application of Ibragimov theorem to such equation gives the components

$$
\begin{aligned}
C^0 &= \tau \mathcal{L} + W\left[\frac{\partial \mathcal{L}}{\partial u_t} + D_x^2\left(\frac{\partial \mathcal{L}}{\partial u_{txx}}\right)\right] - D_x(W) D_x\left(\frac{\partial \mathcal{L}}{\partial u_{txx}}\right) + D_x^2(W)\frac{\partial \mathcal{L}}{\partial u_{txx}}, \\
C^1 &= \xi \mathcal{L} + W\left[\frac{\partial \mathcal{L}}{\partial u_x} - D_x\left(\frac{\partial \mathcal{L}}{\partial u_{xx}}\right) + D_x^2\left(\frac{\partial \mathcal{L}}{\partial u_{xxx}}\right) + D_x D_t\left(\frac{\partial \mathcal{L}}{\partial u_{xxt}}\right)\right. \\
&\quad \left. + D_t D_x\left(\frac{\partial \mathcal{L}}{\partial u_{xtx}}\right)\right] + D_x(W) D_x\left[\frac{\partial \mathcal{L}}{\partial u_{xx}} - D_x\left(\frac{\partial \mathcal{L}}{\partial u_{xxx}}\right) - D_t\left(\frac{\partial \mathcal{L}}{\partial u_{xxt}}\right)\right] \\
&\quad - D_t(W) D_x\left(\frac{\partial \mathcal{L}}{\partial u_{xtx}}\right) + D_x^2(W)\frac{\partial \mathcal{L}}{\partial u_{xxx}} + D_t D_x(W)\frac{\partial \mathcal{L}}{\partial u_{xtx}} + D_x D_t(W)\frac{\partial \mathcal{L}}{\partial u_{xxt}},
\end{aligned}
\tag{25}
$$

from each Lie point symmetry generator

$$X = \tau\frac{\partial}{\partial t} + \xi\frac{\partial}{\partial x} + \eta\frac{\partial}{\partial u}.$$

We observe that (25) not only depends on u, but also of the new variable v. This is more evident if we find the components (25) using (1). After some calculations, we have

$$\begin{aligned}
C^0 &= \tau\mathcal{L} + \left[W - \frac{1}{3}D_x^2(W)\right]v + \frac{1}{3}D_x(W)v_x - \frac{1}{3}Wv_{xx}\\
C^1 &= \xi\mathcal{L} + W\left[(3u - \alpha - \beta u^2 - \gamma u^3 - u_{xx})v + (\lambda - \Gamma)v_x - \frac{2}{3}v_{tx} - (\Gamma + u)v_{xx}\right]\\
&\quad D_x(W)\left[-(\lambda + u_x)v + (\Gamma + u)v_x + \frac{1}{3}v_t\right] + \frac{1}{3}v_x D_t(W)\\
&\quad -\frac{2}{3}D_x D_t(W)v - (u + \Gamma)D_x^2(W)v.
\end{aligned} \tag{26}$$

Therefore, the conservation laws we are able to obtain are not really conservation laws for the original Eq. (1), but to the system

$$\begin{cases}
u_t - u_{txx} + 3uu_x + \lambda(u - u_{xx})\\
= 2u_x u_{xx} + uu_{xxx} + (\alpha + \beta u^2 + \gamma u^3)u_x + \Gamma u_{xxx},\\
v_t - v_{txx} + 3uv_x - \lambda(v - v_{xx})\\
= u_{xx}v_x + u_x v_{xx} + (\alpha + \beta u^2 + \gamma u^3)v_x + uv_{xxx} + \Gamma v_{xxx}
\end{cases} \tag{27}$$

in which (1) is embedded by construction.

Although we can establish a conservation law for (27), our intention is the construction of conservation laws for (1), and not (27). In order to eliminate the variable v in (27), we use the concept of *equations nonlinearly self-adjoint* [20]. For further details, see [18, 19] and the review [28].

Definition 1 An equation $F = 0$, with one dependent variable u, is said to be nonlinearly self-adjoint if there exists a function $v = \phi(t, x, u)$, called substitution, and a function $\lambda \in \mathcal{A}$ such that

$$F^*\big|_{v=\phi} = \lambda F. \tag{28}$$

We note that if an equation is nonlinearly self-adjoint, then any solution of the original equation provides a solution $v = \phi$ for the adjoint equation to the original one.

Let us investigate if (1) is nonlinearly self-adjoint. To begin with, we assume that $v = \phi(t, x, u)$ and find the derivatives of v appearing in (27), that is:

$$
\begin{aligned}
v_t &= D_t\phi = \phi_t + \phi_u u_t,\\
v_x &= D_x\phi = \phi_x + \phi_u u_x,\\
v_{xx} &= D_x^2\phi = \phi_{xx} + 2\phi_{xu}u_x + \phi_{uu}u_x^2 + \phi_u u_{xx},\\
v_{txx} &= D_t D_x^2\phi = \phi_{txx} + 2\phi_{txu}u_x + \phi_{tuu}u_x^2 + \phi_{tu}u_{xx} + \phi_{xxu}u_t + 2\phi_{xuu}u_x u_t\\
&\quad +2\phi_{xu}u_{tx} + \phi_{uuu}u_x^2 u_t + \phi_{uu}u_t u_{xx} + 2\phi_{uu}u_x u_{tx} + \phi_u u_{txx}\\
v_{xxx} &= D_x^3\phi = \phi_{xxx} + 3\phi_{xxu}u_x + 3\phi_{xuu}u_x^2 + 3\phi_{xu}u_{xx} + 3\phi_{uu}u_x u_{xx}\\
&\quad +\phi_{uuu}u_x^3 + \phi_u u_{xxx}.
\end{aligned}
$$

Substituting these expressions into F^*, we have

$$
\begin{aligned}
F^*\big|_{v=\phi} &= -\phi_u F + \Big[\lambda\phi - \lambda\phi_{xx} - \phi_t + \phi_x(\alpha - 3u + \beta u^2 + \gamma u^3) + \lambda\phi_u u + \phi_{txx}\\
&\quad +\phi_{xxx}u + \Gamma\phi_{xxx}\Big] + (2\phi_{txu} - 2\lambda\phi_{xu} + 3\phi_{xxu}u + \phi_{xx} + 3\Gamma\phi_{xxu})u_x\\
&\quad (\phi_{tuu} - \lambda\phi_{uu} + 3\phi_{xuu}u + 2\phi_{xu} + 3\Gamma\phi_{xuu})u_x^2 + (\phi_{uuu}u + \phi_{uu} + \Gamma\phi_{uuu})u_x^3\\
&\quad \phi_{xxu}u_t + 2\phi_{xuu}u_x u_t + \phi_{uuu}u_x^2 u_t + 2\phi_{xu}u_{tx} + \phi_{uu}u_t u_{xx} + 2\phi_{uu}u_x u_{tx}\\
&\quad +(\phi_{tu} - 2\lambda\phi_u + 3\phi_{xu}u + \phi_x + 3\Gamma\phi_{xu})u_{xx} + (3\phi_{uu}u + 3\Gamma\phi_{uu})u_x u_{xx}.
\end{aligned}
$$

Substituting the expression above into (29), we conclude that $\lambda = -\phi_u$, $\phi_{xu} = \phi_{uu} = 0$, which implies that $\phi = A(t)u + B(t,x)$. The remaining equations read

$$
\begin{aligned}
&\beta B_x = \gamma B_x = B_{xx} = 0, \quad \lambda B + \alpha B_x - B_t = 0,\\
&2\lambda A - A' - B_x = 0, \quad 2\lambda A - A' - 3B_x = 0.
\end{aligned}
\tag{29}
$$

Solving the system (29), we prove our next result.

Theorem 3 *Equation (1) is nonlinearly self-adjoint. Moreover, its substitution is a linear combination of the functions $\phi_1 = e^{\lambda t}$ and $\phi_2 = e^{2\lambda t}u$.*

Let us now use the results in Theorem 3 and the components (26). We observe that the terms $\tau\mathcal{L}$ and $\xi\mathcal{L}$ vanish on the solutions of (1). These featured terms are called trivial conservation laws (see [26] for further details), meaning that both mathematical and physical relevance of a conservation law is retrieved only from the non-trivial parts of the components.

If we use the translations in space or time in (26) we will obtain components whose divergence vanishes identically and, therefore, no relevant information is provided.

From the remaining generators of symmetries of (1) (see Theorem 1) we are able to find the following conservation laws:

$$D_t(e^{\lambda t}u)+D_x\big[e^{\lambda t}\big(\frac{3}{2}u^2-u_{tx}-uu_{xx}-\frac{1}{2}u_x^2-\alpha u-\frac{\beta}{3}u^3-\frac{\gamma}{4}u^4-\Gamma u_{xx}\big)\big]=0$$

and

$$D_t\big(e^{2\lambda t}\frac{u^2+u_x^2}{2}\big)+D_x\big[e^{2\lambda t}\big(u^3-u^2u_{xx}-uu_{tx}+\Gamma\frac{u_x^2}{2}-\Gamma uu_{xx}-\alpha\frac{u^2}{2}-\beta\frac{u^4}{4}-\frac{\gamma}{5}u^5\big)\big]=0.$$

In the next section, we explore the consequences of the conservation laws and how we can infer qualitative information about the solutions of the equation from them.

4 Conserved Quantities and Its Consequences

Here, we explore some consequences of the conservation laws previously established. First, we proceed a generic analysis, but focused on the Eq. (1). Although we focus on a specific equation, our presentation is easily adapted for any equation/system with time dependence.

Let us start from the divergence

$$D_t C^0 + D_x C^1 = 0, \tag{30}$$

taken on the solutions of the Eq. (1). Such divergence is, as we have already mentioned, called conservation law for the Eq. (1). The vector field $C=(C^0, C^1)$ is called *conserved current*. Its first component, C^0, is known as *conserved density*, while C^1 is its corresponding *conserved flux*.

Let us then define

$$\mathcal{H}(t)=\int_{\mathbb{R}} C^0\,dx. \tag{31}$$

The function $\mathcal{H}(t)$, actually, is a *functional*, associating to each solution $u(t,\cdot)$ of (1) a time function $\mathcal{H}(t)$, that is, we have the relation $u(t,\cdot)\mapsto\mathcal{H}(t)$.

It is also worth mentioning that we obtain the functional (31) by integrating the conserved density over the real line $\mathbb{R}$. However, it would be possible to consider different domains of integration, e.g. over the circle $\mathbb{S}$, which can be identified with the internal $[0, 1)$.

Under very mild conditions, the derivative with respect to t commutes with the integral with respect to x, and then we can measure how $\mathcal{H}(t)$ varies with respect to t, that is,

$$\frac{d}{dt}\mathcal{H}(t) = \int_{\mathbb{R}} D_t C^0 dx \underbrace{=}_{\text{By (30)}} - \int_{\mathbb{R}} D_x C^1 dx = -C^1\Big|_{-\infty}^{+\infty} \tag{32}$$

If we assume that the conserved flux vanishes as long as u and its derivatives vanish, and supposing that this happens whenever $|x| \to \infty$, from (32), we conclude that

$$\frac{d}{dt}\mathcal{H}(t) = 0,$$

meaning that $\mathcal{H}(\cdot)$ is, in fact, constant. Therefore, if we know the solution at a given value of t, then although we may not know the solution for other values, the functional (30) is known for each t as long as the solution exists.

From the results proved in the previous sections, we have the following conserved quantities for (1):

$$\mathcal{H}_1(t) = e^{\lambda t} \int_{\mathbb{R}} u(t,x)dx \tag{33}$$

and

$$\mathcal{H}_2(t) = \frac{e^{2\lambda t}}{2} \int_{\mathbb{R}} \big(u(t,x)^2 + u_x(t,x)^2\big)dx = \frac{e^{2\lambda t}}{2} \|u(t,\cdot)\|^2_{H^1(\mathbb{R})}. \tag{34}$$

The conserved quantity (34) is related with the Sobolev norm of the space $H^1(\mathbb{R})$, as shown in the last equality. For further details about this space, see [22, 27]. On the other hand, if u is either non-negative or non-positive, that is, $u(t,x) = \sigma|u(t,x)|$, where $\sigma = +1$ if u is non-negative, or $\sigma = -1$ if u is non-positive, then the conserved quantity (33) implies on the conservation of the $L^1(\mathbb{R})-$norm of the solution u of (1).

Suppose that we known a function $u_0(x)$ and that it belongs to $H^s(\mathbb{R})$, with $s > 3/2$. Then we can assure, at least at the local level, the existence and uniqueness of solutions of the Eq. (1) satisfying $u(x,0) = u_0(x)$, see [15]. For the particular case $\lambda = 0$, see [9, 10] and references therein. In particular, since we known $u_0(x)$, then we also know

$$\mathcal{H}_0 = \frac{1}{2}\int_{\mathbb{R}} \big(u_0^2 + (u_0')^2\big)dx,$$

and $\mathcal{H}(t)$ as well, in view of the equality $\mathcal{H}(t) = \mathcal{H}_0$.

We note that if we define $m := u - u_{xx}$, then (33) is equivalent to

$$\mathcal{H}_1(t) = \int_{\mathbb{R}} m dx.$$

We recall that ∂_x^k maps a function u from $H^s(\mathbb{R})$ into $\partial_x^k u \in H^{s-k}(\mathbb{R})$. Moreover, if $u_0 \in H^3(\mathbb{R})$, then $m_0 := u_0 - u_0'' \in H^1(\mathbb{R})$. In addition, if we also assume that $m_0 \in L^1(\mathbb{R})$, and m_0 does not change its sign, then the solutions of the Eq. (1), with $\lambda = \alpha = \beta = \gamma = \Gamma = 0$, exists for each $t > 0$, see [6–8, 14].

On the other hand, for different values of the constants in (1), let us define $\kappa = \max\{|\alpha|, |\beta|/3, |\gamma|/4, |\Gamma|\}$, $\theta_0 = \sqrt{2/(1+12\kappa)}$ and let us consider a function $0 \not\equiv u_0 \in H^3(\mathbb{R})$ and define $y(t) := \inf_{x\in\mathbb{R}} u_x(t,x)$.

If there exists $\theta \in (0, \theta_0]$ such that the inequality

$$0 < \lambda < -\frac{y(0)}{4}\frac{\theta^2 u_0'(x_0)^2 - \max\{\|u_0\|_{H^1(\mathbb{R})}, \|u_0\|^4_{H^1(\mathbb{R})}\}}{\theta^2 u_0'(x_0)^2}$$

is satisfied and we can find a point $x_0 \in \mathbb{R}$ such that

$$\theta u_0'(x_0) < \min\{-\|u_0\|^{1/2}_{H^1(\mathbb{R})}, -\|u_0\|^2_{H^1(\mathbb{R})}\},$$

then the corresponding solution of (1) subject to $u(x,0) = u_0(x)$ develop a singularity at a finite time, namely, its derivative with respect to x does not have any lower bound, see [15]. Such phenomenon is better known as *wave breaking*. In the references [6–8, 14] several wave breaking results for equations enclosed in (1) are also reported and, actually, the ideas in [7] are the basic tools for proving blow-up results for equations of the type (1). A common element in them is the fact that the wave breaking appears provided that some relation between the slop of the initial data and its $H^1(\mathbb{R})$ is satisfied.

We close this chapter by giving some words about global solutions of (1). By global, we mean a solution u defined on $[0,\infty) \times \mathbb{R}$. Let us define $u_0(x) := u(0,x)$. If $u_0 \in H^s(\mathbb{R})$, for $s > 3/2$, then we have granted the existence of a unique local solution for (1) with initial data u_0 conserving (34). Then, we have

$$\|u(t,\cdot)\|_{H^1(\mathbb{R})} = e^{-2\lambda t}\|u_0\|_{H^1(\mathbb{R})}. \tag{35}$$

Let us suppose that such a local solution is global (eventually letting s being larger than the value mentioned) and that $\lambda > 0$. It is a foregone conclusion from (35) that $u \to 0$ as long as $t \to +\infty$. On the other hand, in case $\lambda = 0$, if $u_0 \not\equiv 0$, then $\|u(t,\cdot)\|_{H^1(\mathbb{R})} \neq 0$, showing how λ affects the qualitative behaviour of u.

Acknowledgements The authors would like to thank the organisers for their kind invitation to send these notes in honour to the memory of Professor Nail Ibragimov.

References

1. Anco S, da Silva PL, Freire IL (2015) A family of wave-breaking equations generalizing the Camassa-Holm and Novikov equations. J Math Phys 56, paper 091506
2. Bozhkov Y, Freire IL, Ibragimov N (2014) Group analysis of the Novikov equation. Comp Appl Math 33:193–202
3. Camassa R, Holm DD (1993) An integrable shallow water equation with peaked solitons. Phys Rev Lett 71:1661–1664

4. Chen RM, Gui G, Liu Y (2018) On a shallow-water approximation to the Green-Naghdi equations with the Coriolis effect. Adv Math 340:106–137
5. Clarkson PA, Mansfield EL, Pristley TJ (1997) Symmetries of a class of nonlinear third-order partial differential equations. Math Comput Modell 25:195–212
6. Constantin A, Escher J (1998) Global existence and blow-up for a shallow water equation. Ann Sci Norm Sup Pisa 26:303–328
7. Constantin A, Escher J (1998) Wave breaking for nonlinear nonlocal shallow water equations. Acta Math 181:229–243
8. Constantin A, Escher J (1998) Well-Posedness, global existence, and blowup phenomena, for a periodic quasi-linear hyperbolic equation. Commun Pure App Math LI:0475–0504
9. da Silva PL, Freire IL (2019) Well-posedness, travelling waves and geometrical aspects of generalizations of the Camassa-Holm equation. J Diff Equ 267:5318–5369
10. da Silva PL, Freire IL (2020) Integrability, existence of global solutions, and wave breaking criteria for a generalization of the Camassa-Holm equation. Stud Appl Math 145:537–562
11. Dimas S, Tsoubelis D (2004) SYM: a new symmetry-finding package for Mathematica. In: Proceedings of the 10th international conference in modern group analysis, Larnaca, Cyprus, pp 64–70
12. Dimas S, Tsoubelis D (2005) A new heuristic algorithm for solving overdetermined systems of PDEs in Mathematica. In: 6th international conference on symmetry in nonlinear mathematical physics, Kiev, Ukraine, pp 20–26
13. Dullin H, Gottwald G, Holm D (2001) An integrable shallow water equation with linear and nonlinear dispersion. Phys Rev Lett 87, Article 194501
14. Escher J (2016) Breaking water waves. In: Constantin A (eds) Nonlinear water waves. Lecture notes in mathematics, vol 2158. Springer, Cham. https://doi.org/10.1007/978-3-319-31462-4_2
15. Freire IL (2020) Wave breaking for shallow water models with time decaying solutions. J Diff Equ 269:3769–3793
16. Gui G, Liu Y, Sun J (2019) A nonlocal shallow-water model arising from the full water waves with the Coriolis effect. J Math Fluid Mech 21, Article 27
17. Gui G, Liu Y, Luo T (2019) Model equations and traveling wave solutions for shallow-water waves with the Coriolis effect. J Nonlin Sci 29:993–1039
18. Ibragimov NH (2007) A new conservation theorem. J Math Anal Appl 333:311–328
19. Ibragimov NH (2007) Quasi-self-adjoint differential equations. Arch ALGA 4:55–60
20. Ibragimov NH (2011) Nonlinear self-adjointness and conservation laws. J Phys A: Math Theor 44, Paper 432002?00
21. Ibragimov NH, Khamitova RS, Valenti A (2011) Self-adjointness of a generalized Camassa-Holm equation. Appl Math Comp 218:2579–2583
22. Linares F, Ponce G (2015) Introduction to nonlinear dispersive equations. Springer, Berlin
23. Novruzov E (2013) Blow-up phenomena for the weakly dissipative Dullin-Gottwald-Holm equation. J Math Phys 54:092703
24. Novruzov E (2016) Blow-up of solutions for the dissipative Dullin-Gottwald-Holm equation with arbitrary coefficients. J Diff Equ 261:1115–1127
25. Tu X, Liu Y, Mu C (2019) Existence and uniqueness of the global conservative weak solutions to the rotation- Camassa-Holm equation. J Diff Equ 266:4864–4900
26. Olver PJ (1993) Applications of Lie groups to differential equations, 2nd edn. Springer, Berlin
27. Taylor ME (2011) Partial differential equations I, 2nd edn. Springer, Berlin
28. Sampaio JCS, Freire IL (2014) Nonlinear self-adjoint classification of a Burgers-KdV family of equations. Abs Appl Anal 2014, Article 804703
29. Wei L, Wang Y (2019) Symmetry analysis, conserved quantities and applications to a dissipative DGH equation. J Diff Equ 266:3189–3208
30. Wu S, Yin Z (2006) Blow-up, blow-up rate and decay of the solution of the weakly dissipative Camassa-Holm equation. J Math Phys 47, Paper 013504
31. Wu S, Yin Z (2009) Global existence and blow-up phenomena for the weakly dissipative Camassa-Holm equation. J Diff Equ 246:4309–4321

Partial Invariance and Problems with Free Boundaries

V. V. Pukhnachev

Abstract The foundations of group analysis of differential equations were laid by S. Lie. This theory was essentially developed in works of L. V. Ovsiannikov, N. H. Ibragimov, their students, and followers. Notion of the partially invariant solution to the system of differential equations (Ovsiannikov 1964) substantially extended possibilities of exact solutions construction for multidimensional systems of differential equations admitting the Lie group. It is important to note that fundamental equations of continuum mechanics and physics fall in this class a priori as invariance principle of space, time, and moving medium there with respect to some group (Galilei, Lorenz, and others) are situated in the base of their derivation. It should be noticed that classical group analysis of differential equations has a local character. To apply this approach to initial boundary value problems, one need to provide the invariance properties of initial and boundary conditions. Author (1973) studied these properties for free boundary problems to the Navier–Stokes equations. Present chapter contains an example of partially invariant solution of these equations describing the motion of a rotating layer bounded by free surfaces.

1 Introduction

I had occasion to work with N. H. Ibragimov for 12 years before he moves from Novosibirsk to Ufa. We are of the same age, and we both are pupils of L. V. Ovsyannikov. Lev Vasil'evich supervised the theoretical department of our Institute, which included several research aspects, such as the group analysis of differential equations, mathematical problems of gas dynamics, and theory of problems with free boundaries. I managed to work in all these research fields, but the last one became the main aspect for me.

V. V. Pukhnachev (✉)
Lavrentyev Institute of Hydrodynamics, Siberian Branch, Russian Academy of Sciences,
Novosibirsk State University, Novosibirsk 630090, Russia

A. C. J. Luo and R. K. Gazizov (eds.), *Symmetries and Applications of Differential Equations*, Nonlinear Physical Science,
https://doi.org/10.1007/978-981-16-4683-6_9

The Navier–Stokes equations are used as the basic mathematical model in fluid dynamics. By 1970, the main boundary value and initial boundary value problems for these equations with a fixed flow domain have been studied. However, there was not a single result for problems where the domain boundary or some part of this boundary is free. (Here, I mean problems with a free boundary in the exact formulation. Approximate models of motions with free boundaries have been developed since the paper published by Stokes in 1848. The problem of justification of approximate models had also to be solved.)

Recognizing that my efforts were insufficient to develop a general theory for three-dimensional problems, I decided to consider problems of smaller dimensions. A helpful fact was that the Navier–Stokes equations admit a wide Lie group G_∞. Let the equation of the free boundary Γ_t have the form $F(x, t) = 0$. If Γ_t is an invariant manifold of the group $H \subset G_\infty$, then the conditions on this surface are written in terms of invariants of the group H. My proof of this theorem was rather cumbersome, but Nail made it significantly simpler, and I am extremely grateful for that. This statement allows one to construct invariant solutions of the Navier–Stokes equations, which are preliminary matched with the conditions on the free boundary.

In early 1970s, N. Kh. Ibragimov and I worked on our doctor's dissertations, whereas V. M. Men'shchikov, one more pupil of L. V. Ovsyannikov, prepared his candidate's thesis. Lev Vasil'evich posed the following problem for him: Is it possible to continue the invariant solution of gas-dynamic equations through the shock wave? Men'shchikov answered this question positively under the condition that the equation of the strong discontinuity surface is an invariant manifold of the corresponding group in the space x, t.

Now I return to the year 1964, when L. V. Ovsyannikov made one of his main mathematical discoveries: he introduced the notion of a partially invariant solution of a system of differential equations. This notion is specific for systems and arises in a situation where the subgroup H of the group G admitted by the system has too few invariants for the invariant H-solution to exist. The procedure of constructing a partially invariant solution was described in detail in Ovsyannikov's book entitled *Group Analysis of Differential Equations* [1]. The procedure consists in splitting the original system into the resolving and automorphic subsystems. The first one relates only the invariants of the group H and contains a smaller number of independent variables than the original system. If the solution of this system is known, then the automorphic system is integrated in quadratures.

This discovery of L. V. Ovsyannikov significantly extended the possibility of constructing exact solutions of multidimensional systems of differential equations admitting the Lie group. It is important to note here than the fundamental equations of mechanics and physics of continuous media are *a priori* included into this class because their derivation is based on the principles of invariance of space and time and the moving medium with respect to a certain group (Galileo, Lorentz, and other groups).

The first examples of partially invariant solutions were obtained for gas-dynamic equations, and it seemed that their existence is a privilege of hyperbolic systems. In 1972, however, V. O. Bytev (our common pupil with L. V. Ovsyannikov) found an example of a partially invariant solution of unsteady two-dimensional Navier–Stokes equations, which do not have any particular kind at all. Later on, partially invariant solutions of systems of boundary layer equations, gravity-induced thermal convection equations, and other equations were obtained.

In 1973, I found that the theorem of invariance of conditions on the free boundary is also valid if the solution of the Navier–Stokes equations is only partially invariant. It is sufficient that the unknown boundary with the equation $F(x, t) = 0$ defines an invariant manifold in the space x, t. This fact made it possible to obtain new solutions of problems with a free boundary and with an interface of immiscible fluids.

2 Definition of the Partially Invariant Solution

The notion of a partially invariant solution of a system of differential equations was introduced by Ovsyannikov [2, 3]. The theory of partial invariance is described in Chap. VI of his monograph [1]. L. V. Ovsyannikov demonstrated that the possibility of constructing partial solutions of differential equations can be extended by eliminating the property of full invariance of the solution.

Let $G = G'(f)$ be a local r-parameter Lie group of transformations of the n-dimensional space X generated by the mapping $f : V \times O \to X$ of the product of the open set $V \in X$ and the neighborhood of zero of the parametric space of this group.

The orbit of the manifold $\Psi \in V$ is understood as a set $f(\Psi, O)$ of all possible points $x \in \Psi$. In other words, the orbit $f(\Psi, O)$ of the manifold Ψ is the sum of the orbits of all points of this manifold. There is an alternative: the orbit $f(\Psi, O)$ either contains a certain open set of the space X, or is a manifold in this space with a dimension smaller than $\dim X = n$. If

$$\dim f(\Psi, O) < n, \tag{1}$$

then Ψ is called the proper subspace of the space X. If the manifold orbit satisfies inequality (1), then the group G is intransitive. The following inclusion is valid for any invariant manifold Φ of the group G containing the manifold $\Psi : f(\Psi, O)$. Therefore, the orbit $f(\Psi, O)$ is the smallest invariant manifold of the group G containing Ψ. Clearly, if Ψ itself is an invariant manifold of the group G, then $f(\Psi, O) = \Psi$.

The rank of the manifold Ψ with respect to the group G is understood as the rank of its orbit $f(\Psi, O)$. This rank is considered as a function of the pair (Ψ, G) and is denoted by $\rho(\Psi, G)$. The defect of the manifold Ψ with respect to the group G is the difference between the dimension of its orbit $f(\Psi, O)$ and the dimension of the manifold Ψ. Being considered as a function of the pair (Ψ, G), this defect is denoted by $\delta(\Psi, G)$, so that

$$\delta(\Psi, G) = \dim f(\Psi, O) - \dim \Psi. \tag{2}$$

The number $\delta(\Psi, G)$ shows how far the manifold Ψ is from the invariant manifold. The equality $\delta(\Psi, G) = 0$ is a criterion of invariance of the manifold Ψ. The manifold Ψ for which $\delta(\Psi, G) > 0$ is called a partially invariant manifold of the group G with the invariance defect equal to $\delta(\Psi, G)$. It is inconvenient to use formula (2) for calculating the defect because it implies that either the orbit dimension dim $f(\Psi, O)$, or the rank of this orbit is known. Ovsyannikov [1] derived a formula for defect calculation, where the defect is expressed via the mapping $\psi(x) = 0$ defining the manifold Ψ in the space X.

Let us consider a system of differential equations SE. The solution $u \in SE$ is called a partially invariant solution if the manifold U defined by the relations $u = u(x)$ is a partially invariant manifold of the group H admitted by the system SE. In this case, the rank $\rho = \rho(U, H)$ and the defect $\delta = \delta(U, H)$ are called the rank and defect (of invariance) of this partially invariant solution U, respectively .

The algorithm of constructing the partially invariant solution was described in [1]. It consists of constructing two systems (resolving and automorphic) based on the system SE. The resolving system relates the invariants of the group H. It is simpler than the original system because it contains a smaller number of independent variables and sought functions. In turn, the automorphic system is a (consistent) overdetermined system and can be easily solved in most cases. It should be emphasized that the concept of a partially invariant solution is specific for system of differential equations. Examples of partially invariant solutions of gas-dynamic equations can be found in [1, 4, 5].

Let $X = \mathrm{R}^2\ (x, t)$ and let the variables t and x be treated as the time and distance, respectively. The solution $u = v(x - ct)$, where $c = const$, is called a traveling wave, and c is the traveling wave velocity. The existence of such solutions of the system of gas-dynamic equations is caused by the group of translations in terms of the variables and admitted by this system. Obviously, the traveling wave is an invariant solution of these equations. In general three-dimensional case, the system of gas-dynamic equations admits a group of translations along the coordinate axes and time. The notion of a double wave is an extension of the notion of the traveling wave to the case of two-dimensional motions. The system of equations of two-dimensional isentropic motion of a gas relates its velocity components $u(x, y, t)$, $v(x, y, t)$ and density $\rho(x, y, t)$. The solution of this system is called a double wave if $\rho = r(u, v)$. This solution is a partially invariant solution of this system of equations of rank 2 and defect 1 with respect to the group of translations over the axes x, y, t. Finding this solution is reduced to integrating the system of equations with two independent variables u, v.

Yanenko [6] started systematic investigations of the notion of multiple, in particular, double and triple traveling waves. Such waves are described by partially invariant solutions of gas-dynamic equations [1]. The theory of multiple waves is a significant part of the monograph [7].

3 Equations of Hydrodynamics and Their Group Properties

The main mathematical model in hydrodynamics is the Navier–Stokes equations. For an incompressible fluid moving in a potential field of external forces, these equations have the following form [8, 9]:

$$\mathbf{v}_t + \mathbf{v} \cdot \nabla \mathbf{v} = -\rho^{-1} \nabla p + \nu \Delta\, \mathbf{v}, \quad \nabla \cdot\, \mathbf{v} = 0. \tag{3}$$

In system (3), $\mathbf{v}(x, t) = (v_1, v_2, v_3)$ is the fluid velocity vector in the initial inertial coordinate system, $x = (x_1, x_2, x_3)$, t is the time, and $p(x, t)$ is a modified pressure related to the true pressure p_g by the equality $p = p_g - \rho G(x, t)$, where G is the potential of acceleration of external forces. It should be noted that the most important fields from the viewpoint of applications, i.e., the gravity field and the field of centrifugal forces, are potential fields. The fluid density $\rho > 0$ is assumed to be constant, as well as the kinematic viscosity coefficient $\nu > 0$. The gradient over the variables x_1, x_2, x_3 is denoted by ∇, so that $\nabla\, \mathbf{v}$ is the tensor with the elements $(\nabla \mathbf{v})_{jk} = \partial v_k / \partial x_j$, and $\nabla \cdot \mathbf{v}$ is the divergence of the vector $\mathbf{v}$.

The widest group G_∞ admitted by system (3) was calculated by Yu.A. Danilov [10]. However, his paper was published as a preprint, which has limited access, and the result obtained by Danilov was repeated twice [11, 12]. The Lie algebra L_∞ corresponding to the group G_∞ is generated by the infinitesimal operators

$$Z = 2t\partial_t + \sum_{i=1}^{3} \left(x_i \partial_{x_i} - v_i \partial_{v_i} \right) - 2p\partial_p, \quad X_0 = \partial_t, \tag{4}$$

$$X_{kl} = x_k \partial_{x_l} - x_l \partial_{x_k} + v_k \partial_{v_l} - v_l \partial_{v_k}; \quad k, l = 1, 2, 3; \quad k < l,$$

$$\Phi = \varphi \partial_p, \quad \Psi_k = \psi_k \partial_{x_k} + \dot{\psi}_k \partial_{v_k} - \rho x_k \ddot{\psi}_k \partial_p; \quad k = 1, 2, 3.$$

Here, ψ_i and φ are arbitrary (of class C^∞) functions of time, and the dot means differentiation with respect to t. Thus, the group admitted by system (3) is infinite-dimensional.

Assuming that $\nu = 0$ in system (3), we obtain a system of the Euler equations

$$\mathbf{v}_t + \mathbf{v} \cdot \nabla \mathbf{v} = -\rho^{-1} \nabla p, \quad \nabla \cdot\, \mathbf{v} = 0, \tag{5}$$

which describes the motion of an ideal incompressible fluid. The group $\hat{G}_\infty$ admitted by this system is a direct product of the group G_∞ and the dilation group with the operator

$$\hat{Z} = t\partial_t + \sum_{i=1}^{3} x_i \partial_{x_i}. \tag{6}$$

The algebra corresponding to the group $\hat{G}_\infty$ is denoted by $\hat{L}_\infty$.

The presence of the dilation operator Z in the algebra L_∞ means scaling invariance of Eq. (3). This important property forms the basis of physical modeling of viscous fluid flows. The set of three operators X_{kl} generates a group of consistent rotations in the space of coordinates and in the space of velocities admitted by system (3). This property reflects the absence of preferential directions in the spaces mentioned above. It should be noted that the existence of axisymmetric solutions of the Navier–Stokes equations is directly related to the fact that the algebra L_∞ contains rotation operators, as well as the existence of steady solutions of these equations is related to the presence of the translation operator in terms of time X_0 in L_∞.

The operators Φ, Ψ_i are specific for equations of incompressible fluid dynamics. The first of them implements the possibility of adding an arbitrary function of time to the pressure without changing the equations of motion. This fact is consistent with the statement that the pressure in an incompressible medium is not a thermodynamic variable [8]. The operator Ψ_i $(i = 1, 2, 3)$ corresponds to the transformation of the transition to a new coordinate system (which is non-inertial, generally speaking), which moves along the axis x_i with a velocity $\dot{\psi}(t)$ with respect to the initial system. In this case, there appears an additional term ψ_i in the ith momentum Eq. (3), i.e., acceleration of the inertia force, which is compensated by adding the function $-\rho x_i \ddot{\psi}_i$ to the pressure.

Assuming if $\Psi_i = 1$ $(i = 1, 2, 3)$ and $\Psi_i = t$ in system (4), we obtain the operators

$$X_i = \partial_{x_i}, \quad Y_i = t\partial_{x_i} + \partial_{v_i} \quad (i = 1, 2, 3). \tag{7}$$

The set of the operators X_0, X_i, Y_i, X_{jk} forms a ten-parameter Lie algebra L_{10}. The corresponding group G_{10} is called the Galileo group. The presence of operators of translations along the coordinates x_i in the algebra L_{10} is a consequence of space homogeneity. The presence of Galileo translation operators Y_i reflects the fact that the fluid motion laws are independent of the choice of the inertial coordinate system. Supplementing the operators of L_{10} with the dilation operator Z, we obtain an eleven-parameter Lie algebra L_{11}. The corresponding Lie group G_{11} is called the extended Galileo group. Supplementing the algebra L_{11} with the operator $\hat{Z}$ (6), we obtain a twelve-parameter group G_{12}. The groups G_{10} and G_{11} (G_{10} and G_{12}) play an important role in studying invariant and partially invariant solutions of problems with a free boundary for the Navier–Stokes (Euler) equations.

4 Problems with Free Boundaries

The notion of the free boundary of the fluid is an idealized interface of two immiscible fluids if the density of one fluid is much smaller than the density of the other fluid. A typical example of such a situation is the water-air interface. At small velocities of air, it is possible to neglect its dynamic action on water, and the atmospheric pressure p_a can be imposed on the free surface Γ_t. The subscript $t$in Γ_t characterizes the dependence of the free surface shape on time.

Let the free boundary Γ_t be defined by the equation $F(x, t) = 0$. The conditions for the Navier–Stokes equations (3) on this surface have the form

$$F_t + \mathbf{v} \cdot \nabla F = 0 \text{ for } F = 0, \tag{8}$$

$$(p_a - p_g)\,\mathbf{n} + 2\rho\nu D \cdot \mathbf{n} = -2\sigma K\,\mathbf{n} \text{ for } F = 0. \tag{9}$$

Here, p_g is the pressure in the fluid, D is the strain rate tensor, $2D_{ij} = \partial v_i/\partial x_j + \partial v_j/\partial x_{ij}$ $(i, j = 1, 2, 3)$, K is the mean curvature of the surface Γ_t, $\mathbf{n}$ is the unit vector of the external normal to this surface, and $\sigma \geq 0$ is the surface tension coefficient. Condition (8) means that the surface Γ_t is a Lagrangian surface so that the velocity of its motion in the direction $\mathbf{n}$ coincides with the normal component of the fluid velocity. This condition is called the kinematic condition. Condition (9) is called the dynamic condition. It reflects the fact that the shear stress on the free surface of the fluid is equal to zero, and the difference between the normal stress and atmospheric pressure is equal to the capillary pressure.

Then it is assumed that $\sigma = const$. This assumption is valid for isothermal motions in the absence of surfactants. The conditions on the interface of immiscible fluids in the case of their non-isothermal motion on the basis of thermodynamics of the Gibbs surface were derived in [13, 14] (Chap. II). Conditions (8) and (9) are obtained from the general conditions as a result of a limiting transition. In what follows, the motion is assumed to be isothermal.

Let us now consider the system of the Euler equations (5). For this system, the conditions on the free boundary have the form of Eq. (8) and

$$p_g - p_a = 2\sigma K, \quad x \in \Gamma_t. \tag{10}$$

Now we pass to considering the motion of two immiscible viscous incompressible fluids. The motion occurs in the domain $\Omega_t \subset \mathrm{R}^3$, which is divided by a smooth surface Γ_t into two subdomains Ω_{1t} and Ω_{2t}. In each subdomain, let the functions $\mathbf{v}_1$, p_1 and $\mathbf{v}_2$, p_2 satisfying Eq. (3) with replacement of the coefficients ν, ρ by ν_1, ρ_1 and ν_2, ρ_2, respectively, be defined. At each point Γ_t at any time, we assume that there exist the limiting values of the functions $\mathbf{v}_i$, p_i and their first derivatives with respect to all variables from the subdomains Ω_{1t} and Ω_{2t} It turns out that these set of functions cannot be arbitrary: they have to be related by appropriate expressions following from conservation laws and thermodynamic postulates.

The first relations have a kinematic character. They are based on the fact that the surface Γ_t is the Lagrangian (or material) surface. Thus, we avoid considering such processes as dissolving of one of the contacting fluids in the other, condensation, and evaporation, i.e., mass transfer through the interface is prohibited.

Let us use $\mathbf{n}$ to denote the unit vector of the normal to the surface Γ_t directed to the domain Ω_{2t} and V_n to denote the velocity of motion of the surface Γ_t in the direction of the normal $\mathbf{n}$. The fact that this surface is material is expressed by the following equalities [13, 14]:

$$\mathbf{v}_1 \cdot \mathbf{n} = \mathbf{v}_2 \cdot \mathbf{n} = V_n, \quad x \in \Gamma_t. \tag{11}$$

Equalities (11) and continuity equation (the second equation of system (3)) ensure the validity of the integral law of mass conservation in an arbitrary material subdomain of the domain $\Omega_{1t} \cup \Omega_{2t}$.

The integral law of momentum conservation across the interface yields the following expression [13, 14]:

$$(-p_1 + p_2)\,\mathbf{n} + 2(\rho_1 \nu_1 D_1 - \rho_2 \nu_2 D_2) \cdot \mathbf{n} = -2\sigma K \mathbf{n}, \quad x \in \Gamma_t. \tag{12}$$

Here, D_i $(i = 1, 2)$ is the strain rate tensor corresponding to the velocity vector $\mathbf{v}_i$, and K is the mean curvature of the surface Γ_t (it is assumed that $K > 0$ if this surface is convex outward of the domain Ω_{2t}).

To conclude, we postulate the condition of continuity of the total velocity vector across the interface:

$$\mathbf{v}_1 = \mathbf{v}_2, \quad x \in \Gamma_t. \tag{13}$$

In fact, conditions (13) contain two additional scalar conditions because the continuity of the normal component is already implied in conditions (11).

5 Theorems of Invariance of Conditions on the Free Boundary

This paragraph deals with the properties of invariance of conditions (8)–(13) with respect to transformations that ensure conservation of the Navier–Stokes equations (3). For simplicity, we consider a situation without external forces. Then the function p_g involved into the dynamic condition (9) coincides with the function p involved into the momentum equation (the first equation of system (3)).

Let us consider the Euclidean space R^8 with the coordinates of its points being $x_1, x_2, x_3, t, v_1, v_2, v_3, p$. This space is subjected to the action of the Galileo group G_{10} with the basis operators X_0, X_j, Y_j, X_{ij} $(i, j = 1, 2, 3;\ i < j)$ defined by formulas (4). It is admitted by system (3). Let us consider a certain k–parameter subgroup H of the group G_{10}. Let $l \leq k$ be the maximum number of operators of this group that are not linearly related. It should be noted that $l < k$ only if $k \geq 3$ and H contains a group of rotations $\langle X_{12}, X_{13}, X_{23} \rangle$. We are interested only in intransitive groups G_{10}; in this case, $l < 8$.

Let I_α $(\alpha = 1, \ldots, 8 - l)$ be a complete set of functionally independent invariants of the group H. Let us use $m$to denote the rank of the matrix $(\partial I_\alpha / \partial v_\beta)$, where $\beta = 1, 2, 3, 4$, and it is assumed that $v_4 = p$. Clearly, $m \leq \min(4,\ 8 - l)$. In what follows, we consider only such groups H where $m < 8 - l$. Then there exist $n = 8 - l - m$ invariants of H that do not contain $\mathbf{v}$, p, which are the sought functions in system (3). Without loss of generality, we can assume that these invariants are $I_{m+1}, \dots, I_{m+n}$. The condition $m < n$ is a necessary condition for the existence of an invariant H-solution of the above-mentioned system, and the number $n - m$ is the rank of this solution.

Let us now assume that the equation $F(x, t) = 0$ defines a non-singular invariant manifold of the group H. This means that F can be written in the form

$$F = Q[I_{m+1}(x, t), \dots, I_{m+n}(x, t)]$$

with a certain function Q.

Theorem 1 *Let the free boundary $F(x, t) = 0$ be a non-singular invariant manifold of the subgroup $H \in G_{10}$. Then conditions (8) and (9) satisfied on this surface are also invariant with respect to H.*

Theorem 1 was put forward by Pukhnachev [15]. The proof of this theorem can be also found in [16]. At $\sigma \neq 0$, it turned out that the group G_{10} in the condition of Theorem 1 cannot be replaced by a wider subgroup of the infinite-dimensional group G_∞ admitted by system (3). This extension is possible if $\sigma = 0$. In this case, the statement of Theorem 1 remains valid if G_{10} is replaced with an extended Galileo group G_{11} by means of supplementing the generators of the group G_{10} with the dilation operator Z.

Let us now consider the conditions on the free boundary (8) and (10) for the Euler equations (7). If there are no external forces, it may be assumed that $p_g = p$. Here, we have an analog of Theorem 1; the proof is skipped.

As was noted earlier, the notion of the free surface is understood as an idealized interface of two immiscible fluids. Below, we formulate the properties of invariance of the conditions at the interface (11)–(13). Let us use G_{10}^{Φ} to denote the direct product of the Galileo group G_{10} and the group generated by the operator $\Phi = \varphi\partial_p$, where $\varphi(t) \in C^\infty$ is an arbitrary function.

Theorem 2 *Let the interface of immiscible fluids $F(x, t) = 0$ be a non-singular invariant manifold of the subgroup $H \in G_{10}^{\Phi}$. Then conditions (11)–(13) satisfied on this surface are also invariant with respect to H.*

If $\sigma = 0$ in condition (16), then the group G_{10}^{Φ} in the formulation of Theorem 2 can be replaced by the direct product of the group G_Φ and the extended Galileo group G_{11}. The maximum extension of the admitted group is observed at $\sigma = 0$ and identical densities of the contacting fluids.

Theorem 3 *Let us assume that $\rho_1 = \rho_2$ and $\sigma = 0$ in conditions (12). Moreover, let the interface $F(x, t) = 0$ be a non-singular invariant manifold of the subgroup $H \in G_\infty$. Then conditions (11)–(13) satisfied on this surface are also invariable with respect to H.*

Theorems 2 and 3 are new. They show that the invariant properties of the conditions at the interface of immiscible fluids are richer than similar properties of the conditions on the free boundary. This is caused by the fact that the concept of the interface is more natural from the physical viewpoint than the concept of the free surface. Theorems 2 and 3 are proved in accordance with the scheme used to prove Theorem 1; the proof is not provided here.

Theorem 1 was used to construct invariant solutions of the Navier–Stokes equations, which were preliminary matched with the conditions on the free boundary, which is an invariant manifold of the corresponding group [15, 16]. However, retaining the invariance of the free boundary, it is possible to alleviate the requirement to the solution of system (3): it can be partially invariant. Examples of such solutions of problems with a free boundary are provided in the next paragraphs.

6 Partially Invariant Solutions of the Navier–Stokes Equations

It is not an exaggeration to say that the initial trend of studying the Navier–Stokes equations was to obtain their exact solutions. Here, we should mention the solution of Hiemenz [17], which describes the flow near the stagnation point, and also the solution of Karman, which describes the motion in a half-space induced by plane rotation [18]. There is a popular opinion that these both steady solutions are self-similar solutions of system (3). In reality, these solutions have a group-theoretical nature, but it is more complicated. Petrova et al. [19] considered a problem of unsteady motion of a fluid near the stagnation point. It turned out that the solution of this problem is a partially invariant solution of system (3) of rank 2 and defect 2 with respect to the group generated by the operators $X_1 = \partial_{x_1}$ and $Y_1 = t\partial_{x_1} + \partial_{v_1}$. The corresponding resolving system inherits some of the group properties of system (3). The solution obtained by Hiemenz is an invariant solution of the resolving system with respect to the operator $X_0 = \partial_t$.

A similar situation is observed for the Karman solution. To write down this solution, we use system (3) in cylindrical coordinates $r = (x_1^2 + x_2^2)^{1/2}$, $\phi = \arctan(x_2/x_1)$, $z = x_3$:

$$\frac{dv_r}{dt} - \frac{v_\phi^2}{r} = -\frac{1}{\rho}\frac{\partial p}{\partial r} + \nu\Big(\Delta v_r - \frac{2}{r^2}\frac{\partial v_\phi}{\partial \phi} - \frac{v_r}{r^2}\Big), \tag{14}$$

$$\frac{dv_\phi}{dt} + \frac{v_r v_\phi}{r} = -\frac{1}{\rho r}\frac{\partial p}{\partial \phi} + \nu\Big(\Delta v_\phi + \frac{2}{r^2}\frac{\partial v_r}{\partial \phi} - \frac{v_\phi}{r^2}\Big),$$

$$\frac{dv_z}{dt} = -\frac{1}{\rho}\frac{\partial p}{\partial z} + \nu\Delta v_z, \quad \frac{\partial v_r}{\partial r} + \frac{v_r}{r} + \frac{1}{r}\frac{\partial v_\phi}{\partial \phi} = 0,$$

$$\frac{dv_z}{dt} = -\frac{1}{\rho}\frac{\partial p}{\partial z} + \nu\Delta v_z, \quad \frac{\partial v_r}{\partial r} + \frac{v_r}{r} + \frac{1}{r}\frac{\partial v_\phi}{\partial \phi} + \frac{\partial v_z}{\partial z} = 0.$$

Here

$$\frac{d}{dt} = \frac{\partial}{\partial t} + v_r \frac{\partial}{\partial r} + \frac{v_\phi}{r}\frac{\partial}{\partial \phi} + v_z \frac{\partial}{\partial z}, \quad \Delta = \frac{\partial^2}{\partial r^2} + \frac{1}{r}\frac{\partial}{\partial r} + \frac{1}{r^2}\frac{\partial^2}{\partial \phi^2} + \frac{\partial^2}{\partial z^2}.$$

The motion is called rotationally symmetric if the sought functions in system (14) are independent of the variable ϕ. The Karman solution refers to this class. It is described by the formulas

$$v_r = r\Omega F(\varsigma), \quad v_\phi = r\Omega G(\varsigma), \quad v_z = (\nu\Omega)^{1/2} H(\varsigma), \quad p = \rho\nu\Omega Q(\varsigma), \tag{15}$$

where $\varsigma = (\Omega/\nu)^{1/2} z$. Substitution of Eq. (15) into Eq. (14) yields a system of ordinary differential equations for the functions F, G, H, Q:

$$F^2 - G^2 + F'H = F'', \quad 2FG + G'H = G'', \quad HH' = Q' + H'', \quad 2F + H' = 0. \tag{16}$$

Let us impose the following boundary conditions on the solution of system (16):

$$F = 0, \quad G = 1, \quad H = 0 \text{ at } \varsigma = 0, \quad F \to 0, \quad G \to 0 \text{ as } \varsigma \to \infty. \tag{17}$$

Then solution (15) describes the fluid motion in the half-space $z > 0$ induced by rotation of the bounding solid plane around the axis of symmetry with an angular velocity Ω.

Following [20], we demonstrate how the Karman solution can be obtained on the basis of group considerations. Let us consider a five-parameter subgroup H of the group C_∞ with the basis operators $X_1, X_2, Y_1, Y_2, X_{12}$. This subgroup corresponds to the partially invariant solution of system (3) of rank 2 and defect 2. Its invariant part in cylindrical coordinates has the form $v_z = h(z, t), \quad p = q(z, t)$. By virtue of the continuity equation, there exists a relationship between the radial and axial velocity components: $\partial v_r/\partial r + v_r/r + \partial v_z/\partial z = 0$. By requiring that the function v_r is bounded as $r \to 0$, we have $v_r = rf(z, t)$, where $f = -h/2$.

Now we substitute the expressions for v_r, v_z, p into the first equation of system (14) and consider the fact that the sought functions are independent of ϕ. As a result, we obtain the presentation of the circumferential velocity $v_\phi = rg(z, t)$, where g is expressed via the function h and its derivatives. Thus, the general presentation of the partially invariant solution of system (3) with respect to the group H, which is regular on the axis of symmetry, is

$$v_r = rf(z, t), \quad v = rg(z, t), \quad v_z = h(z, t), \quad p = q(z, t). \tag{18}$$

The functions f, g, h, q satisfy the system of equations

$$\frac{\partial f}{\partial t}+h\frac{\partial f}{\partial z}+f^2-g^2=\nu\frac{\partial^2 f}{\partial z^2},\quad \frac{\partial g}{\partial t}+h\frac{\partial g}{\partial z}+2fg=\nu\frac{\partial^2 g}{\partial z^2},$$

$$\frac{\partial h}{\partial t}+h\frac{\partial h}{\partial z}=-\frac{1}{\rho}\frac{\partial q}{\partial z}+\nu\frac{\partial^2 h}{\partial z^2},\quad 2f+\frac{\partial h}{\partial z}=0. \tag{19}$$

System (19) inherits some part of the group properties of the original system (3), in particular, the translation with respect to time. The corresponding steady solution of system (19) coincides with the Karman solution (15) with accuracy to notations. We can say that the Karman solution is an invariant solution of a certain partially invariant sub-model of the Navier–Stokes equations, whereas a solution of the form of Eq. (18) is an unsteady analog of the Karman solution. It turns out that this solution can describe the process of layer spreading on a rotating plane [21] and [16], Chap. VII.

We require that conditions (10) and (11) should be satisfied on the invariant manifold $z = s(t)$ of the group H for solution (18). For this purpose, the unknown functions in system (19) have to be subjected to the boundary conditions

$$\frac{\partial f}{\partial z}=\frac{\partial g}{\partial z}=0,\quad q-2\rho\nu\frac{\partial h}{\partial z}=0\ \text{ for }\ z=s(t),\quad \frac{ds}{dt}=h[s(t),t]\ \text{ for }\ t>0. \tag{20}$$

In addition, we impose the boundary conditions

$$f=0,\quad g=\omega(t),\quad h=0\ \text{ for }\ z=0,\quad t>0. \tag{21}$$

Here, $\omega(t)$ is a specified function, $\omega(0)=0,\ \ \omega'(0)=0$. The formulation of the problem with an unknown boundary for system (19) is closed by setting the initial conditions

$$f=g=0,\quad h=0,\quad s=s_0>0\ \text{ for }\ t=0. \tag{22}$$

Thus, we obtain the following problem: we have to find a function $s(t)$ and a solution of system (19) in the domain $S_T=\{z,t:\ \ 0<z<s(t),\ \ 0<t<T\}$ so that conditions (20)–(22) are satisfied. Let us assume that the function $\omega(t)$ belongs to the Hölder class $C^{1+\alpha/2}[0,T]$, where $0<\alpha<1$. Then there exists a unique solution of problem (19)–(22) for any $T>0$ [16].

The resultant solution is interpreted as follows. At the initial time, the quiescent fluid occupies an infinite layer $0<z<s_0$ whose lower boundary is a solid surface, whereas the upper boundary is free. Then the plane starts smooth rotation around the z axis with an angular velocity $\omega(t)$ and sets the fluid into the corresponding motion. The characteristic feature of the problem implies that the free boundary remains flat for all $t>0$. This property was used to develop a technology of depositing coatings onto a flat disk (see [22] and the references therein). As the problem solution exists for all $t>0$, it is of interest to study its behavior as $t\to\infty$. This was made in [23], where the asymptotic behavior of the solution was found for the case with the function $\omega(t) = At^n$ for large values of t (A and n are constants). In the same paper, results of the numerical solution of problem (19)–(22) were reported for several typical values of $\omega(t)$ (see also Chap. VII of the monograph [16]).

Now we construct an example of a partially invariant solution of system (3), which describes plane motion with an internal interface. In what follows, x and y are the Cartesian coordinates on the plane, while u and v are the corresponding components of the velocity vector. It is assumed that both fluids have an identical density, $\rho_1 = \rho_2 = \rho$, but different viscosities, ν_1 and ν_2. The boundaries of the flow domain $\Pi_T = \{x, y, t : x \in \mathrm{R}, \ 0 < y < l(t), \ 0 < t < T\}$ are solid impermeable walls. One of them, $y = 0$, is stationary, while the other one, $y = l(t)$, moves along the y axis. The line $y = s(t)$ is the interface between the fluids. The band $0 < y < s(t)$ is occupied by the fluid indicated by the subscript 1, and the band $s(t) < y < l(t)$ is occupied by the fluid indicated by the subscript 2. At the initial time, both fluids are at rest.

Let us consider a subgroup of the group G_∞ generated by the operators $X = \partial_x$ and $Y = t\partial_x + \partial_u$. It corresponds to a partially invariant solution of system (3) of rank 2 and defect 2 of the form

$$u_i = -x\frac{\partial v_i}{\partial y}, \quad v_i = v_i(y, t), \quad p = \frac{\rho}{2}a(t)x^2 + m(y, t), \quad i = 1, 2, \tag{23}$$

where a is a given function of t. (We confine ourselves to solutions of system (3), where the functions v_i and p are even functions of the variable x, while u_i are odd functions of this variable). Substituting expressions (23) into system (3), we obtain equations satisfied by the functions v_1, v_2, and m:

$$\frac{\partial^2 v_1}{\partial y \partial t} + v_1\frac{\partial^2 v_1}{\partial y^2} - \Big(\frac{\partial v_1}{\partial y}\Big)^2 = \nu_1\frac{\partial^3 v_1}{\partial y^3} + a(t) \ \text{ for } \ 0 < y < s(t), \ \ 0 < t < T, \tag{24}$$

$$\frac{\partial^2 v_2}{\partial y \partial t} + v_2\frac{\partial^2 v_2}{\partial y^2} - \Big(\frac{\partial v_2}{\partial y}\Big)^2 = \nu_2\frac{\partial^3 v_2}{\partial y^3} + a(t) \ \text{ for } \ s(t) < y < l(t), \ \ 0 < t < T,$$

$$\frac{\partial v_1}{\partial t} + v_1\frac{\partial v_1}{\partial y} = -\frac{1}{\rho}\frac{\partial m}{\partial y} + \nu_1\frac{\partial^2 v_1}{\partial y^2} \ \text{ for } \ 0 < y < s(t), \ \ 0 < t < T, \tag{25}$$

$$\frac{\partial v_2}{\partial t} + v_2\frac{\partial v_2}{\partial y} = -\frac{1}{\rho}\frac{\partial m}{\partial y} + \nu_2\frac{\partial^2 v_2}{\partial y^2} \ \text{ for } \ s(t) < y < l(t), \ \ 0 < t < T.$$

The boundary conditions on the solid regions of the boundary of the domain Π_T (no-slip conditions) have the form

$$v_1 = 0 \ \text{ for } \ y = 0, \ \ 0 \le t \le T, \quad v_2 = \frac{dl}{dt} \ \text{ if } \ y = l(t), \ \ 0 \le t \le T. \tag{26}$$

Based on relations (11)–(13), the boundary conditions on the interface between the fluids are written as

$$v_1 = v_2, \quad v_1 \frac{\partial v_1}{\partial y} = v_2 \frac{\partial v_2}{\partial y}, \quad v_1 \frac{\partial^2 v_1}{\partial y^2} = v_2 \frac{\partial^2 v_2}{\partial y^2} \quad \text{for} \quad y = s(t), \quad 0 \le t \le T, \tag{27}$$

$$\frac{ds}{dt} = v_1[s(t), t] = v_2[s(t), t] \quad \text{for} \quad 0 < t < T. \tag{28}$$

They are supplemented with the initial conditions

$$v_1(y, 0) = 0 \quad \text{for} \quad 0 \le y \le s_0, \quad v_2(y, 0) = 0 \quad \text{if} \quad s_0 \le y \le l_0, \, s(0) = s_0, \tag{29}$$

where $l_0 = l(0)$, $s_0 \in (0, l_0)$ is a specified constant.

The problem with an unknown boundary is formulated. We have to determine a function $s(t)$ and a solution v_1, v_2, m of system (24), (25) that satisfies conditions (8)–(29). It should be noted that relations (24), (26)–(29) form a closed system for finding the functions v_1, v_2, and s. If these functions are found, then the remaining sought function m is reconstructed by a quadrature from Eq. (25). The additive function of time arising thereby makes it possible to ensure the continuity of the function m across the interface.

Problem (24)–(29) is rather non-standard. At the moment, the uniqueness of its classical solution can be guaranteed. To prove the existence theorem, it is reasonable to pass in this problem to the Lagrangian coordinates in which the flow domain is fixed. It can be expected that problem (24)–(29) does have a solution, at least, for a sufficiently small value $T > 0$.

7 Example of Partially Invariant Solution of the Euler Equations

Below we study a rotationally symmetric solution of the Euler equations (5) in a cylindrical layer $\Omega_T = \{r, z, t : \; b < r < s(t), \; z \in \mathrm{R}, \; 0 < t < T\}$. The equations of motion are obtained from Eq. (14) by assuming that $\nu = 0$ and taking into account that the sought functions are independent of the variables ϕ:

$$\frac{\partial v_r}{\partial t} + v_r \frac{\partial v_r}{\partial r} + v_z \frac{\partial v_r}{\partial z} - \frac{v_\phi^2}{r} = -\frac{1}{\rho} \frac{\partial p}{\partial r}, \quad \frac{\partial v_\phi}{\partial t} + v_r \frac{\partial v_\phi}{\partial r} + v_z \frac{\partial v_\phi}{\partial z} + \frac{v_r v_\phi}{r} = 0,$$
$$\frac{\partial v_z}{\partial r} + v_r \frac{\partial v_z}{\partial r} + v_z \frac{\partial v_z}{\partial z} = -\frac{1}{\rho} \frac{\partial p}{\partial z}, \quad \frac{\partial v_r}{\partial r} + \frac{v_r}{r} + \frac{\partial v_z}{\partial z} = 0. \tag{30}$$

System (30) admits a group with operators ∂_z and $t\partial_z + \partial_{v_z}$, which corresponds to its partially invariant solution of rank 2 and defect 1 of the form

$$v_r = u(r,t), \quad v_\phi = v(r,t), \quad v_z = -z\Big(\frac{\partial u}{\partial r} + \frac{u}{r}\Big), \quad p = p(r,t). \tag{31}$$

Substituting expressions (31) into Eq. (30), we obtain the resolving system of equations for the sought invariant functions, which can be conveniently written in the form

$$\frac{\partial u}{\partial r} + \frac{u}{r} = L, \quad \frac{\partial L}{\partial t} + u\frac{\partial L}{\partial r} - L^2 = 0, \quad \frac{\partial v}{\partial t} + u\frac{\partial v}{\partial r} + \frac{uv}{r} = 0, \tag{32}$$

$$\frac{\partial u}{\partial t} + u\frac{\partial u}{\partial r} - \frac{v^2}{r} + \frac{1}{\rho}\frac{\partial p}{\partial r} = 0.$$

Let us assume that the layer boundary $r = b$ is a permeable surface. It is subjected to the boundary condition

$$u = c(t), \quad r = b, \quad z \in \mathrm{R}, \quad 0 < t < T, \tag{33}$$

where $c(t)$ is a prescribed function. The boundary $r = s(t)$ is assumed to be free. It is subjected to the conditions

$$p = p_a + \frac{\sigma}{s(t)}, \quad r = s(t), \quad \frac{ds}{dt} = v[s(t), t], \quad 0 < t < T, \tag{34}$$

which follow from conditions (8) and (10). Moreover, the following initial conditions are imposed:

$$u(r,0) = u_0(r), \quad v(r,0) = v_0(r), \quad b \le r \le s_0, \quad s(0) = s_0. \tag{35}$$

Here, $u_0(r)$ and $v_0(r)$ are specified functions, and $s_0 > b$ is a prescribed constant.

System (32) has a recurrent structure: its first two equations are separated from the others. An effective analysis of problem (32)–(35) is reached by means of the transition to the Lagrangian coordinate ξ instead of r. The relationship between r and ξ is determined by solving the Cauchy problem

$$\frac{dr}{dt} = u(r,t), \quad t > 0; \quad r = \xi, \quad t = 0. \tag{36}$$

The following notations are introduced:

$$u[r(\xi,t),t] = U(\xi,t), \quad L[r(\xi,t),t] = \Lambda(\xi,t),$$

$$v[r(\xi,t),t] = V(\xi,t), \quad p[r(\xi,t),t] = P(\xi,t).$$

Here, $r(\xi,t)$ is the solution of the Cauchy problem (36). We denote $u_0'(\xi) + u_0(\xi)/\xi = a(\xi)$. In the new variables, system (32) takes the form

$$\frac{\partial \Lambda}{\partial t} = \Lambda^2, \quad r\frac{\partial^2 r}{\partial r \partial t} + \frac{\partial r}{\partial \xi}\frac{\partial r}{\partial t} = r\frac{\partial r}{\partial \xi}\Lambda, \tag{37}$$

$$\frac{\partial V}{\partial t} + \frac{UV}{r} = 0, \quad \frac{\partial U}{\partial t} - \frac{V^2}{r} + \frac{1}{\rho}\Big(\frac{\partial r}{\partial \xi}\Big)^{-1}\frac{\partial P}{\partial \xi} = 0. \tag{38}$$

The solution of the first equation of (37) with the initial condition $\Lambda(\xi, 0) = a(\xi)$ has the form

$$\Lambda = -\frac{a(\xi)}{a(\xi)t + 1}. \tag{39}$$

The second equation of (37) can be written as

$$\frac{\partial}{\partial t}\left(\frac{\partial r^2}{\partial \xi}\right) = \Lambda\left(\frac{\partial r^2}{\partial \xi}\right),$$

after which it is easily integrated with allowance for equalities (39) and $r = \xi$ at $t = 0$:

$$r(\xi, t) = \Big[\int\limits_{\varsigma(t)}^{\xi} \frac{2\eta d\eta}{a(\eta)t + 1} + b^2\Big]^{1/2}. \tag{40}$$

Here, the function $\varsigma(t)$ determines the image of the fixed boundary $r = b$ of the flow domain in passing to the Lagrangian coordinates. From Eq. (40) and the equalities $r[\varsigma(t), t] = b$, $r_t[\varsigma(t), t] = c(t)$, we obtain the Cauchy problem for finding the function $\varsigma(t)$:

$$2\varsigma\frac{d\varsigma}{dt} = -bc(t)[a(\varsigma)t + 1], \quad t > 0; \quad \varsigma = b, \quad t = 0.$$

In accordance with the second condition of (34), the image of the free boundary $r = s(t)$ on the plane of the Lagrangian coordinates is a segment of the straight line $\xi = s_0$.

Using formulas (40) and $U = r_t(\xi, t)$, we find the functions V and P with the help of quadratures from Eq. (38). In this case, the function P is determined with the accuracy up to the additive function of time. Thus, the first condition of (34) can be satisfied on the free boundary $r = s(t)$. As a result, we obtain a parametric representation of the solution of the problem with a cylindrical free surface for the Euler equations.

References

1. Ovsyannikov LV (1982) Group analysis of differential equations. Academic, New York, p 432
2. Ovsyannikov LV (1964) Partly invariant solutions of the equations admitting a group. In: Proceedings of the 11th international congress of applied mechanics, Münich, pp 868–870
3. Ovsyannikov LV (1969) Partial invariance. Dokl Akad Nauk SSSR 186(1):22–25 (in Russian)
4. Ovsyannikov LV (1994) Program SUBMODELS. Gas dynamics. J Appl Math Mech 58(4):30–55
5. Ovsyannikov LV (1995) Special vortex. J Appl Mech Tech Phys 36(3):45–52
6. Yanenko NN (1956) Traveling waves of the system of differential equations. Dokl Akad Nauk SSSR 109(1):44–47 (in Russian)
7. Sidorov AF, Shapeev VP, Yanenko NN (1984) Method of differential constrains and its applications in gas dynamics. Nauka (Sibirskoe Otdelenie), Novosibirsk, 272 (in Russian)
8. Serrin J (1959) Mathematical principles of classical fluid mechanics. Handbuch der Physik. Band VIII/1. Strömungsmechanik I. Berlin/Göttingen/Heidelberg, pp 125–263
9. Kochin NE, Kibel IA, Rose NV (1963) Theoretical hydrodynamics. Part 2. Interscience Publication Co, New York, p 736
10. Danilov YuA (1967) Group properties of Maxwell and Navier–Stokes equations. Inst Atomic Energy, 1452, Moscow, 15. (in Russian)
11. Bytev VO (1972) Group properties of Navier - Stokes equations. Chislennye Metody Mekhaniki Sploshnoi Sredy, Novosibirsk 3(3):13–17 (in Russian)
12. Lloyd SP (1981) The infinitesimal group of the Navier - Stokes equations. Acta Mech 38(1–2):85–98
13. Napolitano LG (1979) Thermodynamics and dynamics of surface phases. Acta Astron 6(9):1093–1112
14. Andreev VK, Gaponenko YuA, Goncharova ON, Pukhnachev VV (2020) Mathematical models of convection, 2nd edn. De Gruyter, Berlin/Boston, p 418
15. Pukhnachev VV (1972) Invariant solutions of the Navier - Stokes equations, which describe motions with free boundaries. Dokl Akad Nauk SSSR 202(2):302–305 (in Russian)
16. Andreev VK, Kaptsov OV, Pukhnachov VV, Rodionov AA (1998) Applications of group-theoretical methods in hydrodynamics. Kluwer Academic Publishers, Dordrecht, p 396
17. Hiemenz K (1911) Die Grenzschicht an einem in den gleichfoermigen Fluessigkeitstrom eingetauchten geraden Kreiszylinder. Dingl Polytech J 326:321–410
18. Karman Th (1921) Ueber laminare und turbulentne Reibung. Z Angew Mat Mech 1(4):233–252
19. Petrova AG, Pukhnachev VV, Frolovskaya OA (2016) Analytical and numerical investigation of unsteady flow near a critical point. J Appl Math Mech 80(3):215–224
20. Meleshko SV, Pukhnachev VV (1999) One class of partially invariant solutions of the Navier–Stokes equations. J Appl Mech Tech Phys 40(2):208–216
21. Pukhnachev VV (1972) Unsteady motions of a viscous liquid with a free boundary described by partially invariant solutions to the Navier–Stokes equations. Dinamika Sploshnoi Sredy, Novosibirsk 10:125–137 (in Russian)
22. Matsumoto S, Saito K, Takashima Y (1973) Thickness of liquid film on a rotating disk. J Chem Eng Jpn 6(6):503–507
23. Lavrenteva OM, Volkova GB (1996) Limit regimes of the spreading of a layer on a rotating plane. Dinamika Sploshnoi Sredy, Novosibirsk 111:58–74 (in Russian)

Group Properties of the Riemann Function

A. V. Aksenov

Abstract A linear hyperbolic equation of the second order in two independent variables is considered. The Riemann function of the adjoint equation is shown to be invariant with respect to the symmetries of the fundamental solutions. The Riemann function is constructed with the aid of fundamental solutions symmetries. Examples of the application of the algorithm for constructing Riemann function are given.

1 Introduction

In the study [1], B. Riemann suggested the Riemann's method of integrating that was applied to a hyperbolic second-order partial equation with two independent variables. In order to apply the Riemann's method, it is necessary to construct Riemann function that is a solution of Cauchy special characteristic problem [2]. General method for Riemann function construction does not exist.

In [3], an extensive analysis of six certain methods for creation Riemann function of particular types of equations. Ibragimov recommended to find Riemann function with the aid of equation symmetries [4, 5] basing on Ovsyannikov study result [6] in group classification of linear hyperbolic second-order equations.

The most complete review of research on the Riemann function is given in [7].

In the present study, we consider the method for constructing the Riemann function based on the use of the symmetry of fundamental solutions. Examples of using the method are given.

A. V. Aksenov (✉)
Lomonosov Moscow State University, Moscow, Russian Federation
e-mail: aksenov@mech.math.msu.su

A. C. J. Luo and R. K. Gazizov (eds.), *Symmetries and Applications of Differential Equations*, Nonlinear Physical Science,
https://doi.org/10.1007/978-981-16-4683-6_10

2 Riemann's Method

Let us consider the general linear hyperbolic equation of the second order in two independent variables

$$Lu = u_{xy} + a(x, y)u_x + b(x, y)u_y + c(x, y)u = 0 . \qquad (1)$$

Riemann's method is based on the following identity

$$2(vLu - uL^*v) = (vu_y - uv_y + 2auv)_x + (vu_x - uv_x + 2buv)_y ,$$

where $L^*v = v_{xy} - (av)_x - (bv)_y + cv{=}0$ is the adjoint equation. Riemann's method allows us to convert the integration problem of Eq. (1) to construction of the intermediary Riemann function $R(x, y; x', y')$, that obeys the following adjoint equation of variables x, y

$$L^*R = 0$$

and the following conditions at the characteristics:

$$(R_y - aR)\big|_{x=x'} = 0 , \qquad (R_x - bR)|_{y=y'} = 0 , \quad R(x', y'; x', y') = 1 .$$

General solutions of Cauchy problem and Goursat problem are constructed with the aid of Riemann function for Eq. (1) [2].

Riemann function $u = R^*(x, y; x', y')$ of the adjoint equation satisfies the Eq. (1) and following conditions at characteristics:

$$(R^*_y + aR^*)\big|_{x=x'} = 0 , \qquad (R^*_x + bR^*)\big|_{y=y'} = 0 , \qquad R^*(x', y'; x', y') = 1 . \quad (2)$$

Riemann functions R and R^* have properties of reciprocity [2]

$$R(x, y; x', y') = R^*(x', y'; x, y) . \qquad (3)$$

3 Symmetries of Fundamental Solutions

Fundamental solutions of linear partial differential equations are frequently invariant under transformations admitted by the original equation [8]. Below, a fundamental solution is constructed using the algorithm from [9, 10] proposed for finding fundamental solutions of linear partial differential equations. The algorithm makes use of the symmetries admitted by a linear partial differential equation with a delta function on its right-hand side. Let us briefly describe the main result of this work.

Consider the pth-order linear partial differential equation

$$Au \equiv \sum_{\alpha=1}^{p} B_\alpha(x) D^\alpha u = 0\,, \qquad x \in R^m\,. \tag{4}$$

Here, the standard notation is used: $\alpha = (\alpha_1, \ldots, \alpha_m)$ is a multi-index with nonnegative integer components, $\alpha = \alpha_1 + \cdots + \alpha_m$, and

$$D^\alpha \equiv \left(\frac{\partial}{\partial x^1}\right)^{\alpha_1} \cdots \left(\frac{\partial}{\partial x^m}\right)^{\alpha_m}.$$

The fundamental solutions of Eq. (4) are solutions of the equation

$$Au = \delta(x - x_0)\,. \tag{5}$$

It was shown in [11] that Eq. (4) with $p \geqslant 2$ and $m \geqslant 2$ can admit only symmetry operators of the form (the finite-dimensional part of Lie algebra)

$$Y = \sum_{i=1}^{m} \xi^i(x)\,\frac{\partial}{\partial x^i} + \eta(x, u)\,\frac{\partial}{\partial u}, \qquad \frac{\partial^2 \eta}{\partial u^2} = 0\,.$$

The basic Lie algebra of symmetry operators of Eq. (4) regarded as a vector space is the direct sum of two subalgebras: one consisting of operators of the form

$$X = \sum_{i=1}^{m} \xi^i(x)\,\frac{\partial}{\partial x^i} + \zeta(x)\,u\,\frac{\partial}{\partial u}\,, \tag{6}$$

and the infinite-dimensional subalgebra generated by the operators

$$X_\infty = \varphi(x)\,\frac{\partial}{\partial u}\,, \tag{7}$$

where $\varphi(x)$ is an arbitrary solution of Eq. (4). Note that operators (7) are symmetry operators of Eq. (5). In what follows, we consider only symmetry operators of form (6).

Let denote $\underset{p}{X}$ an extension of order p of symmetry operator (6).

Proposition 1 *The infinitesimal operator given by* (6) *is a symmetry operator of Eq.* (4) *if and only if there exists a function* $\lambda = \lambda(x)$ *satisfying the identity*

$$\underset{p}{X}(Au) \equiv \lambda(x)\,Au \tag{8}$$

for any function $u = u(x)$ *from the domain of Eq.* (4).

Let us formulate the main result [9, 10].

Theorem 1 *The Lie algebra of symmetry operators of Eq.* (5) *is a subalgebra of the Lie algebra of symmetry operators of Eq.* (4) *and is defined by the relations*

$$\begin{aligned} &\xi^i(x_0) = 0, \qquad i = 1, \dots, m\,, \\ &\lambda(x_0) + \sum_{i=1}^{m} \frac{\partial \xi^i(x_0)}{\partial x^i} = 0\,. \end{aligned} \tag{9}$$

Let us describe an algorithm for finding fundamental solutions by applying symmetries [9, 10]:
1. Find a general symmetry operator of Eq. (4) and the corresponding function $\lambda(x)$ satisfying identity (8).
2. Use this operator and relations (9) to obtain the basis for the Lie algebra of symmetry operators of Eq. (5).
3. Construct invariant fundamental solutions with the help of the symmetries of Eq. (5).
4. Obtain new fundamental solutions from the known ones with the help of the symmetries of Eq. (5) (production of solutions).

Remark 1 To find generalized invariant fundamental solutions, we need to search for invariants in the class of generalized functions.

Remark 2 In works [12, 13], instead of the second relation from (9) another relation was proposed.

4 Method for Constructing of the Riemann Function

Symmetry operator of the Eq. (1) has a form

$$X = \xi^1(x)\,\frac{\partial}{\partial x} + \xi^2(y)\,\frac{\partial}{\partial y} + \zeta(x, y)\,u\,\frac{\partial}{\partial u} \tag{10}$$

and as this takes place the following relations must be hold

$$\begin{aligned} &\frac{\partial \zeta}{\partial x} + \frac{\partial (b\,\xi^1)}{\partial x} + \xi^2\,\frac{\partial b}{\partial y} = 0\,, \\ &\frac{\partial \zeta}{\partial y} + \frac{\partial (a\,\xi^2)}{\partial y} + \xi^1\,\frac{\partial a}{\partial x} = 0\,, \\ &\frac{\partial^2 \zeta}{\partial x \partial y} + a\,\frac{\partial \zeta}{\partial x} + b\,\frac{\partial \zeta}{\partial y} + \frac{\partial (c\,\xi^1)}{\partial x} + \frac{\partial (c\,\xi^2)}{\partial y} = 0\,. \end{aligned} \tag{11}$$

Function $\lambda = \lambda(x, y)$ that satisfies the identity law $\underset{2}{X}(Lu) \equiv \lambda\, Lu$ has the form

$$\lambda = \zeta - \frac{d\,\xi^1}{d\,x} - \frac{d\,\xi^2}{d\,y}. \tag{12}$$

Let us consider the equation

$$Lu = \delta(x - x')\,\delta(y - y'), \tag{13}$$

that describes fundamental solutions of homogeneous Eq. (1). So the symmetry operators of fundamental solutions (or the symmetries of Eq. (13)) satisfy following additional conditions as Theorem 1 takes place

$$\begin{aligned} &\xi^1(x') = 0, \qquad \xi^2(y') = 0, \\ &\lambda(x', y') + \frac{d\,\xi^1(x')}{d\,x'} + \frac{d\,\xi^2(y')}{d\,y'} = 0. \end{aligned} \tag{14}$$

Show that conditions at characteristics (2) are invariant under symmetry operator (10) at (11), (12) and (14). Note that characteristics $x = x'$, $y = y'$ are invariant under operators of the symmetry of the fundamental solutions. $\zeta(x', y') = 0$ results from relations (12) and the second one of (14). This implies that the latest one of relations (2) is invariant.

Write the invariance condition at the characteristic $x = x'$

$$\underset{1}{X}(u_y + au)\Big|_{\substack{x = x' \\ u = R^*}} = 0,$$

$$\left\{\left(\zeta - \frac{d\xi^2}{dy}\right)(u_y + au) + u\left[\frac{\partial\zeta}{\partial y} + \frac{\partial(a\,\xi^2)}{\partial y} + \xi^1\frac{\partial a}{\partial x}\right]\right\}\Big|_{\substack{x = x' \\ u = R^*}} = 0. \tag{15}$$

Invariance condition (15) is realized owing to (11) and (2) at the characteristic $x = x'$. Invariance of the condition at the characteristic $y = y'$ is convinced analogous.

The normalization condition $R^*(x', y'; x', y') = 1$ gives the additional relation

$$\eta(x', y') = 0. \tag{16}$$

Thus, we have proved the following theorem.

Theorem 2 *The symmetries of the fundamental solutions of the second-order linear hyperbolic equation with the additional relation* (16) *leave the Riemann function* $R^*(x, y; x', y')$ *of the adjoint equation invariant.*

It follows from Theorem 2 that the Riemann function $R^*(x, y; x', y')$ of the adjoint equation is an invariant of the symmetries of fundamental solutions of the original equation. The Riemann function $R(x, y; x', y')$ of the original equation is determined from the reciprocity relation (3).

Formulate the algorithm of Riemann function construction based on the usage of the fundamental solutions symmetries [14, 15]:

1. Solving for symmetries of linear homogeneous equation (1).
2. Computation of the fundamental solutions symmetries.
3. Construction of invariant solutions with the aid of fundamental solutions symmetries.
4. Extraction of Riemann function $R^*(x, y; x', y')$ from the computed invariant solutions invoking continuity condition of Riemann function and its first derivatives at the point (x', y') and the condition $R^*(x', y'; x', y') = 1$.

Remark 3 This algorithm allows one to find the Riemann function of a hyperbolic equation without passing to characteristic variables. This stresses the invariant nature of this method of constructing of the Riemann function.

Remark 4 Let us consider the equation

$$Lu = \delta(x - x')\,\delta(y - y'). \tag{17}$$

As Adamar noticed in [16], the fundamental solution of Eq. (4) or the solution of the Eq. (17) can be written as

$$u_f = R^*\theta(x - x')\theta(y - y')\,.$$

Next, as Theorem 2 takes place, its right part is invariant of fundamental solutions symmetries.

5 Examples

Consider examples of the method application.

Example 1 Consider an equation

$$u_{xy} + u = 0\,. \tag{18}$$

The symmetries of Eq. (18) can be found using the symmetry-finding algorithm from [8]. Symmetry operators basis of the finite-dimensional part of Lie algebra of Eq. (18) has the form

$$X_1 = \frac{\partial}{\partial x}, \qquad X_2 = \frac{\partial}{\partial y}, \qquad X_3 = x\frac{\partial}{\partial x} - y\frac{\partial}{\partial y}, \qquad X_4 = u\frac{\partial}{\partial u}\,.$$

The general form of the coordinates of the symmetry operator of Eq. (18) is as follows

$$\xi^1 = a_1 + a_3 x, \qquad \xi^2 = a_2 - a_3 y, \qquad \eta = a_4 u.$$

Therefore, function λ is equal $\lambda = a_4$ (from the relation (16), it follows that a_4 is zero) and the equation

$$u_{xy} + u = \delta(x - x')\delta(y - y')$$

admits the symmetry operator

$$Y = (x - x')\frac{\partial}{\partial x} - (y - y')\frac{\partial}{\partial y}\,. \tag{19}$$

The solution that is invariant under symmetry operator (19) is found as

$$u = f(z)\,, \qquad z = (x - x')(y - y')\,.$$

Here, function $f = f(z)$ is a solution of the ordinary differential equation

$$zf'' + f' + f = 0\,. \tag{20}$$

The general solution of Eq. (18) has the form

$$f = C_1 J_0(2\sqrt{z}) + C_2 Y_0(2\sqrt{z})\,,$$

where C_1, C_2 are arbitrary constants; $J_0(z)$, $Y_0(z)$ are Bessel functions [17]. The condition $f(0) = 1$ implies that

$$R^*(x, y; x', y') = R(x, y; x', y') = J_0(2\sqrt{(x - x')(y - y')})\,.$$

Example 2 Consider an equation

$$u_{xy} + \frac{1}{4(x + y)^2}u = 0\,. \tag{21}$$

The symmetries of Eq. (21) can be found using the symmetry-finding algorithm from [8]. Symmetry operators basis of the finite-dimensional part of Lie algebra of (21) has the form

$$X_1 = \frac{\partial}{\partial x} - \frac{\partial}{\partial y}\,, \qquad X_2 = x\frac{\partial}{\partial x} + y\frac{\partial}{\partial y}\,, \qquad X_3 = x^2\frac{\partial}{\partial x} - y^2\frac{\partial}{\partial y}\,, \qquad X_4 = u\frac{\partial}{\partial u}\,.$$

The general form of the coordinates of the symmetry operator of the Eq. (21) is as follows

$$\xi^1 = a_1 + a_2 x + a_3 x^2, \qquad \xi^2 = -a_1 + a_2 y - a_3 y^2, \qquad \eta = a_4 u.$$

Therefore, function λ is equal $\lambda = -2a_2 - 2a_3(x - y) + a_4$ (from the relation (16), it follows that a_4 is zero) and the equation

$$u_{xy} + \frac{1}{4(x+y)^2} u = \delta(x - x')\delta(y - y')$$

admits the symmetry operator

$$Y = (x - x')(x + y')\frac{\partial}{\partial x} - (y - y')(y + x')\frac{\partial}{\partial y}. \tag{22}$$

The solution that is invariant under symmetry operator (22) is found as

$$u = f(z), \qquad z = \frac{(x - x')(y - y')}{(x' + y')(x + y)}.$$

Here function $f = f(z)$ is a solution of the ordinary differential equation

$$(z^2 + z) f'' + (2z + 1) f' + \frac{1}{4} f = 0. \tag{23}$$

The general solution of Eq. (23) has the form

$$f = C_1 EllipticK(i\sqrt{z}) + C_2 EllipticCK(i\sqrt{z}),$$

where C_1, C_2 are arbitrary constants; $EllipticK(z)$ is complete elliptic integral of the first kind and $EllipticCK(z)$ is complementary complete elliptic integral of the first kind [17]. From the condition $f(0) = 1$ follows

$$R^*(x, y; x', y') = R(x, y; x', y') = \frac{2}{\pi} EllipticK\Big(i\sqrt{\frac{(x - x_0)(y - y_0)}{(x_0 + y_0)(x + y)}}\Big).$$

Example 3 Let us consider Euler–Poisson–Darboux equation

$$\frac{\partial^2 u}{\partial r^2} + \frac{2\alpha}{r}\frac{\partial u}{\partial r} - \frac{\partial^2 u}{\partial z^2} = 0. \tag{24}$$

The symmetries of Eq. (24) can be found using the symmetry-finding algorithm from [8]. In case of $\alpha \neq 0$ Eq. (24) admits the following basis of the finite part of Lie algebra symmetry operators

$$Y_1 = \frac{\partial}{\partial z}, \qquad Y_2 = r\,\frac{\partial}{\partial r} + z\,\frac{\partial}{\partial z}, \qquad Y_3 = u\,\frac{\partial}{\partial u},$$
$$Y_4 = 2rz\,\frac{\partial}{\partial r} + (r^2 + z^2)\,\frac{\partial}{\partial z} - 2\alpha z u\,\frac{\partial}{\partial u}.$$

The general form of the coordinates of the symmetry operator of Eq. (24) is as follows

$$\xi^1 = r(a_2 + 2a_4 z), \qquad \xi^2 = a_1 + a_2 z + a_4(r^2 + z^2), \qquad \eta = (a_3 - 2\alpha a_4 z)u.$$

Therefore, function λ is equal $\lambda = -2(a_2 + 2a_4 z)$ (from the relation (16) it follows that $a_3 = 2\alpha a_4 z'$) and the equation

$$\frac{\partial^2 u}{\partial r^2} + \frac{2\alpha}{r}\,\frac{\partial u}{\partial r} - \frac{\partial^2 u}{\partial z^2} = \delta(r - r')\delta(z - z')$$

admits the symmetry operator

$$Y = 2r(z - z')\,\frac{\partial}{\partial r} + \left[r^2 + (z - z')^2 - r'^2\right]\frac{\partial}{\partial z} - 2\alpha(z - z')u\,\frac{\partial}{\partial u}. \tag{25}$$

The symmetry operator (25) has two functionally independent invariants

$$\xi = \frac{r^2 - (z - z')^2 + r'^2}{2rr'}, \qquad \tau = r^\alpha\, u.$$

Invariant solutions are sought in the form

$$\tau = f(\xi),$$

or

$$u = r^{-\alpha}\, f(\xi).$$

Euler–Poisson–Darboux solutions (18), that are invariant under symmetry operator (20), have the following form

$$u = r^{-\alpha}\left[C_1\, P_{-\alpha}(\xi) + C_2\, Q_{-\alpha}(\xi)\right],$$

where C_1, C_2 are arbitrary constants; $P_{-\alpha}(\xi)$, $Q_{-\alpha}(\xi)$ are Legendre functions of the first and second kinds [17]. The condition $f(0) = 1$ gives us Riemann function

$$R^*(r, z; r', z') = \left(\frac{r}{r_0}\right)^{-\alpha} P_{-\alpha}(\xi). \tag{26}$$

Remark 5 The Riemann function (26) describes, up to a constant factor, a solution to the characteristic problem of interpenetration of two centered rarefaction waves [18].

Example 4 Let us consider the equation

$$\frac{\partial^2 u}{\partial x \partial y} + \frac{\beta}{x+y} u = 0 . \tag{27}$$

The symmetries of Eq. (27) can be found using the symmetry-finding algorithm from [8]. In case of $\beta \neq 0$, Eq. (27) admits the following basis of the finite part of Lie algebra symmetry operators

$$X_1 = \frac{\partial}{\partial x} - \frac{\partial}{\partial y}, \qquad X_2 = u .$$

The general form of the coordinates of the symmetry operator of the Eq. (27) is as follows:

$$\xi^1 = a_1, \qquad \xi^2 = -a_1, \qquad \eta = a_2 u.$$

Therefore, function λ is equal $\lambda = a_2$ (from the relation (16), it follows that $a_2 = 0$) and the equation

$$\frac{\partial^2 u}{\partial x \partial y} + \frac{\beta}{x+y} u = \delta(r - r')\delta(z - z')$$

no admits a symmetry operator.

6 Conclusion

The main result of the study is the consideration of the method for constructing the Riemann function based on the use of symmetries of fundamental solutions. Its effectiveness is performed by examples.

References

1. Riemann B (1860) Ueber die fortpflanzung ebener luftwellen von endlicher schwingungsweite. Abhandlungen der Königlichen Gesellschaft der Wissenschaften in Göttingen 8:43–66
2. Darboux G (1889) Leçons sur la théorie générale des surfaces et les applications géométriques du calcul infinitésimal. Dèuxieme Partie. Paris, Gauthier-Villars et Fils
3. Copson ET (1957/1958) On the Riemann-Green function. Arch Rat Mech Anal 1(1):324–348
4. Ibragimov NH (1992) Group analysis of ordinary differential equations and the invariance principle in mathematical physics (for the 150th anniversary of Sophus Lie). Russ Math Surv 47(4):89–156
5. Ibragimov NH (1994) Sophus Lie and harmony in mathematical physics, on the 150th anniversary of his birth. Math Intell 1(1):20–28

6. Ovsiannikov LV (1960) Group properties of S. A. Chaplygin's equation. J Appl Mech Tech Phys 1(3):126–145 (in Russian)
7. Zeitsch PJ (2018) On the Riemann function. Mathematics 6(12):316
8. Ovsiannikov LV (1982) Group analysis of differential equations. Academic, New York
9. Aksenov AV (1994) Symmetries of linear partial differential equations and fundamental solutions. Uspekhi Matematitcheskih Nauk 49(4):143–144 (in Russian)
10. Aksenov AV (1995) Symmetries of linear partial differential equations and fundamental solutions. Dokl Math 51(3):329–331
11. Bluman G (1990) Simplifying the form of Lie groups admitted by a given differential equation. J Math Anal Appl 145(1):52–62
12. Berest YuYu (1993) Group analysis of linear differential equations in generalized functions and the construction of fundamental solutions. Diff Equ 29(11):1700–1711
13. Berest YuYu, Ibragimov NH (1994) Group theoretic determination of fundamental solutions. Lie Groups Appl 1(1):65–80
14. Aksenov AV (2017) Method of construction of the Riemann function for a second-order hyperbolic equation. J Phys: Conf Ser 937:012001
15. Aksenov AV (2018) Symmetries of fundamental solutions and their application in continuum mechanics. Proc Steklov Inst Math 300:1–12
16. Hadamard J (1952) Lectures on Cauchy's problem in linear partial differential equations. Dover Publications, New York
17. Abramowitz M, Stigan IA (1964) Handbook of mathematical functions with formulas, graphs, and mathematical tables. National Bureau of Standards, Tenth Printing, Washington
18. Courant R, Friedrichs KO (1948) Supersonic flow and shock waves. Interscience Publishers, New York

The manufacturer's authorised representative in the EU is Springer Nature Customer Service Centre GmbH, Europaplatz 3, 69115 Heidelberg, Germany. If you have any concerns regarding our products, please contact ProductSafety@springernature.com

Printed and bound by CPI Group (UK) Ltd, Croydon, CR0 4YY
24/07/2026
02175916-0002